The State of Art of Nuclear Chemistry: Theoretical and Practical Aspects

The State of Art of Nuclear Chemistry: Theoretical and Practical Aspects

Edited by:

Maria Emilova Velinova

ARCLER PRESS

www.arclerpress.com

The State of Art of Nuclear Chemistry: Theoretical and Practical Aspects

Maria Emilova Velinova

Arcler Press
224 Shoreacres Road
Burlington, ON L7L 2H2
Canada
www.arclerpress.com
Email: orders@arclereducation.com

ISBN: 978-1-77407-770-2 (Hardcover)

DECLARATION

Some content or chapters in this book are open access copyright free published research work, which is published under Creative Commons License and are indicated with the citation. We are thankful to the publishers and authors of the content and chapters as without them this book wouldn't have been possible.

ABOUT THE EDITOR

Maria Velinova is Ph.D. holder in Quantum chemistry at the University of Sofia since April 2012. Her major research experience is in the field of Computational Chemistry, especially in statistical mechanics methods applied to different sorts of biomolecules. Member of the Laboratory of Quantum and Computational Chemistry at the University of Sofia.

TABLE OF CONTENTS

LIST OF CONTRIBUTORS

Julia Even
KVI – CART, University of Groningen, 9747 AA Groningen, The Netherlands

Christoph E. Düllmann
Institute of Nuclear Chemistry, Johannes Gutenberg University, 55128 Mainz, Germany
GSI Helmholtzzentrum für Schwerionenforschung, 64291 Darmstadt, Germany
Helmholtz Institute Mainz, 55099 Mainz, Germany

Valeria Pershina
GSI Helmholtzzentrum für Schwerionenforschung GmbH, 64291 Darmstadt, Germany

Giancarlo Cavazzini
CNR Istituto di Geoscienze e Georisorse, Padova, Italy

Matthew Jennings
School of Chemistry & Physics, University of Adelaide, Adelaide, SA 5000, Australia
Department of Medical Physics, Royal Adelaide Hospital, Adelaide, SA 5000, Australia

Loredana G. Marcu
School of Chemistry & Physics, University of Adelaide, Adelaide, SA 5000, Australia
Faculty of Science, University of Oradea, 410087 Oradea, Romania

Eva Bezak
School of Chemistry & Physics, University of Adelaide, Adelaide, SA 5000, Australia
Department of Medical Physics, Royal Adelaide Hospital, Adelaide, SA 5000, Australia

Francesca Sammarruca
Physics Department, University of Idaho, Moscow, ID, United States

Randy Millerson
Physics Department, University of Idaho, Moscow, ID, United States

E. Ruiz Arriola
Departamento de Física Atómica, Molecular y Nuclear and Instituto Carlos I de Física Teórica y Computacional, Universidad de Granada E-18071 Granada, Spain

J. E. Amaro
Departamento de Física Atómica, Molecular y Nuclear and Instituto Carlos I de Física Teórica y Computacional, Universidad de Granada E-18071 Granada, Spain

R. Navarro Perez
Nuclear and Chemical Science Division, Lawrence Livermore National Laboratory Livermore, California 94551, USA

H.B. Yang
Institute of Modern Physics, Chinese Academy of Sciences, 730000 Lanzhou, China
University of Chinese Academy of Sciences, 100049 Beijing, China
School of Nuclear Science and Technology, Lanzhou University, 730000 Lanzhou, China

Z.Y. Zhang
Institute of Modern Physics, Chinese Academy of Sciences, 730000 Lanzhou, China

J.G. Wang
Institute of Modern Physics, Chinese Academy of Sciences, 730000 Lanzhou, China

Z.G. Gan
Institute of Modern Physics, Chinese Academy of Sciences, 730000 Lanzhou, China

L. Ma
Institute of Modern Physics, Chinese Academy of Sciences, 730000 Lanzhou, China
University of Chinese Academy of Sciences, 100049 Beijing, China
School of Nuclear Science and Technology, Lanzhou University, 730000 Lanzhou, China

L. Yu
Institute of Modern Physics, Chinese Academy of Sciences, 730000 Lanzhou, China
University of Chinese Academy of Sciences, 100049 Beijing, China

J. Jiang
Institute of Modern Physics, Chinese Academy of Sciences, 730000 Lanzhou, China
University of Chinese Academy of Sciences, 100049 Beijing, China

Y.L. Tian
Institute of Modern Physics, Chinese Academy of Sciences, 730000 Lanzhou, China

B. Ding
Institute of Modern Physics, Chinese Academy of Sciences, 730000 Lanzhou, China

S. Guo
Institute of Modern Physics, Chinese Academy of Sciences, 730000 Lanzhou, China

Y.S. Wang
Institute of Modern Physics, Chinese Academy of Sciences, 730000 Lanzhou, China

T.H. Huang
Institute of Modern Physics, Chinese Academy of Sciences, 730000 Lanzhou, China

M.D. Sun
Institute of Modern Physics, Chinese Academy of Sciences, 730000 Lanzhou, China
University of Chinese Academy of Sciences, 100049 Beijing, China
School of Nuclear Science and Technology, Lanzhou University, 730000 Lanzhou, China

K.L. Wang
Institute of Modern Physics, Chinese Academy of Sciences, 730000 Lanzhou, China
University of Chinese Academy of Sciences, 100049 Beijing, China

S.G. Zhou
Institute of Theoretical Physics, Chinese Academy of Sciences, 102413 Beijing, China

Z.Z. Ren
Department of Physics, Nanjing University, 210093 Nanjing, China

X.H. Zhou
Institute of Modern Physics, Chinese Academy of Sciences, 730000 Lanzhou, China

H.S. Xu
Institute of Modern Physics, Chinese Academy of Sciences, 730000 Lanzhou, China

G.Q. Xiao
Institute of Modern Physics, Chinese Academy of Sciences, 730000 Lanzhou, China

I. Silisteanu
Horia Hulubei National Institute of Physics and Nuclear Engineering, Bucharest-Magurele, RO-077125, Romania

C. I. Anghel
Horia Hulubei National Institute of Physics and Nuclear Engineering, Bucharest-Magurele, RO-077125, Romania
University of Bucharest, Faculty of Physics, Bucharest-Magurele, RO-077125, Romania

Magali Estienne
Subatech (CNRS/IN2P3), IMT Atlantique, Université de Nantes, 4 Rue Alfred Kastler, 44307 Nantes, France

Muriel Fallot
Subatech (CNRS/IN2P3), IMT Atlantique, Université de Nantes, 4 Rue Alfred Kastler, 44307 Nantes, France

Lydie Giot
Subatech (CNRS/IN2P3), IMT Atlantique, Université de Nantes, 4 Rue Alfred Kastler, 44307 Nantes, France

Loïc Le Meur
Subatech (CNRS/IN2P3), IMT Atlantique, Université de Nantes, 4 Rue Alfred Kastler, 44307 Nantes, France

Amanda Porta
Subatech (CNRS/IN2P3), IMT Atlantique, Université de Nantes, 4 Rue Alfred Kastler, 44307 Nantes, France

N.A. Fedorov
Joint Institute for Nuclear Research, Dubna, Russia
Lomonosov Moscow State University, Moscow, Russia

D.N. Grozdanov
Joint Institute for Nuclear Research, Dubna, Russia
Institute for Nuclear Research and Nuclear Energy of the Bulgarian Academy of Sciences, Sofia, Bulgaria

V.M. Bystritskiy
Joint Institute for Nuclear Research, Dubna, Russia

Yu.N. Kopach
Joint Institute for Nuclear Research, Dubna, Russia

I.N. Ruskov
Joint Institute for Nuclear Research, Dubna, Russia
Institute for Nuclear Research and Nuclear Energy of the Bulgarian Academy of Sciences, Sofia, Bulgaria

V.R. Skoy
Joint Institute for Nuclear Research, Dubna, Russia

T.Yu. Tretyakova
Skobeltsyn Institute of Nuclear Physics, Lomonosov Moscow State University, Moscow, Russia

N.I. Zamyatin
Joint Institute for Nuclear Research, Dubna, Russia

D. Wang
Joint Institute for Nuclear Research, Dubna, Russia
Xi'an Jiao Tong University, Xi'an, China

F.A. Aliev
Joint Institute for Nuclear Research, Dubna, Russia
Institute of Geology and Geophysics of Azerbaijan National Academy of Sciences, Baku, Azerbaijan

C. Hramco
Joint Institute for Nuclear Research, Dubna, Russia
Institute of Chemistry of the Academy of Sciences of Moldova, Chisinau, Republic of Moldova

A. Gandhi
Banaras Hindu University, Varanasi, India

A. Kumar
Banaras Hindu University, Varanasi, India

S. Dabylova
Joint Institute for Nuclear Research, Dubna, Russia

E.P. Bogolubov
All-Russia Research Institute of Automatics (VNIIA), Moscow, Russia

Yu.N. Barmakov
All-Russia Research Institute of Automatics (VNIIA), Moscow, Russia

F. Tárkányi
Institute for Nuclear Research, Hungarian Academy of Sciences (ATOMKI), Debrecen, Hungary

F. Ditrói
Institute for Nuclear Research, Hungarian Academy of Sciences (ATOMKI), Debrecen, Hungary

S. Takács
Institute for Nuclear Research, Hungarian Academy of Sciences (ATOMKI), Debrecen, Hungary

A. Hermanne
Cyclotron Laboratory, Vrije Universiteit Brussel (VUB), Brussels, Belgium

A. V. Ignatyuk
Institute of Physics and Power Engineering (IPPE), Obninsk, Russia 249020

Katharina Breunig
Forschungszentrum Julich, Institute of Neuroscience and Medicine, INM-5: Nuclear Chemistry, 52428 J ¨ ulich, Germany ¨

Bernhard Scholten
Forschungszentrum Julich, Institute of Neuroscience and Medicine, INM-5: Nuclear Chemistry, 52428 J ¨ ulich, Germany ¨

Ingo Spahn
Forschungszentrum Julich, Institute of Neuroscience and Medicine, INM-5: Nuclear Chemistry, 52428 J ¨ ulich, Germany ¨

Alex Hermanne
Vrije Universiteit Brussel (VUB), Cyclotron Laboratory, 1050 Brussels, Belgium

Stefan Spellerberg
Forschungszentrum Julich, Institute of Neuroscience and Medicine, INM-5: Nuclear Chemistry, 52428 J ¨ ulich, Germany ¨

Heinz H. Coenen
Forschungszentrum Julich, Institute of Neuroscience and Medicine, INM-5: Nuclear Chemistry, 52428 J ¨ ulich, Germany ¨

Bernd Neumaier
Forschungszentrum Julich, Institute of Neuroscience and Medicine, INM-5: Nuclear Chemistry, 52428 J ¨ ulich, Germany ¨

Donatella Puglisi
Department of Electronics, Information and Bioengineering, Politecnico di Milano, Campus Como, 22100 Como, Italy
Department of Physics, Chemistry and Biology, Sensor and Actuator Systems, Linköping University, 58183 Linköping, Sweden

Giuseppe Bertuccio
Department of Electronics, Information and Bioengineering, Politecnico di Milano, Campus Como, 22100 Como, Italy
Italian National Institute of Nuclear Physics (INFN), Section Milano, 20133 Milan, Italy

F.M. Brunbauer
CERN, 385 Route de Meyrin 1217 Meyrin, Geneva, Switzerland
European Spallation Source (ESS AB), SE-22100 Lund, Sweden

M. Lupberger
University of Bonn, Regina-Pacis-Weg 3, 53113 Bonn, Germany

H. Müller
CERN, 385 Route de Meyrin 1217 Meyrin, Geneva, Switzerland

E. Oliveri
CERN, 385 Route de Meyrin 1217 Meyrin, Geneva, Switzerland

D. Pfeiffer
European Spallation Source (ESS AB), SE-22100 Lund, Sweden
CERN, 385 Route de Meyrin 1217 Meyrin, Geneva, Switzerland

L. Ropelewski
CERN, 385 Route de Meyrin 1217 Meyrin, Geneva, Switzerland

L. Scharenberg
CERN, 385 Route de Meyrin 1217 Meyrin, Geneva, Switzerland
University of Bonn, Regina-Pacis-Weg 3, 53113 Bonn, Germany

P. Thuiner
CERN, 385 Route de Meyrin 1217 Meyrin, Geneva, Switzerland

M. van Stenis
CERN, 385 Route de Meyrin 1217 Meyrin, Geneva, Switzerland

Pavel P. Povinec
Department of Nuclear Physics and Biophysics, Faculty of Mathematics, Physics and Informatics, Comenius University, Mlynska dolina F1, 84248 Bratislava, Slovakia

Artem Lebedev
UCLA Department of Molecular and Medical Pharmacology, David Geffen School of Medicine, Los Angeles, California, United States of America
Traceability, Culver City, California, United States of America

Jing Jiao
UCLA Department of Molecular and Medical Pharmacology, David Geffen School of Medicine, Los Angeles, California, United States of America

Jason Lee
UCLA Department of Molecular and Medical Pharmacology, David Geffen School of Medicine, Los Angeles, California, United States of America
UCLA Crump Institute for Molecular Imaging, Los Angeles, California, United States of America

Fan Yang
The Department of Chemistry at The Scripps Research Institute, La Jolla, California, United States of America

Nathanael Allison
UCLA Biomedical Physics Interdepartmental Program, Los Angeles, California, United States of America

Harvey Herschman
UCLA Department of Molecular and Medical Pharmacology, David Geffen School of Medicine, Los Angeles, California, United States of America
UCLA Crump Institute for Molecular Imaging, Los Angeles, California, United States of America
UCLA Ahmanson Translational Imagining Division, Los Angeles, California, United States of America

Saman Sadegh
UCLA Department of Molecular and Medical Pharmacology, David Geffen School of Medicine, Los Angeles, California, United States of America
UCLA Crump Institute for Molecular Imaging, Los Angeles, California, United States of America
UCLA Biomedical Physics Interdepartmental Program, Los Angeles, California, United States of America
UCLA Ahmanson Translational Imagining Division, Los Angeles, California, United States of America

LIST OF ABBREVIATIONS

AMS	Accelerator Mass Spectrometry
AOs	atomic orbitals
CENTA	Centre for Nuclear and Accelerator Technologies
CN	compound nucleus
DS	delta-shells
DFT	density functional theory
ECP	Effective core potentials
EFT	Effective Field Theory
EA	electron affinities
FLNR	Flerov Laboratory for Nuclear Reactions
FWHM	full width at half maximum
FFF	Fused Filament Fabrication
GaAs	gallium arsenide
GEMs	Gaseous Electro n Multipliers
HFIR	High Flux Isotope Reactor
HU	Hounsfield unit
ICPMS	Inductively Coupled Plasma Mass Spectrometry
ICVF	intracellular volume fraction
IBA	ion beam analysis
IBM	ion beam modification
IP	ionization potentials
LBNL	Lawrence Berkeley National Laboratory
LEBIT	Low Energy Beam and Ion Trap
MRI	magnetic resonance imaging
MBPT	manybody perturbation theory
MPGDs	MicroPattern Gaseous Detectors
MP	model potentials
NSCL	National Superconducting Cyclotron Laboratory

NRA	nuclear reaction analysis
PWA	partial wave analysis
PET	Positron emission tomography
PCB	Printed Circuit Board
PP	pseudo potentials
QED	Quantum electrodynamic
RFQ	radio-frequency quadrupole
SRS	Scalable Readout System
SR	scalar relativistic
SAS	semiconductor alpha-spectrometry
SMFA	shell model formation amplitude
SUV	standardized uptake value
SHE	superheavy elements
THGEM	Thick Gaseous Electron Multiplier

PREFACE

Nuclear chemistry concerns the study of the properties of the chemical elements at varying the structure of the atomic nucleus. In the last decades, it has become very interdisciplinary because its applications range from the study of the formation of the elements in the universe to the design of radioactive drugs for diagnostic medicine: radiotracers.

Section 1 of *The State of Art of Nuclear Chemistry: Theoretical and Practical Aspects* book focuses on the nuclear structure and proprieties of superheavy elements (SHE). In particular, it discusses of techniques of chemical separation, of the search for elements characterized by atomic number greater than 118, of the importance of relativistic and quantum electrodynamic effects for the study of SHE.

Section 2 treats the radiative decay as a second-order transformation process, discussing its consequences on radiometric dating, and lastly, it focuses on the use of radiotracers in PET and theoretical tumor modeling.

Section 3 concentrates on the nuclear forces in the medium, investigating the nuclear equation of state and highlighting some applications in neutron-rich nuclei and neutron stars. Moreover, it discusses the validity conditions of the Effective Field Theory based on chiral expansions.

Section 4 takes into consideration the three kinds of decay, alpha, beta and gamma, focusing on their sources and their measurement. Moreover, it presents the determination of the gamma emission probabilities of a particular isotope.

Section 5 focuses on the nuclear reactions of fission. In particular, it presents the experimental and theoretical study of deuteron induced nuclear reactions on platinum and the production of medically useful bromine isotopes via alpha-particle induced nuclear reactions.

Section 6 treats two kinds of radiation detectors, a semiconductor microstrip radiation detector in silicon carbide, and the 3D printed gaseous radiation detectors.

Finally, the last Section 7 presents recent works about radiochemical techniques. In detail, it describes the new developments in ultra-sensitive radioanalytical technologies for environmental, life and space sciences; and the synthesis of an in vivo stable radiotracer that is radiolabeled directly on a heteroaromatic ring.

SECTION 1:
NUCLEAR STRUCTURE AND PROPRIETIES OF SUPERHEAVY ELEMENTS

1

Chemistry Aided Nuclear Physics Studies

Julia Even

KVI – CART, University of Groningen, 9747 AA Groningen, The Netherlands

ABSTRACT

Studies of the superheavy elements bring several challenges through low production yields, short half-lives, and high background rates. This paper describes the possibilities of chemical separations as techniques to overcome the background problematic and to investigate the nuclear properties of the heaviest nuclides.

INTRODUCTION

The transuranium elements up to nobelium (No, Z = 102) have been discovered applying radiochemical separation methods (see, for example, reference [1]). The discoveries of the heaviest actinoide element lawrencium (Lr, Z = 103) and the first transactinoide rutherfordium (Rf, Z = 104) have been achieved through the application of nuclear physics techniques and

Citation: Julia Even. "Chemistry aided nuclear physics studies" EPJ Web Conf., 131 (2016) 07008 DOI: https://doi.org/10.1051/epjconf/201613107008.

followed shortly after by radiochemical separations. With increasing proton number, the production cross sections and the half-lives decrease. Thus, from the discovery of dubnium (Db, Z = 105) on, physics techniques were the driving forces for the discovery of new elements. In the early eighties, the area of superheavy elements at electromagnetic separators started with the discovery of bohrium (Bh, Z = 107), hassium (Hs, Z = 108), and meitnerium (Mt, Z = 109) at SHIP at GSI, Darmstadt (see contribution of S. Hofmann in this proceedings, and reference [2]). Although physics techniques dominate the research on heavy elements in our days, chemists are still actively involved in this field. Besides studies of the chemical properties, chemical separations play a role in nuclear physics studies. In this paper, examples will be given how the combination of chemistry and physics can overcome recent challenges.

IDENTIFICATION OF THE PROTON NUMBER

The elements flerovium (Fl, Z = 114) up to element Z = 118 have been so far only synthesized in ^{48}Ca-induced fusion reactions with actinoide targets [3]. However, an unambiguous identification of the proton number of these elements could still not be achieved. All these elements decay into previously unknown spontaneous fission isotopes. Thus, there is no direct link to the well-known part of the nuclear chart. In case of 288115, the α-decay chains end in a long-lived nuclide which decays through spontaneous fission. This long-lived fission nucleus was assigned to ^{268}Db or after decay through electron capture to ^{268}Rf. One option to identify the proton number of ^{288}115 is the chemical isolation and identification of the daughter nuclide ^{268}Db or ^{268}Rf [4, 5]. For this approach, ^{288}115 was produced in the reaction ^{243}Am(48Ca,3n) and implanted in a copper block. After irradiation, the upper copper layer was dissolved, elements of group four and group five were separated through ion-exchange chromatography. Samples of the chromatography fractions were prepared for spontaneous fission and α spectroscopy. Each sample was measured for 960 hours. In the group four and group five fraction, several fission events were observed, which were assigned to the decay of long-lived Rf or Db isotopes [4, 5]. These results were yet not included in the assignment of element 115 by the joint working party of IUPAP and IUPAC [6]. Their report notes that a new chemical separation scheme for Db and Rf was applied. In the past, Db and Rf chemistry had already been studied with the isotopes ^{262,263}Db and ^{257,261}Rf, which were directly produced in fusion-evaporation reactions studies strongly relies on the support from theoretical predictions and studies of the lighter homologue elements. The structure of the chemical compounds cannot unambiguously be determined. Thus, the combination of gas-chemistry with more sophisticated methods such as mass spectrometry would be beneficial. The approach of chemical reactions in a gas catcher combined with mass measurements appears feasible. A good system to start with are the transition metal carbonyl complexes. In mass measurements of Fe and Co at LEBIT, it was observed that Fe and Co formed mono-, bi-, and even tri-carbonyl complexes with small impurities of carbon monoxide in the buffer gas [24]. Controlling the amount of carbon monoxide in the buffer gas will open the door for kinetic studies of the carbonyl complex formation of even the transition metals of the 7th period of the periodic table.

CONCLUSION AND OUTLOOK

Although the period of chemically driven element discovery passed, nuclear physics experiments can still benefit from chemical techniques. Chemical studies can be used to identify the proton number of an isotope. Furthermore, fast gas-phase transport of volatile species can be applied for background free decay spectroscopy. Group 8 tetroxide and the volatile transition metal carbonyl complexes are powerful chemical separators. Currently, new detectors are under development to improve the detection efficiency and allow α, β, and γ coincidence measurements aided by gas-phase chemistry [26]. This new detector type will have a sandwich structure. The inner chromatography channel consists of an α-detector array followed by layers of detectors sensitive for electrons and photons.

However, also superheavy element chemistry can benefit from developments in experimental nuclear physics. Ion-chemistry combined with single molecule mass spectrometry will open the door to unambiguously identify the structure of chemical compounds of the heaviest elements. It will open the door for studies of chemical reaction kinetics. A wise choice of combination of the different experimental methods will lead to many interesting discoveries in the future.

The author acknowledges support through the Rosalind Franklin fellowship program of the University Groningen co-funded by the EU.

REFERENCES

1. Y. Nagame and M. Hirata, Radiochim. Acta 99, 377 (2011)
2. G. Münzenberg and K. Morita, Nucl. Phys. A944, 3 (2015)
3. Y.T. Oganessian and V.K. Utyonkov, Nucl. Phys. A944, 62 (2015)
4. D. Schumann *et al.*, Radiochim. Acta 93, 727 (2005)
5. N.J. Stoyer *et al.*, Nucl. Phys. A787, 388 (2007)
6. P.J. Karol *et al.*, Pure and Appl. Chem. 88, 139 (2016)
7. J.V. Kratz, Radiochim. Acta 99, 477 (2011)
8. M. Schädel, Phil. Trans. R. Soc. A 373, 20140191 (2015)
9. C.E. Düllmann *et al.*, Nature 418, 859 (2002)
10. J. Dvorak *et al.*, Phys. Rev. Lett. 97, 242501 (2006)
11. J. Dvorak *et al.*, Phys. Rev. Lett. 100, 132503 (2008)
12. J. Dvorak *et al.*, Phys. Rev. C 79, 037602 (2009)
13. R. Graeger *et al.*, Phys. Rev. C 81, 061601 (2010)
14. J. Even *et al.*, Inorg. Chem. 51, 6431 (2012)
15. J. Even *et al.*, Science 345, 1491 (2014)
16. J. Even *et al.*, Radiochim. Acta 102, 1093 (2014)
17. J. Even *et al.*, J. Radioanal. Nucl. Chem. 303, 2457 (2015)
18. I. Usoltsev *et al.*, Radiochim. Acta 104, 141 (2016)
19. H. Haba *et al.*, Phys. Rev. C 83, 034602 (2011)
20. H. Haba *et al.*, Phys. Rev. C 85, 0246411 (2012)
21. M. Block *et al.*, Eur. Phys. J. A 50, 49 (2005) [EDP Sciences]
22. P. Schury *et al.*, Nucl. Instr. Meth. B317, 537 (2013)
23. J.M. Gates, these proceedings
24. R. Ferrer *et al.*, Phys. Rev. C 81, 044318 (2010)
25. A.A. Kwiatkowski *et al.*, Phys. Rev. C 80, 051302 (2009)
26. A. Di Nitto *et al.*, in *GSI Scientific Report 2014*, GSI Report 2015-1, p. 184

2

On The Search for Elements Beyond Z =118. An Outlook Based on Lessons from the Heaviest Known Elements

Christoph E. Düllmann[1,2,3]

[1] Institute of Nuclear Chemistry, Johannes Gutenberg University, 55128 Mainz, Germany
[2] GSI Helmholtzzentrum für Schwerionenforschung, 64291 Darmstadt, Germany
[3] Helmholtz Institute Mainz, 55099 Mainz, Germany

ABSTRACT

Recently, IUPAC approved all elements up to Z = 118 as discovered. Search experiments for the heavier elements with Z = 119 and 120 have been performed in recent years, but have so far not led to their discovery. I will review some aspects associated with the study and identification of the heaviest known elements that are relevant for future search experiments for elements beyond Z = 118 and highlight pressing issues that should be addressed, both on the experimental as well as on the theory side, to allow for performing these future experiments under improved and better informed conditions.

Citation: Christoph E. Düllmann. "On the search for elements beyond Z =118. An outlook based on lessons from the heaviest known elements" EPJ Web Conf., 131 (2016) 08004 DOI: https://doi.org/10.1051/epjconf/201613108004

INTRODUCTION

At the time of the Nobel Symposium NS160, held in beautiful Bäckaskog Castle, Sweden, in May/June 2016, all elements up to Z = 118 have been officially accepted by IUPAC as discovered [1, 2]. One week later, the recommended names for the four most recent elements were announced. Figure 1 shows the corresponding cut-out of the chart of nuclei (with decay properties updated according to [3]), where all known or claimed nuclei are indicated in coloured boxes, overlaying the calculated shell-correction energy landscape according to a macroscopic-microscopic model [4]. Also indicated (by white boxes for yet unknown isotopes) are the hypothetical decay chains originating from the 3n and 4n compound nucleus (CN) evaporation channels following complete fusion of $^{50}Ti + ^{249}Bk$ leading to $^{299}119*$ (red bordered boxes), $^{50}Ti + ^{249}Cf$ leading to $^{299}120*$ (light blue bordered boxes), and $^{54}Cr + ^{248}Cm$, $^{58}Fe + ^{244}Pu$, and $^{64}Ni + ^{238}U$, which all lead to $^{302}120*$ (dark blue bordered boxes). Decay modes of unknown isotopes were assumed based on [5–7].

On the first day of this Nobel Symposium, the introductory speaker for the physics of the superheavy elements, Matti Leino, listed among his conclusions that it is "difficult to increase Z, and N. How about deepening our understanding of already synthesized species?" I follow this recipe and devote Sect. 2 to lessons from known elements that bear relevance to the search for new elements. Section 3 will be devoted to an outlook, going beyond Z = 118, picking up on the reactions mentioned above.

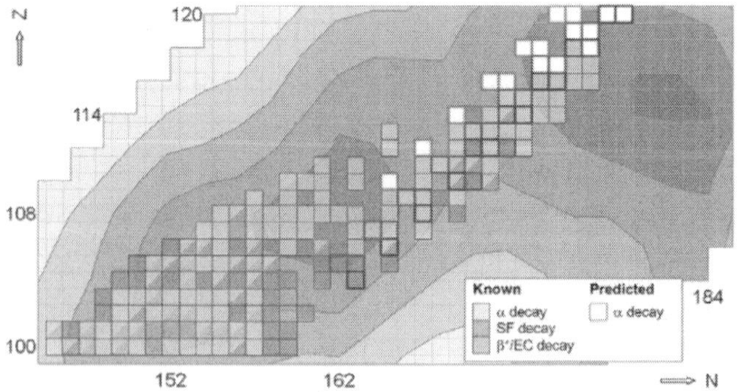

Figure 1: Chart of nuclei of the heaviest elements. The underlying contour plot shows calculated shellcorrection energies [4]. Known nuclei are indicated by coloured boxes, new nuclei occurring in the hypothetical 3n and 4n decay chains produced in complete fusion of different target and projectile combina-

tions leading to Z = 119 and Z = 120 are indicated by white boxes. See text for more details.

LESSONS FROM KNOWN ELEMENTS

For this, I will look at the case of the element 115. In experiments performed at the Flerov Laboratory for Nuclear Reactions (FLNR) in Dubna, Russia [8, 9], at the GSI Helmholtzzentrum in Darmstadt, Germany [3, 10], and at the Lawrence Berkeley National Laboratory (LBNL) in Berkeley, CA, USA [11], over 100 decay chains have been observed following the irradiation of ^{243}Am targets with ^{48}Ca beams and subsequent separation of evaporation residues in gas-filled recoil separators. All were assigned to originate from element 115, produced in complete fusion, neutron-evaporation reactions. Indirect arguments on the reaction mechanism and nuclear structure of the produced nuclei were invoked to support this assignment [8, 9], along with the IUPAC-preferred tool of "crossbombardments", in which a certain nucleus is produced in multiple pathways; for example, in the case of element 115, the claimed production of ^{289}115 via ^{243}Am(^{48}Ca,2n) and via ^{249}Bk(^{48}Ca,4n) $^{293}117 \rightarrow \alpha$ decay [1]. None of the experiments performed to date were able to directly identify the Z or the A of any nucleus ascribed to belong to any element 115 decay chain (or, in a broader perspective, of any nucleus produced in a ^{48}Ca + actinide target reaction). While the current assignment appears correct based on indirect arguments, K.E. Gregorich argues in his contribution to these proceedings that the assignment appears "good, but not good enough" [12]. Still, IUPAC approved these data as originating from element 115 [1], and especially highlighted the cross-bombardment aspect in its assessment, which formed the basis for their approval. The IUPAC-approved cross bombardment case of elements 115 and 117, originally invoked by the FLNR-led collaborations has since been challenged [13], and the identification of scientifically sound connections of element 117 and element 115 chains is ongoing [14]; even if found, they will be unable to prove the Z of the originating element, though. The safe, proven assignment of the α decay of the first member of any decay chain to element 115 is only possible if the nuclear reaction can be shown to lead to an isotope of element 115 (i.e., if complete fusion is achieved, followed by the evaporation exclusively of neutrons), and if its decay is uniquely detected.

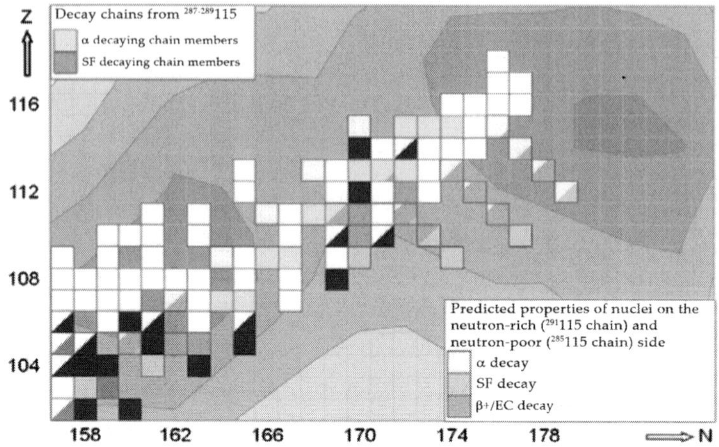

Figure 2: Black-outlined coloured boxes: decay chains populated in ^{243}Am(^{48}Ca,2-4n) reactions, as given in "scenario 2" in [3]. White (black) boxes: other known α -decaying (SF-decaying) nuclei. Redbordered coloured boxes: predicted decay properties [16] of element 115 decay chains located on the neutron-rich (^{291}115) and neutron-deficient (^{285}115) side of those from the ^{243}Am(^{48}Ca,2-4n) reactions. The endpoints of the ^{285}115 chain (^{261}Lr and ^{265}Rf) are known and decay by SF.

I will focus here on the second aspect. For this, it is illustrative to review predictions on partial half-lives of different decay modes in the relevant region of the nuclear chart. The growing importance of β^{+}/electron capture (EC) decay, especially towards the neutron-rich region in the isotopic chains of element 113 and 115, has been noted, e.g., in [15, 16], see Fig. 2.

In none of the presently reported decay chains assigned to proceed through element 115 are there members that were originally (neither by IUPAC) assigned to possess a β^{+}/EC branching. To fully appreciate the relevance of this, one should bear in mind that the detection techniques used for the study of the ^{48}Ca +243Am reaction at the recoil separators in Dubna, Darmstadt, and Berkeley, are sensitive to α as well as to spontaneous fission (SF) decay; however, the direct detection of EC decay is not possible. The presence of an EC-decaying member in the chain could only be inferred indirectly by comparing decay properties of later chain members with known data – which is only possible if decay properties of (i) nuclei populated if EC decays are present, as well as of (ii) those populated in the absence of

EC decay, are known. None of the nuclei populated by the decay chains obtained in the ^{48}Ca +^{243}Am reaction have been known before, though.

Chemical approaches able to separate individual elements have the power to achieve Z identification. For this, chemical properties of the element need to be established and exploited. The heaviest element, which had its chemical properties reproducibly studied, is Cn (Z = 112) using the isotope ^{283}Cn produced as α -decay daughter in the reaction $^{242}Pu(^{48}Ca,3n)^{287}Fl$, see review in [17]. Z identification is still not fully direct, but relies on chemical similarities with the lighter members in group 12 of the Periodic Table [18]. For element 113, first attempts for its chemical study have been performed, and for Fl, conflicting results on its chemical behaviour are reported [19, 20], see discussion in [17]. The power of chemical separators can only be fully employed, if the chemical properties of the element to be studied are already established. Once this is achieved, chemical studies of isotopes with unknown Z will allow verifying its atomic number.

Coming back to the decay chains allegedly originating from element 115, more specifically to the 96 five-α long chains from $^{288}115$, we note that there are arguments supporting that no chain member is present which decays (exclusively or in large fractions) by an intermediate β^+/EC decay, which would render the chain to continue through eveneven isotopes: first, statistical analyses of lifetime distributions of each individual generation within these chains showed these distributions to be consistent with what would be expected if every specific decay step occurred from the same nuclear state in all chains. Significant abundance of branched decay in any member is thus unlikely and would only go undetected by this method if a partially β^+/EC-decaying member with a lifetime that is very long or very short compared to that of its daughter were present. Only in these two extreme cases would the distribution of apparent lifetimes (which would then sometimes be the sum of lifetimes of β^+/EC decay and subsequent α decay) at this position in the decay chain fit with what is expected for a decay of a single nuclear state. Secondly, β^+/EC decay in this chain would populate isotopes known or expected to be fast fissioning even-even nuclei. Therefore, experimental evidence suggests that no decay branches other than α (and terminating SF, potentially preceded by β^+/EC decay) are present in these long chains.

For $^{289}115$, data is scarcer, precluding as definite statements. Interestingly, isotopes of elements 115, 113, and 111 that are only slightly more neutron-rich than those produced in the ^{48}Ca+^{243}Am reaction are predicted to possess sizable β^+/EC branches. As follows from Fig. 4 of [15], this decay mode is expected also in lighter isotopes of these elements, also in those populated in the decay chains of ^{288},^{289}115. On the neutron-deficient side, taking the ^{285}115 chain as an example, β^+/EC branches are furthermore predicted for the element 107 and 105 members. The α and +/EC daughters of the element 105 member, i.e., ^{261}Lr and ^{265}Rf, are known to decay by SF, in agreement with the calculations. The question thus arises how safe a Z assignments can be if β^+/EC decay as a prominent possibility is neglected (or, more accurately: not detectable). The problem is illustrated in Fig. 3. Due to the multiple decay modes with significant branches as they are predicted to occur in several of the populated nuclei (assuming decay properties as predicted in [16]) each decay chain from ^{291}115 will pass through one out of in total 19 different pathways. In Fig. 3, just five out of these pathways are highlighted. All include two α decays and different numbers of β^+/ ECdecaying members at different positions in the chains, and are terminated by SF. Table 1 presents these five pathways in more detail.

As can be seen, all five pathways lead to the detection of α - α -SF decay chains in the absence of sensitivity to β^+/EC decays. None of the three detected nuclei can be fixed in A or Z. Focusing on the Z of the first detected decaying nucleus, we see that this will be 115 in three of the five pathways, 114 in another one, and 113 in the last discussed one!

A safe assignment of the first α particle to element 115 is thus impossible. It thus appears crucial to expand current techniques to be sensitive also to -decay processes. Detection of this decay mode and its detailed study are routinely performed in other scientific communities. As just one example, neutrino mass determination via a "missing-mass" type of approach, using EC decaying isotopes – e.g., ^{163}Ho, which is used due to its very low Q_{EC} value [21] – is mentioned, with the "Electron Capture in Holmium-163" (ECHo) collaboration [22, 23] being one of the groups working in this field. Metallic magnetic microcalorimeters [24] are used. In an exploratory study, similar detectors have been applied for studies of fusion-evaporation reaction products behind the SHIP separator at GSI [25], but the technique has not yet reached maturity to be directly applicable in superheavy element research. Approaches based on more traditional techniques may appear better suited to achieve direct EC observation [26], including delayed X-ray coincidence observation, which is possible if the detection system is operated in a low-background environment as present, e.g., at the SHIP [27] or the planned BGS-FIONA setup [28].

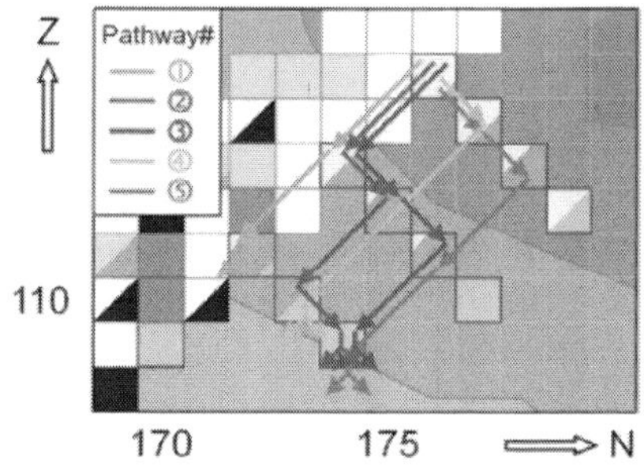

Figure 3: Cut-out from Fig. 2, indicating predicted decay modes and their relative importance for $^{291}115$ and its daughters. Among all possible 19 decay pathways, five are selected; see text for more details.

Table 1: Five decay pathways given in Fig. 3, originating from ²91115

#	Decay pathway	Apparent chain[1]	α_1	α_2	SF
1	α–α–SF	α–α–SF	^{291}**115**	^{287}**113**	^{283}**111**
2	α–β^+/EC–α–β^+/EC–SF	α–α–SF	^{291}**115**	^{287}**112**	^{283}**109**
3	α–β^+/EC–β^+/EC–α–SF	α–α–SF	^{291}**115**	^{287}**111**	^{283}**109**
4	β^+/EC–α–α–SF	α–α–SF	^{291}**114**	^{287}**112**	^{283}**110**
5	β^+/EC–β^+/EC–α–α–SF	α–α–SF	^{291}**113**	^{287}**111**	^{283}**109**

[1] In case of non-detection of β^+/EC decay.

Having discussed Z assignment and the problem of detecting β^+/EC-decaying isotopes, it is worthwhile to discuss the assignment of A to observed decay chains. This is discussed in [3, 13, 14] for chains produced in ^{48}Ca +^{243}Am reactions. Important aspects include that nuclei having significant branches for different decay modes are prevalent also among the heaviest known elements. Therefore, the length of a decay chain (i.e., the number of chain members) cannot be used to judge which set of chains originates from the same nucleus. Not even elaborate statistical analyses as presented in [3] are sufficient to judge whether the A of one type of chain is identical or different to that of a distinctly different set of chains, not least because the occurrence of isomeric states, which decay, e.g., by α decay to analog states in the daughter rather than to their own ground state, is expected according to theoretical calculations (see Sect. 6 in [3]).

To conclude, despite all progress made in recent years, culminating in the IUPACapproved discovery and naming of all elements up to Z = 118, the direct identification of A or Zof any nucleus produced in ^{48}Ca + actinide reactions has not been achieved yet. Detection to date relies exclusively on the observation of α and SF decay. β^+/EC decay, expected to be present also in the region that is already being explored, cannot be detected experimentally.

Table 2: Fusion reactions employed to search for elements beyond Z = 118

	Compound nucleus	Beamtime invested	Cross section limit[1]	Ref.
$^{50}Ti + ^{249}Bk$	$^{299}119^*$	4.5 months	2	[30]
$^{64}Ni + ^{238}U$	$^{302}120^*$	4 months	≥ 90 fb	[31]
$^{58}Fe + ^{244}Pu$	$^{302}120^*$	2 months	≥ 400 fb	[32]
$^{54}Cr + ^{248}Cm$	$^{302}120^*$	> 1 month	≥ 580 fb	[33]
$^{50}Ti + ^{249}Cf$	$^{299}120^*$	> 1 month	2	[30]

[1] 63.2% confidence level ("one-event limit").

[2] Under final analysis.

THE WAY TO ELEMENTS BEYOND Z =118

Recent work suggests fusion reactions with actinide targets to provide the highest cross sections for the synthesis of elements beyond Z = 118. To proceed on the successfully established ^{48}Ca beam route, targets beyond $_{98}Cf$ would be necessary. Current technology for the production of transuranium isotopes, e.g., at the ORNL's High Flux Isotope Reactor (HFIR) is capable of producing g and pg amounts of long-lived $_{99}Es$ and $_{100}Fm$ isotopes, respectively [29]. This is orders of magnitude less than the experiments discovering and studying the heaviest known elements require. Thus, reactions induced by beams heavier than ^{48}Ca will be needed. The most commonly suggested reactions giving access to Z ≥ 119 include ^{48}Ca $+^{249}Bk$ (leading to Z = 119) and $^{64}Ni + ^{238}U$, $^{58}Fe + ^{244}Pu$, $^{54}Cr + ^{248}Cm$, and ^{50}Ti $+^{249}Cf$ (all four leading to Z = 120). All five have been employed in search experiments, but none has led to the conclusive discovery of a new element. 2 summarizes the efforts invested to date to find elements beyond Z = 118.

As can be seen, more than one year of beamtime has been devoted to search experiments, which were carried out at GSI Darmstadt (^{50}Ti+^{249}Bk, ^{64}Ni+^{238}U, ^{54}Cr+^{248}Cm, and ^{50}Ti+^{249}Cf) and FLNR Dubna (^{58}Fe+^{244}Pu). Most likely, non-observation of new elements was due to insufficient sensitivity. It is noteworthy that some of the predictions of the nuclear half-lives of the produced isotopes, especially in the case of Z = 120, suggest these to be in the order of microseconds, i.e., comparable to the flight-time through the separator. Hence, losses due to decay-in-flight might have negatively affected these experiments. Preparations for new, improved experiments are under way at several laboratories. Thus, it appears appropriate to reflect on which aspects should get most attention to perform the best possible experiments, thus maximizing the likelihood of success. Obvious, purely technical aspects include higher primary beam intensities, targets that can withstand such intensities over extended periods [34, 35], more efficient separators, more sensitive detection systems that allow safe identification of single decay chains from new elements, and longer irradiation times. I will focus here on a different aspect: numerous theoretical calculations of maximum cross sections of the reactions presented in Table 2 have been published in recent years (see overview in [30] for the reactions leading to Z = 120), which span ranges of several orders for each reaction. This translates to unknown minimum irradiation times needed to guarantee the discovery of a new element with high confidence. In general, maximum cross sections scale with reaction asymmetry, but theoretical uncertainties are comparable to differences between different reactions in many cases. ^{54}Cr+^{248}Cm is favoured among the reactions leading to ^{302}120∗. The more asymmetric ^{50}Ti +^{249}Cf leads to ^{299}120∗. The influence of the three extra neutrons in the former is often assumed to be favourable, as the reaction product is situated closer to the next spherical shell closure expected at N = 184. The opposite conclusion can also be reached, e.g., considering the fission barrier landscape as depicted in Fig. 6 of [36], which shows fission barrier heights calculated in the macroscopicmicroscopic approach, which is very successful in describing experimental data in the region of the heaviest known elements. Attempts to extract fission barrier heights from experimental observables have been performed [37] but are inherently model-dependent. Thus, no safe guidance on the optimum reaction, based purely on experimental data, exists.

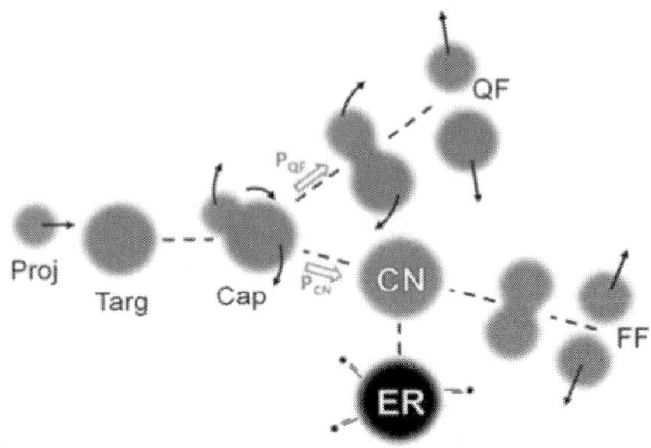

Figure 4: Schematic of the relevant processes in nucleus-nucleus collisions aimed at synthesizing the heaviest elements. Proj: projectile; Targ: target; Cap: captured system; CN: compound nucleus; ER: evaporation residue; QF: quasifission; FF: fusion-fission; P_{CN} : probability for the captured system to form a compound nucleus; P_{QF} : probability for the captured system to fission. Obviously, P_{QF} =1-P_{CN} .

Another aspect that is at least as critical as the proper choice of the nuclear reaction concerns the proper choice of the irradiation energy, as excitation functions are only a few MeV wide. Reliable experimental systematics exists for the ^{48}Ca + actinide reactions (e.g., Fig. 4 in [38]). However, the situation is dramatically different as soon as the reactions given in Table 2 are concerned. As detailed in [30], where the system ^{50}Ti +^{249}Cf is employed to illustrate the current situation, optimum beam energies suggested by different theoretical frameworks differ by more than 20 MeV! Recalling that typical energy ranges covered in a single experiment are roughly 5 MeV highlights the need for a well-informed choice of the irradiation energy. Performing at multiple beam energies searches that are significantly more sensitive than the results shown in Table 2 (which were all obtained at one single energy) would consume years of beamtime. In this situation, performing complementary experiments studying these fusion reactions, which yield information that guides future new element search efforts, appears well justified. For this, we recall that the formation of a superheavy nucleus is parametrized as the product of three individual steps, each characterized with its own probability: $\sigma_{\mathrm{ER}} = \sigma_{\mathrm{cap}} \cdot P_{CN} \cdot W_{sur}$ with σ_{ER} the evaporation residue (ER) cross section, σ_{cap} the capture cross section, P_{CN} the compound nucleus formation probability, and W_{sur} the exit channel survival probability. The formation of superheavy elements is significantly suppressed by the quasifission (QF) process, in which two colliding nuclei reseparate on a fast time-scale before coalescing into a compound nucleus. See [39] for a detailed account, and Fig. 4 for a schematic of the relevant processes.

QF is the complementary process to compound nucleus formation, thus $P_{QF} = 1 - P_{CN}$, where P_{QF} can reach large values. In-depth studies of superheavy element formation reactions relying on ER observation are hampered by the tiny σ_{ER}, which renders such studies very (beam)time consuming. The competing QF process profits from much higher rates for systems where P_{CN} is small. The understanding of QF processes has matured to a point where its detailed study, e.g., by recording mass angle distributions, provides significant insight. Thus, it appears timely to perform such studies of the reactions given in Table 2, and to gain an experimental data set of P_{CN} values, and of the heights of the Coulomb barrier in these reactions. Currently, theoretical frameworks only rarely predict observables of the QF process, but mostly focus on ER cross sections. Ideally, these frameworks will be extended to deliver predictions on observables of this "most likely outcome" of nucleus-nucleus collisions, i.e., QF (at least in cases where the underlying physics is contained in the framework). Such observables include mass angle distributions [40] or mass energy distributions [41]. Combined with the corresponding experimental data, which can be recorded in short experiments, this will allow testing different models, thus providing guidance concerning which ones deliver the most trustworthy predictions. Better informed selection of, e.g., the optimum irradiation energy will be invaluable. First steps in this direction have been made recently [39, 42]. A combined effort encompassing theoretical as well as experimental work will provide an improved basis for the certainly coming future efforts to discover new elements with atomic numbers Z > 118 and to thus open a new row in the Periodic Table of the Elements.

REFERENCES

1. P.J. Karol et al., Pure Appl. Chem. 88, 139 (2016).
2. P.J. Karol et al., Pure Appl. Chem. 88, 155 (2016).
3. U. Forsberg et al., Nucl. Phys. A953, 117 (2016).
4. Z. Patyk, R. Smolanczuk, A. Sobiczewski, Nucl. Phys. A626, 337c (1997).
5. I. Muntian, Z. Patyk, A. Sobiczewski, Phys. At. Nucl. 66, 1015 (2003).
6. I. Muntian et al., Acta Phys. Pol. B34, 2073 (2003).
7. R. Smolanczuk, J. Skalski, A. Sobiczewski, Phys. Rev. C ′ 52, 1871 (1995).
8. Yu.Ts. Oganessian et al., Phys. Rev. C 69, 021601 (2004).
9. Yu.Ts. Oganessian et al., Phys. Rev. C 87, 014302 (2013).
10. D. Rudolph et al., Phys. Rev. Lett. 111, 112502 (2013).
11. J.M. Gates et al., Phys. Rev. C 92, 021301(R) (2015).
12. K.E. Gregorich, EPJ Web Conf., these proceedings (2016).
13. U. Forsberg et al., Phys. Lett. B760, 293 (2016).
14. D. Rudolph et al., (in preparation) (2016).
15. A.V. Karpov et al., Int. J. Mod. Phys. E 21, 1250013 (2012).
16. V.I. Zagrebaev, A.V. Karpov, W. Greiner, Phys. Rev. C 85, 014608 (2012).
17. A. Türler, R. Eichler, A. Yakushev, Nucl. Phys. A944, 640 (2015).
18. R. Eichler et al., Angew. Chem. Intl. Ed. 47, 3262 (2008).
19. R. Eichler et al., Radiochim. Acta 98, 133 (2010).
20. A. Yakushev et al., Inorg. Chem. 53, 1624 (2014).
21. S. Eliseev et al., Phys. Rev. Lett. 115, 062501 (2015).
22. L. Gastaldo et al., J. Low Temp. Phys. 176, 876 (2014).
23. L. Gastaldo et al., Eur. Phys. J. Special Topics, (in preparation) (2016).
24. L. Gastaldo et al., Nucl. Instrum. Methods A 711, 150 (2013).
25. S. Kraft-Bermuth, PhD Thesis, Fachbereich Physik, JGU University Mainz, 2004.
26. F.P. Heßberger et al., Eur. Phys. J. A 52, 192 (2016).
27. F.P. Heßberger et al., Eur. Phys. J. A, (submitted) (2016).
28. J.M. Gates, EPJ Web Conf., these proceedings (2016).

29. J.B. Roberto et al., Nucl. Phys. A944, 99 (2015).
30. Ch.E. Düllmann et al., (in preparation) (2016).
31. S. Hofmann, J. Phys. G 42, 114001 (2015).
32. Yu.Ts. Oganessian et al., Phys. Rev. C 79, 024603 (2009).
33. S. Hofmann et al., Eur. Phys. J. A 52, 180 (2016).
34. S.N. Dmitriev and A.G. Popeko, J. Radioanal. Nucl. Chem. 305, 927 (2015).
35. E. Jäger et al., J. Radioanal. Nucl. Chem. 299, 1073 (2014).
36. M. Kowal, P. Jachimowicz, A. Sobiczewski, Phys. Rev. C 82, 014303 (2010).
37. S. Hofmann et al., Eur. Phys. J. A 52, 116 (2016).
38. Yu.Ts. Oganessian, and V.K. Utyonkov, Nucl. Phys. A944, 62 (2015).
39. D.J. Hinde et al., EPJ Web Conf., these proceedings (2016).
40. R. du Rietz et al., Phys. Rev. C 88, 054618 (2013).
41. M.G. Itkis et al., Nucl. Phys. A944, 204 (2015).
42. H.M. David et al., (in preparation) (2016).

3

Theoretical Chemistry of Superheavy Elements: Support for Experiment

Valeria Pershina

GSI Helmholtzzentrum für Schwerionenforschung GmbH, 64291 Darmstadt, Germany

ABSTRACT

Spectacular developments in the relativistic quantum theory and computational algorithms in the last few decades allowed for accurate calculations of properties of the superheavy elements (SHE) and their compounds. Often conducted in a close link to the experimental research, these investigations helped predict and interpret an outcome of sophisticated and expensive experiments with single atoms and to reveal the magnitude and importance of relativistic effects.

IMPORTANCE OF CHEMICAL THEORY FOR SHE

Conceptually, it is the atomic number Z and ground state electronic configuration of an element that define its position in the Periodic Table.

Citation: Valeria Pershina. "Theoretical chemistry of superheavy elements: Support for experiment" EPJ Web Conf., 131 (2016) 07002 DOI: https://doi.org/10.1051/epjconf/201613107002

However, information on chemical behaviour of a new element with respect to its homologs often serves as an additional proof of the proper assignment [1, 2]. This is particularly important for superheavy elements (SHE): It is meanwhile well established that relativistic effects on their electron shells are so strong that deviations from trends in chemical properties observed within lighter homologs in the groups can occur [3]. Important properties to assess similarity of SHE with the lighter congeners are atomic/ionic radii, redox potentials and complex formation.

The atomic number of many SHE can be determined by physical methods, while the electron configuration can currently be provided only by theory. Experimental studies of chemical properties of short-lived SHE isotopes are also limited to measurements of very few properties, mainly, volatility of atoms or molecules using gas-phase chromatography [4], or complex formation using liquid-phase separation techniques [5]. (See also the contribution of A. Türler in this volume). Knowledge of many other important properties, such as chemical composition, stability, geometrical configuration, ionization potentials (IP), or electron transitions, can only be provided by theory. Thus, in the area of SHE, theoretical studies become extremely important and are often the only source of useful chemical information [6–9]. They are also invaluable for predicting and/or interpreting the outcome of sophisticated and demanding experiments with single atoms. Moreover, since nature is relativistic, it is only theory that can demonstrate the importance and magnitude of relativistic effects by comparing the observed behaviour with that predicted on the basis of relativistic vs. non-relativistic calculations.

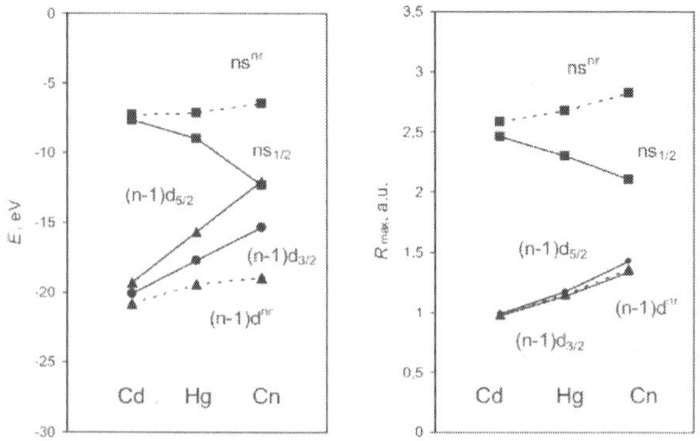

Figure 1: Relativistic (solid line) and nonrelativistic (dashed line) energy, E, and the maximum of the radial charge density, R_{max}, of the group-12 valence ns and (n-1)d AOs [6].

RELATIVISTIC AND QED EFFECTS FOR SHE

With increasing Z, the velocity of an electron increases. The Bohr model for a hydrogenlike species gives it as $v = (2\pi e^2/nh)Z$. As a result, the relativistic mass of an electron also increases

$$m = m_0\left[(1 - (v/c)^2\right]^{-1/2}, \tag{1}$$

where m_0 is the mass at zero velocity (rest mass) and c is the speed of light. From the sixth row onwards, the m/m0 ratio exceeds by 10%, so that relativistic effects cannot be neglected anymore. For example, for Fl, $m/m_0 = 1.79$, and it is 1.95 for element 118. As a result of the relativistic mass increase, the electron (negative) energy increases and the orbital radius shrinks according to

$$E = -(2\pi^2 e^4/n^2h^2)mZ^2, \tag{2}$$

$$r = Ze^2/mv^2, \tag{3}$$

where n is the principal quantum number, e is the charge of the electron, and h is Planck's constant. The contraction (Eq. (3)) and stabilization (Eq. (2)) of the hydrogen-like s and $p_{1/2}$ electrons is a direct relativistic effect and it was shown to originate from the inner K- and Lshell regions. The second (indirect) relativistic effect causes an expansion and destabilization of the $p_{3/2}$, d, f , and g atomic orbitals (AOs) due to increased screening of the nucleus by the relativistically contracted s and $p_{1/2}$ AOs. The third relativistic contribution comes from the spin-orbit (SO) splitting of the AOs with $l > 0$. All the three effects change approximately as Z^2 for the valence shells down a column of the Periodic Table.

Figure 1 shows, e.g., relativistic effects on group-12 elements: the ns AO stabilization reaches 5.8 eV for Cn, which is the maximum in the 7th row of the Periodic Table, and its contraction is 25%; the SO splitting of the 5d AOs is also large, e.g., 3.3 eV for Cn. The SO splitting of the 7p AOs reaches 11.8 eV at element 118. For the heavier elements, relativistic effects on their valence orbitals are even more pronounced and could lead to properties very different from those of the lighter homologs.

Breit effects (accounting for magnetic and retardation interactions) on valence orbital energies and IPs of the heaviest elements are small, for example, only 0.02 eV for element 121. Quantum electrodynamic (QED) effects such as vacuum polarization and electron self-energy can reach a few % in IP, electron affinities (EA), or transition energies for SHE (see, for instance, Refs. [10, 11]).

METHODS TO CALCULATE ELECTRONIC STRUCTURES OF SHE

There are several issues devoted to the relativistic methods (see, for example, [12]). It is meanwhile established that for highly relativistic SHE systems, the most appropriate approaches are those that treat both relativity and electron correlation at the highest possible level of theory: both effects were shown to contribute to more than 50% to binding energies and other properties of the 7th-row elements. Presently the highest theoretical level for the many-body methods is the Dirac-Coulomb-Breit (DCB) Hamiltonian

$$h_{DCB} = \sum_i h_D(i) + \sum_{i<j}(1/r_{ij} + B_{ij}) \tag{4}$$

where B_{ij} is the Breit term and the Dirac-Coulomb one-electron Hamiltonian is

$$h_D(i) = c\vec{\alpha}_i \vec{p}_i + c^2(\beta_i - 1) + V^n(i). \tag{5}$$

It contains the one-electron Dirac Hamiltonian plus the nuclear potential, V^n, and the operator $1/r_{ij}$ for the Coulomb interaction between electrons. The wave function is a four-component (4c) vector (spinor). The V^n includes the effect of the finite nuclear size, while some finer effects, like QED, can be added to h_{DCB} self-consistently for atoms, or perturbatively for molecules. Correlation effects are taken into account by configuration interaction (CI), manybody perturbation theory (MBPT) or, presently at the highest level of theory, coupled cluster theory with single-double (and perturbative triple) excitations (CCSD(T)) or Fock-Space (FS) CC theory.

Wavefunction based (ab-initio) methods solving the Dirac many-electron Eq. (4) are the most accurate. Nevertheless, the problems of electron correlation and proper basis sets make the usage of the 4c-DF methods very limited in molecular calculations. These methods are still too computer time intensive. Therefore, almost equally accurate 2c-methods have recently been developed.

Effective core potentials (ECP) allow for more economic calculations within the DHF schemes by replacing the inner core orbitals, which do not take part in the bond formation, by a special (effective core) potential. In this way, the number of basis functions and, therefore, two-electron integrals, is drastically diminished. There are two main types of the ECPs, pseudo potentials (PP) and model potentials (MP).

Relativistic density functional theory (DFT). Due to the high accuracy and efficiency, computational schemes based on the 4c- and 2c-DFT methods are among the most popular in theoretical chemistry, especially for extended systems, large molecules, liquids, or solids. The accuracy depends on the adequate choice of the exchange-correlation potential, E^{ex}, whose exact form is unknown. There is quite a number of the GGA E^{ex} and their choice is dependent on the system. For solid-state and adsorption calculations of SHE, the ADF BAND program is very promising [13].

ATOMIC PROPERTIES OF SHE

Electronic Configurations

Earlier predictions of chemical properties of elements Z = 104 through Z = 172 were made on the basis of Dirac-Fock (DF) and Dirac-Slater (DS) atomic calculations (see [3] for a review). Later, atomic properties of SHE were calculated using the MCDF and DCB FSCC/CCSD(T) methods, with the heaviest element treated in this way being Z = 122 [9]. Table 1 shows the progress in the predictive power of the atomic methods upon their development for ground-state configurations of elements 121, 122, and 123.

Table 1: Ground states predicted using various relativistic atomic methods[a]

Element	Madelung principle	DS	DF	MCDF	DCB CC
	Seaborg (1968)	*Waber et al.* (1968)	*Fricke et al.* (1971)	*Nefedov et al.* (2006)	*Eliav et al.* (1998)
121	$7d$	$7d$	$8p_{1/2}$	$8p_{1/2}$	$8p_{1/2}$
122	$5g^2$	$7d^2$	$7d8p_{1/2}$	$7d8p_{1/2}$	$7d8p_{1/2}$
123	$5g^3$	$6f^1 7d^2$	$6f7d8p_{1/2}$	$6f^2 8p_{1/2}$	...

[a] See [3, 6, 7, 9] for exact references.

One can see that the calculated states deviate from those predicted via the simple Madelung's principle. In the area of Z > 122, the elements are characterized by mixing of configurations coming from the partially filled $8p_{1/2,3/2}$, $7d_{3/2,5/2}$, $6f_{5/2,7/2}$ and $5g_{7/2,9/2}$, etc. shells. The proximity of the valence SO bands makes the search for the correct ground state very difficult. The usual classification on the basis of a simple electronic configuration and the placement of these elements in this part of the Periodic Table becomes, therefore, problematic, so that the shape of the table at these high Z is still under discussion and debate [3, 14, 15].

Spectroscopic Properties and Predictions of SHE Transport

Ionization potentials in the DF and DS approximations are available for SHE up to Z = 166 [3], and in the MCDF and DCB CC approximation till Z = 122 [9]. Multiple IPs were also calculated at the MCDF level of theory for elements Rf through Hs. They were used for estimates of stabilities of various oxidation states of these elements (see [6]).

Electron affinities were also calculated for quite a few of the heaviest elements at the DCB CC level of theory. Fl, has, e.g., no EA. On the contrary, element 118 has a positive EA of 0.058 eV, shown by the DCB FSCC + QED calculations. Both unusual results were shown to be due to the influence of relativistic effects.

Atomic/ionic/covalent radii and polarizabilities of SHE are now also available at various levels of theory [2, 6, 7, 14]. An important finding is smaller covalent radii of the 6d elements than those of the 5d ones in group 9 through 12, called a "transactinide break" [14, 15].

The atomic data were useful in predicting chemical behaviour of SHE [3, 6, 7], such as transport through Teflon capillaries from an accelerator to the chemistry set up, and/or their interaction with inert materials of gas-phase chromatography column. For that purpose, the following adatom-slab model for the adsorption energy, E_{ads}, was used:

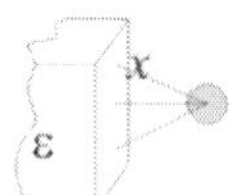

$$E(x) = -\frac{3}{16}\left(\frac{\varepsilon-1}{\varepsilon+2}\right)\frac{\alpha_{mol}}{\left(\frac{1}{IP_{slab}}+\frac{1}{IP_{mol}}\right)x^3} \quad (6)$$

Here, ε is the dielectric constant of the adsorbent material and x is the adatom-surface distance related to the van der Waals radius of the species. Table 2 shows two examples of such predictions for very different cases: for Fl with a small α and a high IP, and for element 120 with a large α and a small IP. The obtained low value of E_{ads} of Fl is indicative of its easy transport through Teflon capillaries to the chemistry set up, which was not obvious for element 120.

Table 2: DC CCSD(T) calculated atomic properties of Fl ($7s^2 7p_{1/2}{}^2$) and of element 120 ($8s^2$), and their adsorption enthalpies, $-\Delta H_{ads}$, on Teflon [7]

Element	α, a.u.	IP, eV	R_{vdW}, Å	$-\Delta H_{ads}$, kJ/mol
Fl	29.5	8.539	3.94	10.4
120	29.5	8.539	3.94	10.4

Using the calculated properties and equation 6, a yield N/N_0 (N_0 is the number of atoms entering the column or capillary, N is the number at the exit) of a hypothetical $^{300}120$ of 90% at the end of a Teflon capillary of 1 m was then determined. Thus, this element should also be transported, and the limiting factor is its half-life [7].

PREDICTIONS OF EXPERIMENTAL BEHAVIOUR OF SHE

Gas-Phase Experiments

Group 4-8 molecules. Gas-phase chromatography experiments provide measurements of volatility of SHE as adsorption temperature, T_{ads}, on a surface of detectors of the chromatography column usually covered with SiO_2 or gold layers [2, 4]. Obtained on their basis the adsorption enthalpy, ΔH_{ads}, is then used to deduce the sublimation enthalpy, ΔH_{sub}, being a measure of volatility in macrochemistry, using an empirical correlation between these quantities. The task of the theory was to predict both ΔH_{ads} and ΔH_{sub} and to see whether such a correlation is valid in the area of the heaviest elements.

For weak interactions of closed-shell SHE atoms or molecules with inert surfaces, no modern DFT methods can provide reliable values of E_{ads}: there are no dispersioncorrected E^{ex} for SHE, as they should be based on empirical data. Thus, for those systems, adatom/molecule-slab models proved to be useful. For molecules without dipole moments, Eq. 6 was used, while for those with dipole moments, μ, the following equation:

$$E(x) = -\frac{2Qe\mu_{mol}{}^2}{x^2} - \frac{Q^2e^2\alpha_{mol}}{2x^4} - \frac{3}{2}\frac{\alpha_{mol}\alpha_{Cl}}{\left(\frac{1}{IP_{mol}} + \frac{1}{IP_{Cl}}\right)x^6}. \quad (7)$$

All the properties of group-4 through 8 molecules used in Eqs. (6) and (7) have accurately been calculated via 4c-DFT and PP CC methods (see Tables 8 and 9 in [6]). Thus, e.g., for the prediction of adsorption of group-7 oxychlorides MO_3Cl (M = Tc, Re, and Bh) on a chlorinated quartz surface, equation 7 was used [6]. For an effective charge Q = −0.4 of the surface chlorine ions, $-\Delta H_{ads}$ = (48 ± 2) kJ/mol and $-\Delta H_{ads}$ = (78 ± 5) kJ/mol for TcO_3Cl and BhO_3Cl, respectively, were obtained with respect to the measured $-\Delta H_{ads}$ (ReO_3Cl) = (61 ± 2) kJ/mol. These $-\Delta H_{ads}$ of the Tc and Bh oxychlorides were confirmed by experiment giving (51 ± 2) kJ/mol and (75+6 −9) kJ/mol, respectively (see [2]). The sequence in volatility, TcO_3Cl > ReO_3Cl > BhO_3Cl, was revealed and explained by increasing μ in this row.

Another example deals with predictions of volatility of group-8 tetroxides, MO_4 (M = Ru, Os, and Hs). Their adsorption on a SiO_2, or a silicon nitride surface was predicted with the use of Eq. (6) and 4c-DFT calculated molecular properties (Table 3).

Thermochromatography gas-phase experiments have, indeed, shown that $-\Delta H_{ads}$ (HsO_4) is only 6 kJ/mol larger than $-\Delta H_{ads}$ (OsO_4) (see [2] for the reference), in excellent agreement with the calculations (Table 3). Such a reversal of the trend in volatility in the group could not be predicted from a simple extrapolation, but came out as a result of the accurate calculations and precise measurements.

Table 3: Ionization potentials, IP (in eV), polarizabilities, α (in a.u.), bond lengths, R_e (in Å), and adsorption enthalpies, $-\Delta H_{ads}$, (in kJ/mol), of MO_4 (M = Ru, Os, and Hs) on quartz[a]

	Meth.	RuO_4	OsO_4	HsO_4
IP	calc.	12.21	12.35	12.29
	exp.	12.19	12.35	–
α	calc.	58.07	55.28	65.99
	exp.	58.64	55.13	–
R_e	calc.	1.712	1.719	1.779
	exp.	1.706	1.711	–
$-\Delta H_{ads}$	calc.	41.0 ± 1	39.0 ± 1	45.4 ± 1
	exp.	–	39 ± 1	46 ± 2

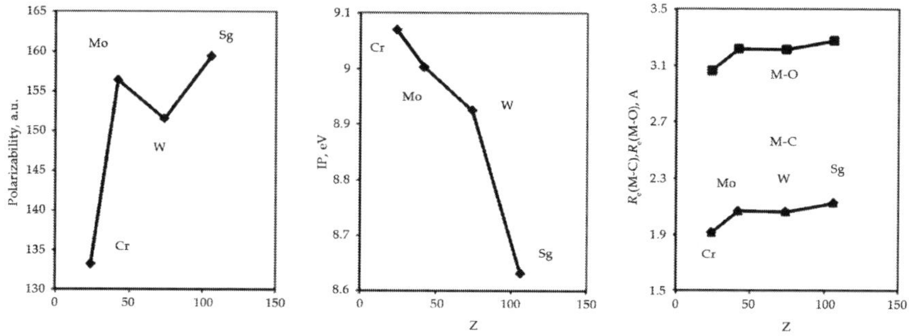

Figure 2: Calculated ionization potentials (IP), polarizabilities (α) and equilibrium bond lengths (R_e) of $M(CO)_6$ (M = Cr, Mo, W, and Sg) [17].

Another example deals with recent successful predictions of volatility of group-6 carbonyls. First of all, stability of $Sg(CO)_6$ had to be known. A work [16] based on the CCSD RECPs calculations have shown that the first bond dissociation energy (FBDE), $M(CO)_6 \rightarrow M(CO)_5 + CO$, should increase in group 6, reaching 2.20 eV for $Sg(CO)_6$. The $Sg(CO)_6$ was, indeed, synthesised and experiments are on the way to measure its FBDE.

To predict volatility of $Sg(CO)_6$, the 4c-DFT calculations were performed for the group-6 carbonyls and Eads were estimated using the adatom-slab model (Eq. (6)) [17]. The trends in the calculated molecular properties are shown in Fig. 2 reflecting, e.g., peculiarities in R_e and α caused by the lanthanide contraction from Mo to W. The $-\Delta H_{ads}$ of (46.2 ± 2.5) kJ/mol of $Sg(CO)_6$ was predicted to be almost equal to the measured $-\Delta H_{ads}$ of $W(CO)_6$ of (46.5 ± 2.5) kJ/mol. The experimental ΔH_{ads} have, indeed, indicated an almost equal volatility of the W and Sg carbonyls giving $-\Delta H_{ads}$ of $Sg(CO)_6$ of (50 ± 4) kJ/mol [18], in very good agreement with the calculations [17].

Elements 112 and 114. Another exciting case was a study of volatility of Cn and Fl, having closed-shell ($6d^{10}7s^2$) and quasi-closed shell ($7s^27p_{1/2}{}^2$) ground states, respectively. Due to the strong relativistic effects on their AOs (Fig. 1), these elements were expected to be extremely inert and volatile. Also their extrapolated ΔH_{sub} are the smallest in the corresponding groups.

Experimentally, volatility of Cn in comparison with that of Hg and Rn was to be studied as adsorption process on a gold surface of detectors of a gas-phase thermochromatography column [19–21]. The questions to the electronic structure theory, therefore, were: are Cn and Fl metallic in the solid state, or are they more like a solid noble gas? How volatile and reactive towards gold are Cn and Fl atoms in comparison with the lighter homologs and with Rn?

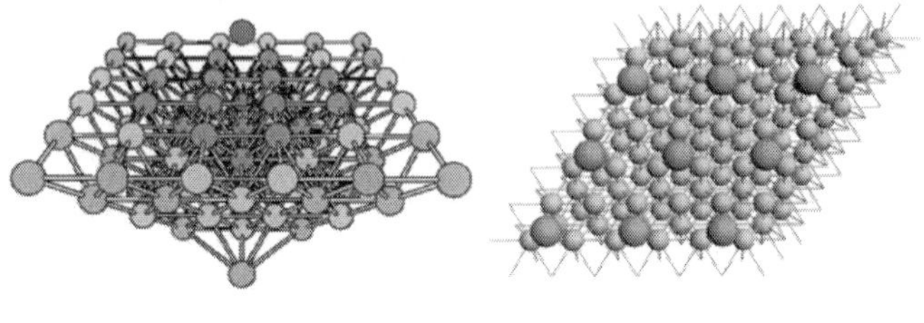

Figure 3: Models simulating adsorption of M on a Au(111) surface. The left panel is a cluster model treated via a molecular code. The right panel is a super-cell one treated via a periodic code.

Table 4: Adsorption enthalpies, $-\Delta H_{ads}$ (in kJ/mol), of Hg, Cn, Pb and Fl on a hydroxylated quartz and a gold (111) surface at very low coverage calculated using the ADF BAND periodic and 4c-DFT cluster approaches for single species in comparison with experiment.

Surface	Hg	Cn	Pb	Fl	Method	References
SiO_2	54	26	152	21	ADF BAND	[25]
	42 ± 2	–	165 ± 4	–	Experiment	see [2]
Au	78[a]	45	240	68	BAND/4c-DFT	[23, 24]
	98 ± 3	52^{+4}_{-3}	234	34^{+54}_{-11}; ≥ 48	Experiment	[19, 21]

[a] Preliminary ADF BAND value for the non-ideal surface.

The 4c-DFT calculations of the binding energies of group-12 homonuclear dimers [6, 7] and the scalar relativistic (SR) DFT periodic ones of cohesive energies [22] have shown Cn-Cn and the bulk of Cn to be more bound than the corresponding Hg systems. Thus, a decrease in ΔH_{sub} in group 12 should not continue with Cn. In contrast to the M-M-bonding, the M-Au one should be weaker for Cn than for Hg due to the stronger relativistic stabilization of the 7s(Cn) AO [23]. This means that $-\Delta H_{ads}$ (Hg) > $-\Delta H_{ads}$ (Cn), which is opposite to ΔH_{sub} (Hg) < ΔH_{sub} (Cn). Thus, there is no correlation between $-\Delta H_{ads}$ and ΔH_{sub} in group 12. The Fl-Au bond was also shown to be weaker than the Pb-Au one, but stronger than the Cn-Au one. Numerous calculations for the Cn/Fl-Au_n systems, where Au_n are clusters simulating a gold surface (Fig. 3, left), have come to the same result, E_b(Fl-Au) > E_b (Cn-Au) (Table 4).

Experiments [19], being in good agreement with the theoretical predictions for adsorption of Cn on gold [23], turned out to be in some disagreement with each other and partially with the theory for adsorption of Fl [20, 21] (Table 4). Therefore, further experiments are required to shed more light on this interesting case.

For element 113, to be experimentally studied next, the calculations have shown that it should be much more reactive than Cn and Fl due to one unpaired $7p_{1/2}$ electron, so that its $-\Delta H_{ads}$ on gold of 150 kJ/mol should be larger than $-\Delta H_{ads}$ of Cn and Fl (Table 4) [7]. In group 13 the trend to an increase in volatility should continue with element 113: its weaker adsorption on gold is a result of the $7p_{1/2}$ AO SO stabilization and, therefore, weaker bonding.

Calculations of E_{ads}(M-gold) using the ADF BAND code are in progress [24] (Fig. 3, right). Modelling non-ideal surfaces (with vacancies, steps) allowed for reaching even better agreement of the calculated E_{ads} of Hg and Pb with their experimental $-\Delta H_{ads}^{Au}$ (see Table 4). Results for SHE will follow.

Assistance was also rendered to experiments on volatility of Hg/Cn and Pb/Fl with the use of SiO_2 covered detectors located ahead of the gold-plated ones in a chromatography column (allowing for differentiating between ΔH_{ads} on surfaces of these different materials).

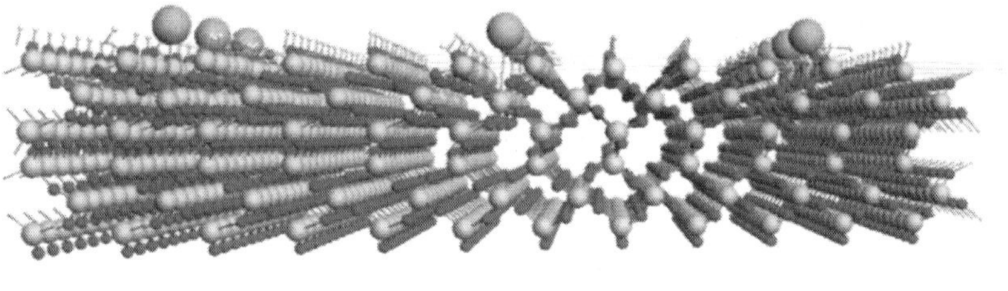

Figure 4: A supercell modelling adsorption of M on vicinal silanols for low adsorbate coverage. Reproduced from [25] with permission from the PCCP Owner Societies.

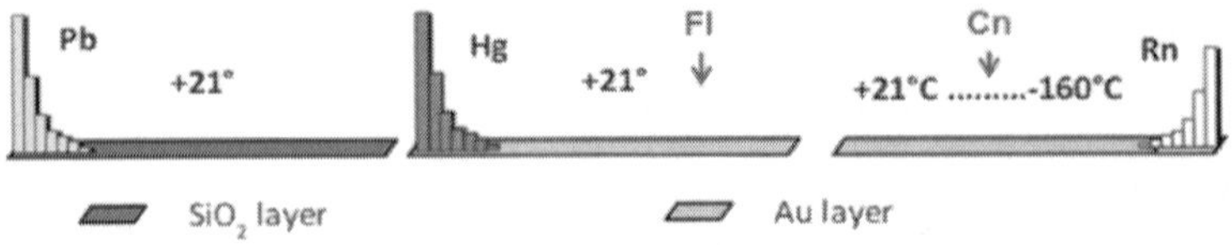

Figure 5: Predicted adsorption of Cn and Fl on quartz and gold surfaces of the chromatography column. Reproduced from [25] with permission from the PCCP Owner Societies.

Accordingly, periodic SR and SO calculations were performed on E_{ads} of these elements on a quartz surface using the ADF BAND code [25]. Two modifications of the hydroxylated SiO_2 surfaces were considered: vicinal (80%) and germinal (20%) silanols. A (4 × 4) super-cell modelling adsorption on a vicinal silanol for very low coverage is shown in Fig. 4.

Thus, according to the calculations [25], Pb should adsorb on the SiO_2 surface at room temperature, while Hg, Cn, and Fl should not. The latter three elements should adsorb, however, on gold, with Hg revealing much stronger interaction (Table 4 and Fig. 5). The reason for the non-interaction of Cn and Fl with quartz and weaker interaction with gold than that of Hg and Pb, respectively, is the strongest relativistic effects on their valence electrons making these elements rather inert. Experiments have, indeed, revealed that Cn and Fl do not adsorb on SiO2 (ΔH_{ads} to be published).

The theory has also considered interaction of even heavier elements such as 115 through 120 with gold and other metals of interest on the basis of Au-containing dimers [6, 7]. These results might be of interest for future vacuum chromatography experiments, provided sufficiently long-living isotopes of these elements are found.

Some molecular species of even heavier elements are suggested in [14], whose stability may be proven by detailed calculations.

Aqueous Chemistry Experiments

Complex formation is known to increase in the groups at the beginning of the 3d-5d- series. It was, therefore, important to establish whether this trend is continued with the 6d-elements. Experimental investigations were accordingly devoted to the study of the complex formation of Rf, Db, and Sg (elements with sufficiently long half-lives) in aqueous solutions of HF, HCl, HBr, and H_2SO_4 acids. Liquid-liquid extraction, or sorption chromatography separations were applied to the studied species [2, 5]. To render assistance to those experiments, predictions of complex-formation constants and distribution coefficients, K_d, between an aqueous and organic phases (or resins) were made with the use of a model that treats a free energy change of a complex formation reaction as a sum of the ionic and covalent constituents. The latter are obtained via 4c-DFT electronic structure calculations of the systems of interest [6, 7].

The results have demonstrated that even though the heaviest elements are homologs of their lighter congeners in the chemical groups, trends in the complex formation are not necessarily continued with them. (See, e.g., predictions of an unexpected trend in the K_d values of the group-5 complexes by extraction from the HCl solutions into amines, confirmed by experiments [5].) The calculations have also shown that simple models based on the ratio of the cation charge to its size, or extrapolations are unreliable in the area of SHE. Only by performing relativistic calculations of the real chemical equilibria, can the complex formation, or hydrolysis be correctly predicted.

A summary of the predicted trends in hydrolysis, complex formation and extraction of the group-4 through 6 elements including the heaviest is given in [6, 7]. As one can see there, most of the predictions have been confirmed by the experiments, while some of them, like predictions for Sg in HF solutions, are still awaiting confirmation.

SUMMARY

A large number of theoretical chemical studies were performed on the heaviest elements and their compounds such as molecules, aqueous complexes, and solid-state systems. They were based on the accurate calculations with the use of the most advanced relativistic methods, such as 4c/2c-DFT, PP and DF CC ones. With many of them carried out in a close link to the experimental research, these studies have contributed to better understanding of the chemistry of these exotic elements, and the role and magnitude of relativistic effects. A synergy between theoretical and experimental research has been achieved.

The results have demonstrated that even though the heaviest elements are homologs of their lighter congeners in the chemical groups, trends in properties are not necessarily continued with them. Thus, in this area of the Periodic Table simple extrapolations or semiempirical models turned out to be unreliable. Relativistic calculations at the highest level of theory are, therefore, mandatory to achieve convincing results.

Relativistic theory will undergo further methodical developments like, e.g., creation of special techniques in atomic CC methods to treat open shell systems (needed for predictions of electronic configurations of elements with $Z > 122$), or inclusion of QED effects on the SCF basis in molecular calculations. Also, dispersive potentials for SHE should be worked out in DFT methods in order to treat more accurately weak interactions.

For elements heavier than $Z = 118$, theoretical investigations will be very exciting, since resemblance of properties with their lighter homologs in the groups is expected to be even less pronounced.

REFERENCES

1. The Chemistry of the Superheavy Elements, 2nd edition, eds. M. Schädel and D. Shaughnessy (Springer, Heidelberg, 2014).
2. A. Türler and V. Pershina, Chem. Rev. 113, 1237 (2013).
3. B. Fricke, Struct. Bond. 21, 89 (1975).
4. H.W. Gäggeler and A. Türler, Gas-Phase Chemistry of Superheavy Elements. In reference 1., p. 415.
5. J.V. Kratz and Y. Nagame, Liquid-Phase Chemistry of Superheavy Elements. In reference 1., p. 309.
6. V. Pershina, Theoretical Chemistry of the Heaviest Elements. In reference 1., p. 135.
7. V. Pershina, Nucl. Phys. A 944, 578 (2015).
8. P. Schwerdtfeger and M. Seth, In: Encyclopedia on Calculational Chemistry (Wiley, New York, 1998) Vol. 4, p. 2480.
9. E. Eliav, S. Fritzsche, U. Kaldor, Nucl. Phys. A 944, 518 (2015).
10. P. Pyykkö, Chem. Rev. 88, 563 (1988).
11. P. Schwerdtfeger et al., Nucl. Phys. A 944, 551 (2015).
12. Relativistic Electronic Structure Theory, Parts I and 2, ed. P. Schwerdtfeger (Elsevier, Amsterdam, 2002).
13. BAND2016, SCM, Theoretical Chemistry, Vrije Universiteit, Amsterdam, The Netherlands, http://www.scm.com.
14. P. Pyykkö, Phys. Chem. Chem. Phys. 14, 14734 (2012).
15. P. Pyykkö, Chem. Rev. 112, 371 (2012).
16. C.S. Nash and B.E. Bursten, J. Am. Chem. Soc. 121, 10830 (1999).
17. V. Pershina and J. Anton, J. Chem. Phys. 138, 174301 (2013).
18. J. Even et al., Science 345, 1491 (2014).
19. R. Eichler et al., Nature, Letters 447, 72 (2007).
20. R. Eichler et al., Radiochim. Acta 98, 133 (2010).
21. A. Yakushev et al., Inorg. Chem. 53, 1624 (2014).
22. N. Gaston et al., Angew. Chem. Int. Ed. 46, 1663 (2007).
23. V. Pershina, J. Anton, T. Jacob, J. Chem. Phys. 131, 084713 (2009).
24. V. Pershina, in progress.
25. V. Pershina, Phys. Chem. Chem. Phys. 18, 17750 (2016).

SECTION 2:
RADIOACTIVE DECAY KINETICS AND RADIOTRACERS

4

Radioactive Decay as A Second-Order Kinetics Transformation Process. Consequences on Radiometric Dating

Giancarlo Cavazzini

CNR Istituto di Geoscienze e Georisorse, Padova, Italy

ABSTRACT

Arguments suggest and recent analysis of experimental work confirm that the current interpretation of the transformation process we call 'radioactive decay' should be revised. The characteristics of this process are better accounted for by re-interpreting it in terms of second-order kinetics. Therefore, the atomic systems of nuclides we observe decay are 'radio-activated', and not, as hitherto believed, 'radio-active'. According to this interpretation, the rate of decay of a radioactive nuclide is at any instant proportional to the concentration of the physical species that determines its activation. The analysis of λ of alfa- and beta-emitting nuclides show the dependence of these parameters from solar activity and distance.

Citation: Giancarlo Cavazzini. "Radioactive Decay as A Second-Order Kinetics Transformation Process. Consequences on Radiometric Dating". Applied Physics Research; Vol. 12, No. 1; 2020. https://doi.org/10.5539/apr.v12n1p26

Therefore, if changes in the emission of energy from the sun occurred over time since the formation of a geological system, changes in the values of λ of the radioactive nuclides would also have occurred, and the calculated radiometric age of the system may differ from the true age. Implications on the science of dating geological samples using parent-daughter decay systematics are investigated.

Keywords: Radioactive decay, Second-order kinetics, Decay constant, Solar activity, Apparent age

INTRODUCTION

Recent careful analysis of measurements of activity of beta-emitting (^{32}Si, ^{36}Cl,) alpha-emitting (^{226}Ra), and electron-capturing (^{54}Mn) nuclides have shown that variations of the respective decay constants occur in relation to sun activity and distance[1-7] (Fischbach et al., 2009; Jenkins & Fischbach, 2009; Jenkins et al., 2009; 2010; 2102; Javorsek II et al., 2010; Sturrock et al., 2011).

A clear periodic oscillation with an approximate period of one year has been observed for the rate of decay of ^{36}Cl for seven successive years (from July 2005 to June 2011), during instrumental calibration at The Ohio State University Research Reactor (OSURR) (Jenkins et al., 2012), and similar oscillations also result from the analysis of half-life measurements taken by two other independent groups: at the Brookhaven National Laboratory (BNL), New York, and at the Physikalisch-Technische Bundesanstalt (PTB) in Braunschweig, Germany.

The BNL group measured the half-life of ^{32}Si for four years (1982-1986), and the PTB group measured the half-life of ^{226}Ra in a standard used for comparison in the measurements of the half-life of ^{152}Eu for fifteen years (1984-1998) (Jenkins et al., 2009; the raw data are from Alburger et al., 1986, and Siegert et al., 1998, respectively). The counting rates of each of these nuclides increase to a maximum in winter, and progressively decrease to a minimum in summer, suggesting a direct relationship between half-lives of nuclides on the Earth and Earth-Sun distance.

In the analysis of the raw data from BNL and PTB, the data sets on ^{32}Si and ^{226}Ra were shown to have not only the same period but, in the two years during which the two data sets overlapped, they also had the same approximate phase and amplitude (Jenkins et al., 2009; Javorsek II et al., 2010). Moreover, the hypothesis of a solar influence on radioactive decays was strengthened by the identification of an additional periodicity in the BNL data at 11.25/year and in the PTB data at 11.21/year. Both of these peaks can be related to the rotation (and probably to the inhomogeneous nature) of the core of the Sun (Sturrock, 2009).

A third periodicity was also identified by Sturrock et al. (2011) in both the BNL and PTB data sets, which is analogous to Rieger periodicity (Rieger et al., 1984).

Jenkins and Fischbach (2009), Jenkins et al. (2009) and Fischbach et al. (2009) also report a relationship between solar activity and the measured value of the decay rate of ^{54}Mn, during a series of solar flares from the beginning of December 2006 to the beginning of January 2007. During some flares, the value of the decay constant of ^{54}Mn strikingly changed by a factor of ~ 2. As correctly suggested by Fishbach et al. (2009), all these observations require a revision of the current model used to describe the process of radioactivity.

Long before these interesting observations, arguments, which we will illustrate in the following, suggested that a revision of the model which describes the process of radioactive decay was needed. The characteristics of this process of transformation are more properly accounted for if re-interpreted in terms of second-order kinetics. Moreover, the relationship between the change in solar activity and the change in the rate of decay of ^{54}Mn raises the question whether the decay constants of the radioactive nuclides were actually constant during the geologic time. For this reason, we will also discuss implications that a second-order kinetics re-interpretation has on the science of dating geological systems using parent-daughter decay systematics. A slow decrease in sun activity over time-spans of the order of 10^9 years may have produced changes in the value of the decay constants which cannot be neglected.

ON THE WAY TO ASCERTAIN IF A SYSTEM IS STABLE/UNSTABLE, AND ON THE INCONSISTENCY OF THE EXPONENTIAL RELATIONSHIP USED TO QUANTITATIVELY DESCRIBE THE DECAY OF THE RADIOACTIVE NUCLIDES WITH AN INTERNAL (SELF-) TRANSFORMATION PROCESS

Stability and Instability

Let us start our critical analysis of the current interpretation of the process of radioactive decay by briefly discussing the method to ascertain if a system is stable or not. First of all, we have to define precisely what we mean by 'unstable': 'unstable' is a system which transforms itself 'spontaneously', i.e. a system which has within itself the cause of its own instability.

We call a system 'unstable' if we observe values of its properties changing over time. We have no other way to ascertain the instability/stability of a system. Obviously, one could correctly argue how long the time-span of observation of a system should be before the stability/instability of the system is ascertained. As a matter of fact, there is not and there cannot be at all any a priori time-span of observation: the system could modify very slowly, so that the longer the time-span of observation, the higher our confidence in believing the system stable; and, on principle, we have no choice but the continuous observation of the properties of the system over an infinite time-span to ascertain if the system is actually a stable one. Thus, we define 'stable' a system whose proprieties change after an infinite time-span of observation.

However, it is obvious that observing a system for an infinite time-span would hinder any decision about the system itself. Thus, we usually call 'stable' a system whose properties do not change over a time-span which is a reasonable (and often quite small) fraction of our biological life-time. If this is the rule, it seems very improper calling unstable the atoms of nuclides such as ^{238}U and ^{232}Th, which were in the solar nebula long before their incorporation in the planet we live, the formation of which is believed to have occurred approximately 4.5 billion years ago.

Internal Instability

In spite of this simple line of reasoning, the transformation process of atoms of nuclides we called 'radio-active' was believed to be due to internal reasons, not related to interactions of these systems with something in their surroundings. This was essentially due to two reasons. The first reason was that experiments showed that, although the atoms of the radioactive nuclides are certainly non-isolated systems and therefore interacting with their surroundings, their rate of transformation did not appreciably change as the external conditions changed.

Fischbach et al. (2009) correctly emphasized that few issues frame the history of natural radioactivity as fundamentally as the question of whether the decay rates of nuclides are constants of nature, unaffected by the external environment. After Henri Becquerel discovered radioactivity, in 1896, intense efforts were made to ascertain whether the decay rates of nuclides could be affected by temperature, pressure, chemical composition, electric and magnetic fields etc., and Rutherford et al. (1930) concluded that the rate of transformation of a radioactive nuclide is a constant under all conditions.

The second reason was that the results of the measurements of the rate of transformation of these atomic systems can be interpreted as if the process occurred according to first-order kinetics, i.e. in terms of probability to decay (Rutherford et al., 1930). According to this interpretation, all the atoms of a radioactive nuclide have equivalent probability to decay in a certain time-span of observation. However, some of them decay whereas the others do not, and the parameter we call the "decay constant" of the transformation, denoted by the Greek letter λ, represents the instantaneous probability one atom has to transform itself in a unit of time:

$$-\frac{dN}{N}\frac{1}{dt}=\lambda \qquad (1)$$

All the atoms in a number of atoms of a radioactive nuclide have the same probability to decay in a time-span of observation because they are equivalent. However, if the atoms are equivalent and the transformation of each of these atomic systems is due to purely internal reasons, why some of them transform and some others do not? Why should the behaviour of systems which are equivalent, and which transform themselves only due to internal reasons, be different?

The Exponential Law of Decay is Inconsistent with Internal Instability

Finally, let us consider the decrease over time in the number of atoms of radioactive nuclides. The law of decrease is the exponential law which results from integration of differential equation (1):

$$N = N_0 e^{-\lambda t}, \qquad (2)$$

where N_0 is the number of atoms at $t = 0$.
According to this equation, the number of atoms of the radioactive nuclide, N, will never be zero, unless an infinite time-span has passed. This is paradoxical, indeed, because, according to the line of reasoning we have illustrated in 2.1, this is absolutely inconsistent with the concept of instability of (some of) the atomic systems, a fortiori with the concept of internal instability. Therefore, relationship (2) cannot describe the change over time in a number of self-unstable systems, and two are the possible reasons for this inconsistency: i) the process of transformation of the atoms of the radioactive nuclides is not due to purely internal reasons; alternatively, ii) relationship (2) is wrong. However, relationship (2) is the result of repeated accurate experimental work on all radioactive species, and we can exclude it is wrong. Therefore, we infer that the processes of decay are not the result of processes of self-transformation, but they result from interactions of atomic systems with something external, as also suggested by lines of reasoning in 2.1 and 2.2: the atomic systems we currently call 'radio-active' are, instead, radio-activated.

"Activation" is not new in physics: processes such as induced nuclear fission and neutron-activation are examples. The latter is a well-known analytical method which is based on transformations of nuclides due to interaction with neutrons. Therefore, there is no conceptual difference between processes as nuclear fission or neutron activation and the process of radioactive decay. The difference is that we know that physical entities we call 'neutrons', by interaction with atoms of some nuclides, cause their fission or transformation, whereas we do not know (because we do not 'see' them) the physical entities which, by interaction with atoms of some nuclides, cause their 'radioactive' transformation.

That is one reason why scholars who investigated this physical process believed that the transformation of these atomic systems was due to internal instability. However, the law of decay eqation (2) they calculated showed that this hypothesis was incorrect, indicating that the concept of probability to decay had to be substituted by the concept of probability to interact with some external physical species.

RADIOACTIVE DECAY AS A SECOND-ORDER KINETICS TRANSFORMATION PROCESS

Relationship (2) can properly describe the decrease over time in the number of systems which transform by interaction with some external entities of matter/energy, i.e. according to second-order kinetics.

In chemistry, second-order kinetics is the kinetic relationship of transformation of molecules of a chemical species X for interaction with a chemical species Y. X and Y are called the 'reactants':

$$\frac{d[X]}{dt} \propto -[X][Y] \ , \qquad (3)$$

where d[X]/dt is the instantaneous rate of decrease of the concentration of species X, [X], and [Y] is the concentration of species Y at the same instant.

Nevertheless, as discussed in 2.2, the results of the measurements of activities over time led to interpreting the process in terms of probability of self-transformation, a process which occurs according to first-order kinetics. Therefore, we also write:

$$\frac{d[X]}{dt} = -\lambda[X] \ . \qquad (4)$$

In order that equations (3) and (4) are simultaneously valid it is necessary that the concentration of the reactant Y is constant over the time-span the measurements of activity are carried out:

$$\lambda \propto [Y] \ . \qquad (5)$$

If this is the case, a process which occurs according to second-order kinetics may be mistaken for a process which occurs according to first-order kinetics; and the value of the decay constant of the transformation contains information on the value of the concentration of reactant Y.

We hypothesize that the value of λ is approximately constant during measurements of activity because the value of the concentration of species Y is approximately constant. If this is correct, how can the concentration of species Y be constant over time?

If we assume that each element of species Y is 'consumed' when activating one atom of species X, a simple explanation to this question is that species Y is immediately and continuously reintroduced in the system as it 'disappears' by interaction with the atoms of the nuclide which is being activated. In this case, the concentration of species Y in a certain volume of matter remains substantially constant over time. This condition is realized if a continuous 'flux' of elements of Y exists, which pervades the matter. This flux must be approximately stable over time-spans which are important fractions of the human biological life, because repeated measurements of activities of nuclides over 30-50 years did not give results which were clearly different. The process of transformation can be described as follows: the atoms of a radioactive nuclide are targets which may be hit or not by elements of the physical species Y which can determine their activation. In a very small time-span dt (an 'instant'), the number of the targets which are hit in the unit of time and which decay to the daughter species are obviously proportional to the total number of targets and to the concentration of species Y at that instant:

$$\frac{dN}{dt} \propto -[Y]N \quad . \qquad (6)$$

Since the rate of transformation depends on the number of targets, the number N of atomic systems which can be activated in a unit of time decreases with time and tends to zero, without ever reaching it, as correctly shown by the exponential relationship (2).

We assume that all the atomic systems of a nuclide species X are equivalent from the point of view of the content in energy. We also assume that the elements of Y are characterized by a continuous spectrum of values of energy in the range $E_1 \div E_2$ ($E_2 > E_1$), and that only the elements which are characterized by a content in energy which is higher than a certain threshold value ε_0 may determine the decay of X. Accordingly, we write the rate of decay of nuclide X as:

$$\frac{dN}{dt} = -\varphi[Y]N = -\frac{\int_{\varepsilon_0}^{E_2} \nu \, dE}{\int_{E_1}^{E_2} \nu \, dE}[Y]N \qquad (7)$$

where υ is the fraction of the elements of Y as a function of energy, and the ratio between the two integrals represents the fraction φ of Y which can determine the activation of nuclide X.

At present, the remarkable fact is that accurate measurements have shown that the decay 'constants' of radioactive nuclides depend on solar 'activity' and distance (Jenkins & Fischbach, 2009; Fischbach et al., 2009; Javorsek II et al., 2010; Sturrock et al., 2011; Jenkins et al., 2009, 2010, 2012). This could mean that the physical species which causes the activation of the nuclide systems are originated in the sun or, alternatively, that the decay process represents the result of an effect of the sun on physical entities which may not strictly originate in the sun, transforming these entities into activators. As regards the nature of these physical entities and of their flux, at the present stage of knowledge any hypothesis is rather speculative, even if neutrinos from the sun have been proposed to play a role (Fischbach et al., 2009).

Whatever the case, this fact has the following important consequence: in space and time the same radioactive nuclides in planetary and sub-planetary matter orbiting around the same star or around different stars generally decay according to different rates. This has implications on the meaning of the age values we calculate from parent-daughter nuclide systematics to study the history and evolution of that planetary matter, which we discuss in the following.

CHANGE OVER TIME OF THE DECAY 'CONSTANTS' OF RADIOACTIVE NUCLIDES IN PLANETARY MATTER AROUND A STAR: CONSEQUENCES ON CALCULATED RADIOMETRIC DATES

It is believed that stars undergo processes of evolution. An initial time-span, in which the production, and therefore the emission of energy increases over time (when the star begins to bright) is followed by a time-span in which the emission of energy is thought to be rather stable. After this time-span, the production/emission of energy progressively decreases to certain levels. The energy radiation per unit of time of a star which lies on the 'main sequence' (like our Sun), for a given chemical composition and stellar age, depends on its mass.

The observed relationship between activity in the sun and the value of the decay 'constant' of a nuclide raises the question whether the value of this parameter was actually constant during the stages of life of the sun and of the solar system. If the value of the decay constant of a radioactive nuclide depends on the concentration of the physical species that determines its activation, and if the concentration of the physical species which activates the nuclide changed over time, then the value of the decay 'constant' changed over time, and the value which we measure presently is only a point above a long variation curve.

This is an important question, because we calculate the age of terrestrial (and extra-terrestrial) geological events – i.e. the formation of rock and mineral systems - by the well-known equation (in the simplified form):

$$t_a = \frac{1}{\lambda}\ln\left(1+\frac{D}{N}\right) , \qquad (8)$$

where D, N represent, respectively, the number of atoms of daughter nuclide and parent nuclide at time t after the formation of the system, and λ is the value of the decay 'constant' we measure presently.

This shows that the calculated age ta may significantly differ from the true age t due to hyperbolic dependence on the value of the 'constant', and this is the reason why we have added symbol a to symbol t, indicating that the calculated age is an apparent one. If the value of λ changed during the stages of life of the Sun, we calculate only apparent ages, and how much the apparent age deviates from the true age must be discussed.

Linear Decrease of the Decosy 'Constant'

If we assume the value of λ to decrease over the true age t of a geological system according to linear relationship

$$\lambda = \lambda_0 + mt , \qquad (9)$$

where λ_0 is the value of the decay 'constant' of the nuclide at the time of formation of the geological system, λ is the value of this parameter we measure presently, and m is the angular coefficient of the evolution straight line. The true age of the geoògical system is one of the two solutions (the positive one) of the following second-degree equation:

$$-\frac{m}{2}t^2 + \lambda t - \ln\left(1+\frac{D}{N}\right) = 0 . \qquad (10)$$

Eq. (10) reduces to eq. (8) in case mt^2 is so small to be negligible. The true age t is obtained by the equivalence between the value of t in eq. (9) and the value of t as solution of eq. (10), assuming the initial value λ0 of the decay 'constant'. This calculation is articulated in three steps: first we assume λ_0 and calculate mt = $\lambda - \lambda_0$ (eq. (9)). Then we replace mt into eq. (10) obtaining an equation of first degree in variable t. Finally, we solve this equation using the measured value of the ratio D/N.

According to this scheme of calculation, the true age of the system is given by the formula:

$$t = \frac{\lambda}{m}\left\{1 - \sqrt{1 - \frac{2}{\lambda}\left(\frac{m}{\lambda}\right)\ln\left(1 + \frac{D}{N}\right)}\right\}, \tag{11}$$

or

$$t = \frac{2}{\lambda_0 + \lambda}\ln\left(1 + \frac{D}{N}\right). \tag{12}$$

For example, let us consider ^{87}Rb-^{87}Sr dating systematics, and consider D/N such that eq. (8) gives t_a = 4.5 Ga (D/N ~ 0.06599, since the present value of λ ^{87}Rb is $1.42 \cdot 10^{-11}$ a^{-1}; Steiger and Jager, 1977). Assuming $\lambda_0 = 1.50 \cdot 10^{-11}$ a^{-1} (a deviation of 5.6% from the present value, i.e. a change of $1.25 \cdot 10^{-9}$ % per year), we calculate m = $-1.775 \cdot 10^{-22}$ a^{-2} and t = 4.38 Ga. If we assume λ_0 = $1.55 \cdot 10^{-11}$ a^{-1} we calculate m = $-2.982 \cdot 10^{-22}$ a^{-2} and t approximately 4.30 Ga.

Therefore, the ratio r between the true age (12) and the apparent age (8) has the same value for systems which have the same age (i.e. the same value of λ_0):

$$r = \frac{2\lambda}{\lambda_0 + \lambda}. \tag{13}$$

Exponential Decrease of the Decay 'Constant'

We can alternatively assume that the value of λ decreased over time according to an exponential function. We write:

$$\lambda = \lambda_0 e^{-ct}, \tag{14}$$

where λ_0 is the value of the decay constant at the instant of formation of the system, and c is a positive constant which represents the ratio between the rate of change with time of the concentration of the physical species that determines the activation of the parent nuclide and the concentration of this physical species at any instant.

In this case, differential equation (1) becomes:

$$-\frac{dN}{N}=\lambda_0 e^{-ct}\,dt \quad , \tag{15}$$

which is integrated to:

$$N=N_0\exp\left[\frac{\lambda}{c}\left(1-e^{ct}\right)\right] \quad . \tag{16}$$

Eq. (16) correctly reduces to eq. (2) in the case the quantity ct is small, so that the Mc Laurin's series expansion of the exponential within the argument of the exponential can be truncated to first-order term with small deviations from the exact value. This occurs a fortiori during laboratory measurements of activity of nuclides, because the time-span over which the measurements are carried out is small.

From eq. (16), we calculate the true age as:

$$t=\frac{1}{c}\ln\left[1+\frac{c}{\lambda}\ln\left(1+\frac{D}{N}\right)\right] \tag{17}$$

or

$$t=-\frac{1}{c}\ln\left[1-\frac{c}{\lambda_0}\ln\left(1+\frac{D}{N}\right)\right] \quad . \tag{18}$$

The Relationship between the True Age and the Apparent Age

It is important to determine the relationship existing between the apparent age and the true age of the geological system. In the case of linear decreasing of λ, re-arranging of eq. (10) gives:

$$t_a=t-\frac{m\,t^2}{2\lambda} \quad , \tag{19}$$

which shows that the difference between the apparent age and the true age is proportional to the square of the true age. In the case of exponential decrease of λ, we calculate the relationship between the two age values by substituting eq. (8) into eq. (17) and re-arranging. We obtain:

$$t_a = \frac{1}{c}\left(e^{ct} - 1\right) . \quad (20)$$

The relationship between the apparent age and the true age depends on the values of the parameters m/λ and c, respectively. In the case of linear decrease of λ, the apparent age and the true age approximately coincide if the ratio m/λ is sufficiently small with respect to true age t. In the case of exponential decrease of λ, the two age values approximately coincide if parameter c is sufficiently small with respect to true age t so that $e^{ct} \sim 1 - ct$.

In the other cases, the apparent age is higher than the true one - the apparent time and the true time running differently, the former running faster than the latter - and the deviation between the two age values increases as the true age increases. Since the relationships between the two age values are not linear, the deviation between the two age values does not increase linearly as the true age increases, i.e. the apparent time runs progressively faster than the true time as the true age increases (Fig. 1).

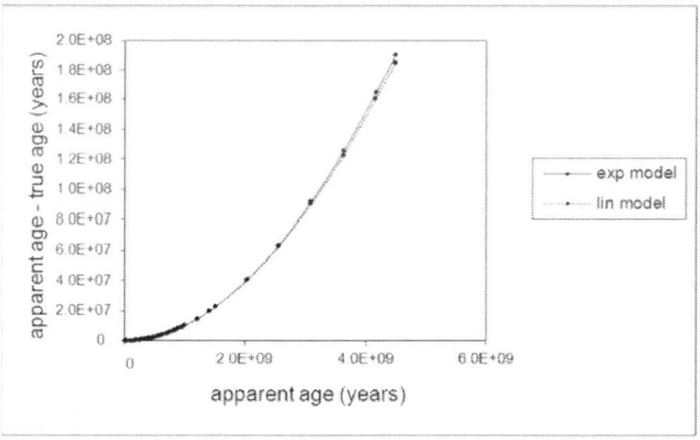

Figure 1: Difference between the apparent age and the true age of a geological system as a function of the apparent age in the case of exponential and linear models of decrease of λ. The two curves are calculated assuming $c = -m/\lambda = 2.1 \cdot 10^{-11}\ a^{-1}$. See text for further explanation.

Let us to refocus on the ^{87}Rb-^{87}Sr dating systematics (present value of λ ^{87}Rb = 1.42·10-11 a1 (Steiger and Jager, 1977)). If D/N is such that eq. (8) gives ta = 4.5 Ga (D/N ~ 0.06599), assuming $c = -m/\lambda = 1.25 \cdot 10^{-11}\ a^{-1}$ (i.e. $\lambda_0 = 1.5 \cdot 10^{-11}\ a^{-1}$) we calculate a decrease of λ per year (m) of $-1.775 \cdot 10^{-22}\ a^{-1}$, and a total decrease in 50 years of $8.875 \cdot 10^{-21}\ a^{-1}$, which would be clearly invisible to our instruments. If $c = -m/\lambda = 2.10 \cdot 10^{-11}\ a^{-1}$ (Fig. 1; $\lambda_0 = 1.55 \cdot 10^{-}$

^{11} a^{-1}), we calculate a decrease of λ per year of $-2.982 \cdot 10^{-22}$ a^{-1}, with a total change in 50 years of $-1.491 \cdot 10^{-20}$ a^{-1}, also invisible to our instruments.

The Apparent Age Coincide for Simultaneous Geological Events on Different Planets of a Planetary System

Let us consider the case of geological events (for example formations of igneous rocks) which occur at the same instant in the geological time on different planets of the same planetary system. Let us assume to calculate the apparent age of these geological systems using the same parent-daughter (P-D) systematics. Although λ, and consequently m, depend on the distance of the planetary matter from the star, ratio m/λ does not depend on this distance. Therefore, if the same P-D systematics is used, the apparent ages of events which occurred at the same instant in the geological time on different planets of a planetary syste, coincide.

Instead, even if the same P-D systematics is used to date the geological events, the apparent ages of systems which formed at the same instant in the geological time on planets of different planetary systems, generally differ. This, because the age value is related to ratio m/λ - the ratio between the rate of change over time of the concentration of the physical species which determines the activation of the parent nuclide and the current concentration of this physical species in the surroundings of the respective stars - which is generally different.

The Case of Different P-D Decay Systematics

In the above discussion, we have considered the case of using the same P-D systematics to date the geological systems because it is possible, on principle, that different parent nuclides are activated by different physical species. Nevertheless, it is also possible that different nuclides are activated by the same physical species, i.e. that differences in the values of λ are due to differences in the respective value of fraction φ.

In this case, the apparent ages of a geological system calculated using those different P-D systematics coincide, because ratio m/λ has the same values in the two cases. However, if the values of apparent age obtained by different systematics coincide, it could be also possible that the physical species which activate the respective parent nuclides are different, but the respective λs did not change significantly over the true age of the system. In this case, the calculated apparent ages are (very) close to the true age of the system. This also occurs for young geological systems, even if λ changed significantly since the beginning of the geological time (see Fig. 1 and eqs. (19) and (20)).

IS THE AGE OF THE EARTH AND OF SOLAR SYSTEM ~ 4.50 GA?

We have shown above that if it is assumed that D/N is such that eq. (8) gives t_a = 4.50 Ga, and c = -m/λ is in the range 1.2 – $2.1\cdot10^{-11}$ a^{-1}, the difference between the true age and the apparent age is -120 ÷ -200 Ma. Now, we want to determine the range of values of parameter c if λ is assumed to decrease slowly and roughly gradually over the geological time, as we can suppose if the sun was not in perfect equilibrium.

Eqs. (19) and (20) give the apparent age of a geological system as a function of the true age, in the case, respectively, of linear and exponential decrease of λ over geological time since the formation of the system. We re-arrange these two equations to obtain the true age as a function of the apparent age. In the case of linear decrease of λ we have, from eq. (19):

$$t = \frac{\lambda}{m}\left\{1 - \sqrt{1 - 2\left(\frac{m}{\lambda}\right)t_a}\right\}, \quad (21)$$

and in the case of exponential decrease of λ, from eq. (20), we have:

$$t = \frac{1}{c}\ln\left(1 + ct_a\right) \ . \quad (22)$$

Let us assume that λ decreased slowly and in a roughly gradual way over geological time because the emission of energy from the sun was not perfectly stable (Fig. 2). The true λ = λ(t) distribution is obviously unknown; however, we can consider the linear function of the time which best fits the points of the distribution (the dashed straight line in Fig. 2). We can also consider the exponential function which best fits the distribution (the continuous line in Fig. 2).

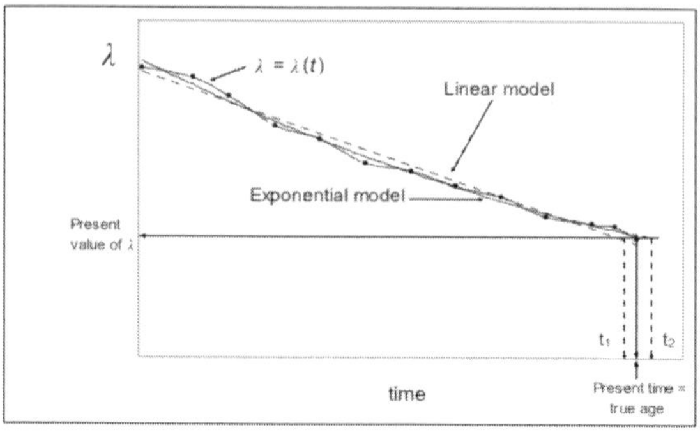

Figure 2: A possible (t, λ) relationship in case the emission of energy from the sun decreased over geological time, and its linear and exponential best-fit models. 'Present time' is the true age of the system, and t1 and t2 are the true ages of the system as calculated by interpreting the distribution in terms of these two models, respectively. See text for further explanations.

Note that we do not know the time-span elapsed since the formation of the sun and of the planetary objects; note, moreover, that the best-fit linear function intersects the time axis (the horizontal line which corresponds to the present value of λ) in a position which is generally different from the position of the intersection of the best-fit exponential function. We call the respective instants of intersection, which are also unknown, t_1 and t_2 (Fig. 2). We have:

$$t_1 = \frac{\lambda}{m}\left\{1 - \sqrt{1 - 2\left(\frac{m}{\lambda}\right)t_a}\right\} \tag{23}$$

and

$$t_2 = \frac{1}{c}\ln(1 + ct_a) \tag{24}$$

If λ decreased in a roughly gradual way, the (t, λ) distribution is approximately fitted by a linear function of the time. In this case, the argument of the exponential function which best fits the distribution is small in absolute value, because, from a geometrical point of view, the exponential function is close to a linear function of the time as well. If it is so, the exponential equation (14):

$$e^{ct} = \frac{\lambda_0}{\lambda}$$

can be written as:

$$1+ct \approx \frac{\lambda_0}{\lambda} \ . \qquad (25)$$

Eq. (25) shows that:

$$c \approx \frac{\lambda_0 - \lambda}{t}\frac{1}{\lambda} = -\frac{m}{\lambda} \qquad (26)$$

and substituting the relationship (26) into eq. (25) we obtain, for the best-fit exponential function:

$$\lambda \approx \lambda_0 + mt \ . \qquad (27)$$

Therefore, if λ decreased slowly and in a roughly gradual way over the whole geological time, parameter c in the exponential model of λ decrease is close to quantity -m/λ in the linear model of λ decrease.

Eq. (27), represents the 'linearization' of the best-fit exponential function. This 'linear' function is not perfectly equivalent to the best-fit linear function, because quantity m/λ has not exactly the same value in eqs. (26) and (21), and also λ_0 is not exactly the same (Fig.2). However, the respective geometrical representations are certainly close to each other, so that -m/λ in eq. (23) can be substituted by c from eq. (26), resulting in very small deviations in the calculated value of t_1. In this case, eq. (23) gives the true age t_1 as eq. (24) gives the true age t_2 , i.e. in terms of parameter c and apparent age ta:

$$t_1 \approx -\frac{1}{c}\left(1-\sqrt{1+2ct_a}\right) \ . \qquad (28)$$

Since the two geometrical representations (of the best-fit linear function and of the best-fit exponential function) are close, the true ages t_1 and t_2 are also close to one another. Thus, if we assume a maximum value for the difference t_1 - t_2 we can calculate the respective values of parameter c. We assume a limit value for this difference of approximately 10 Ma.

In eqs. (24) and (28), ta is the whole apparent geological time-span (4.5 Ga), and we calculate values of t_1 and t_2 assuming progressively higher values of c. If c is in the range $1.0\cdot10^{-20}$ -$1.0\cdot10^{-13}$ a^{-1}, the calculated true ages t_1 and t_2 are substantially undistinguishable from one another (the difference between t_1 and t_2 is always < 1 Ka), as well as from the apparent age (the deviation is $\leq \sim 1$ Ma).

If $c = 1.0 \cdot 10^{-12}$ a^{-1}, t1 - t2 is ~ 15 Ka, and if $c = 1.0 \cdot 10^{-11}$ a^{-1} this difference is ~ 1.4 Ma, which is also negligible. The difference between t_1 and t_2 increases to about 10 Ma if $c = 3.0 \cdot 10^{-11}$ a^{-1}, and to approximately 24 and 68 Ma if $c = 5.0 \cdot 10^{-11}$ a^{-1} and $1.0 \cdot 10^{-10}$ a^{-1}, respectively (if $c > 1.0 \cdot 10^{-10}$ a^{-1} the difference between t_1 and t_2 increases rapidly). Thus $3.0 \cdot 10^{-11}$ a^{-1} can be considered as the approximate upper limit of the range of 'possible' values of c. Fig. 3 shows the change of the values t_1 and t_2 as c increases, and their progressive divergence.

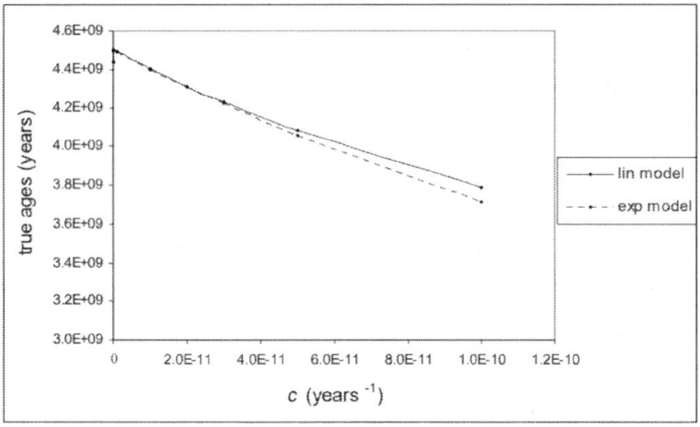

Figure 3: Relationship between parameter c and true ages t_1 and t_2 calculated from linear and exponential best-fit of true (t, λ) distribution, respectively, in the case λ decreased slowly and in a roughly gradual way over the whole geological time

If c is in the range $1.0 \cdot 10^{-20}$ - $3.0 \cdot 10^{-11}$ a^{-1}, the true age is in the range 4.50 – 4.23 Ga. Therefore, in a scenario of slow and roughly gradual decrease of the emission of energy from the sun, the true age of the Earth and of the solar system could be significantly lower than 4.50 Ga, and as low as 4.23 Ga. In the case of ^{87}Rb-^{87}Sr dating systematics, the age value 4.23 Ga corresponds to $\lambda_0 \sim 1.60 \cdot 10^{-11}$ a^{-1}, so that, in this scenario, the average rate of change of λ of ^{87}Rb would be ~ $4.26 \cdot 10^{-22}$ a^{-1} per year.

CONCLUSIONS

Arguments suggest that the process of radioactive decay is better re-interpreted in terms of second-order kinetics. Accordingly, the atomic systems of parent nuclides are not, as hitherto believed, 'radio-active', but 'radio-activated', and the rate of decay of each of these nuclides depends on the instantaneous concentration of one or more physical species which, by interacting with it, determines its transformation. The observed periodical variations in the value of λ of alfa- and beta-emitting nuclides with the Earth-Sun distance, and the change in λ value with the activity of the sun (Fischbach et al., 2009; Jenkins and Fischbach, 2009; Jenkins et al., 2009, 2010, 2012; Javorsek II et al., 2010; Sturrock et al., 2011) indicate that the process of radioactive decay cannot be longer interpreted in terms of first-order kinetics.

Since the value of λ depends on solar activity, the age values calculated using parent-daughter decay systematics are probably apparent ones. If the Sun was not stable from the point of view of emission of energy, snd the value of parameter λ slowly changed monotonically over the true age of a geological system, the calculated age value may significantly differ from the true age value, due to hyperbolic dependence of the calculated age on the value of the decay 'constant'. This in particular if the system is old.

If λ decreased over the true age of a geological sample, the apparent time ran faster than the true time, and we can calculate the deviation of the apparent age from the true age in case λ decreased according to possible models, for example a linear model and an exponential model. This deviation depends on the true age of the sample and on the relative rate of decrease of λ (parameters $-m/\lambda = -(\lambda-\lambda_0)/(\lambda t)$ in the case of linear decrease, and $c = -d\lambda/(\lambda dt)$ in the case of exponential decrease).

The calculations also show that the apparent ages of coeval samples in different planets which orbit around the same star coincide, if the same P-D decay systematics is used to date the samples. If different parent nuclides are activated by the same physical species, the apparent ages obtained by those radioactive P-D systematics also coincide.

Instead, due to generally different changes over time in the emission of energy from the respective stars, the apparent ages of coeval systems in planets which orbit around different stars generally differ, even if calculated using the same P-D systematics.

If the values of apparent age of a geological system determined using different P-D decay systematics coincide, it is possible that the calculated ages are close to the true age of the system, or, alternatively, that the respective parent nuclides are activated by the same physical species.

If λ decreases slowly and in a roughly gradual way over the true age of a geological sample, the linear function which best fits the true (t, λ) distribution and the exponential function which best fits this distribution are close. In this case, we can write the equation which gives the true age of the sample as given by the linear model of decrease of λ and the equation which gives the true age as given by the exponential model of decrease of λ in terms of the same parameters, i.e. the apparent age of the sample and the parameter of exponential decrease of λ, c.

This permits to calculate the range of possible values of c, and, consequently, the range of possible values of the true age of the sample. If the geological sample is the sun-planets system (apparent age ~ 4.5 Ga), we calculate $0 < c \leq 3.0 \cdot 10^{-11}$ a^{-1}, and the true age of the system in the range 4.5 – 4.23

You may also consider the following issues:

- What is the theoretical, clinical, or practical significance of the outcomes, and what is the basis for these interpretations? If the findings are valid and replicable, what real-life psychological phenomena might be explained or modeled by the results? Are applications warranted on the basis of this research? (Note 1)
- What problems remain unresolved or arise anew because of these findings? The responses to these questions are the core of the contribution of your study and justify why readers both inside and outside your own specialty should attend to the findings. Your readers should receive clear, unambiguous, and direct answers. (Note 2)

ACKNOWLEDGMENTS

This work was financially supported by C.N.R., Istituto di Geoscienze e Georisorse. The author is grateful to B. Zanettin, A. Della Giusta, A. Fasson and D. Roccato for discussions. Ida M. Samperi is very gratefully acknowledged for her generous care in language revision.

REFERENCES

1. Alburger, D. E., Harbottle, G., & Norton, E. F. (1986). Half-life of 32Si. Earth Planet. Sci. Lett. 78, 168-176.
2. Fischbach, E., Buncher, J., Gruenwald, J., Jenkins, J. H., Krause, D., Mattes, J., & Newport, J. (2009). Time-dependent nuclear decay parameters: new evidence for new forces? Space Sci. Rev., 145, 285-335.
3. Javorsek II, D., Sturrock, P. A., Lasenby, R. N., Lasenby, A. N., Buncher, J. B., Fischbach, E., Gruenwald, J.T., … Terry, B. (2010). Power spectrum analyses of nuclear decay rates. Astropart. Phys., 34, 173-178.
4. Jenkins, J. H., & Fischbach, E. (2009). Perturbation of nuclear decay rates during the solar flare of 2006 December 13. Astropart. Phys., 31, 407-411.
5. Jenkins, J. H., Fischbach, E., Buncher, J. B., Gruenwald, J. T., Krause, D. E., & Mattes, J. J. (2009). Evidence of correlations between nuclear decay rates and Earth–Sun distance. Astropart. Phys., 32, 42-46.
6. Jenkins, J. H., Herminghuysen, K. R., Blue, T. E., Fischbach, E., Javorsek II, D., Kauffman, A. C., Mundy, D. W., Sturrock, P. A., & Talnagi, J. W. (2012). Additional experimental evidence for a solar influence on nuclear decay rates. Astropart. Phys., 37, 81-88.
7. Jenkins, J. H., Mundy, D. W., & Fischbach, E. (2010). Analysis of environmental influences in nuclear half-life measurements exhibiting time-dependent decay rates. Nucl.Instr. and Meth. in Phys. Res. Sect. A, 620, 332-342.
8. Malatesta, L., & Cenini, S. (1999). Principi di Chimica Generale. Con esercizi. Milano, Italy: Casa Editrice Ambrosiana.
9. Rieger, E., Share, G. H., Forrest, D. J., Kanbach, G., Reppin, C., & Chupp, E. L. (1984). A 154-day periodicity in the occurrence of hard solar flares? Nature, 312, 623- 625.
10. Rutherford, E., Chadwich, J., & Ellis, C. D. (1930). Radiation from radioactive substance. Cambridge, U.K: Cambridge university Press.
11. Siegert, H., Schrader, H., & Schotzig, U. (1998). Half-life measurements of Europium radionuclides and the long-term stability of detectors. Appl. Radiat. Isot., 49, 1397- 1401.

12. Steiger, R. H., & Jager, E. (1977). Subcommision on Geochronology: Convention on the use of decay constants in eo- and Cosmochronology. Earth Planet. Sci. Lett., 36, 359-362.
13. Sturrock, P. A., Fischbach, E., & Jenkins, J. H. (2011). Further evidence suggestive of a solar influence on nuclear decay rates. Solar Phys., 272 (1), 1-10.

5

PET-Specific Parameters and Radiotracers in Theoretical Tumour Modelling

Matthew Jennings[1,2] , Loredana G. Marcu[1,3] , and Eva Bezak[1,2]

[1]School of Chemistry & Physics, University of Adelaide, Adelaide, SA 5000, Australia
[2]Department of Medical Physics, Royal Adelaide Hospital, Adelaide, SA 5000, Australia
[3]Faculty of Science, University of Oradea, 410087 Oradea, Romania

ABSTRACT

The innovation of computational techniques serves as an important step toward optimized, patient-specific management of cancer. In particular, *in silico* simulation of tumour growth and treatment response may eventually yield accurate information on disease progression, enhance the quality of cancer treatment, and explain why certain therapies are effective where others are not. *In silico* modelling is demonstrated to considerably benefit from information obtainable with PET and PET/CT. In particular, models have successfully integrated tumour glucose metabolism, cell proliferation,

Citation: Jennings, M. Marcu, L. Bezak, E. "PET-specific parameters and radiotracers in theoretical tumour modelling" Computational and Mathematical Methods in Medicine, 2015; 2015:415923-1-415923-11. http://dx.doi.org/10.1155/2015/415923.

and cell oxygenation from multiple tracers in order to simulate tumour behaviour. With the development of novel radiotracers to image additional tumour phenomena, such as pH and gene expression, the value of PET and PET/CT data for use in tumour models will continue to grow. In this work, the use of PET and PET/CT information in *in silico* tumour models is reviewed. The various parameters that can be obtained using PET and PET/CT are detailed, as well as the radiotracers that may be used for this purpose, their utility, and limitations. The biophysical measures used to quantify PET and PET/CT data are also described. Finally, a list of *in silico* models that incorporate PET and/or PET/CT data is provided and reviewed.

INTRODUCTION

Anatomic imaging modalities, particularly X-ray computed tomography (CT) and magnetic resonance imaging (MRI), have long been the standard tools for the accurate localization of organs and lesions in radiation oncology. Today, they play a routine role in three-dimensional treatment planning. However, the effectiveness of structural imaging techniques in determining metabolic or functional tissue information is limited. Functional imaging has been demonstrated to be invaluable for the initial diagnosis and staging of cancer as well as the monitoring of therapy and the detection of cancer recurrence [1]. Metabolic changes in tissue commonly precede the structural changes that are detected via CT and MRI. Thus, imaging of metabolic changes may enable the detection of malignant disease at earlier stages of development [2]. In addition, the availability of functional information is advantageous for cases in which there is poor contrast between normal and malignant tissue when using structural imaging. For example, for the initial staging of lymphomas, the metabolic information provided by positron emission tomography (PET) enables more accurate delineation of the extent of nodal disease as compared with CT and bone scans. Similarly and perhaps most notably, PET demonstrates superior local staging capabilities for head and neck cancers over both CT and MRI. FDG-PET alone has a plethora of indications for a wide variety of malignancies [3]. Consequently, functional imaging modalities such as PET are playing an increasingly important role in the management of malignant disease.

PET scanning has progressed into widespread clinical practice since its first commercialization in the late 1970s. For oncological PET studies, the most utilized and extensively researched radiotracer is ^{18}F-fluorodeoxyglucose (FDG). FDG-PET has demonstrated superior accuracy over conventional imaging modalities in multiple scenarios across both the diagnosis and the staging of cancer [3, 4]. In a variety of clinical settings, FDG-PET exhibits improved values of sensitivity, specificity, or both. Examples of clinical scenarios in which FDG-PET has demonstrated efficacy include the evaluation of mass lesions, the staging and restaging of cancer, the planning of radiotherapy treatments, the monitoring of therapy, and the detection of cancer recurrence [3]. However, the primary drawback of functional imaging is the lack of anatomic information that it provides. This necessitates its accompaniment with structural imaging for application in clinical oncology [1].

In order to take advantage of their inherent benefits, the combined use of both anatomic and physiologic imaging modalities is optimal. The accurate structure localization capabilities of CT complement the mapping of normal and abnormal tissue function performed by PET. Because of this, PET images are routinely read alongside CT images in order to both distinguish and localize metabolic irregularities. In the first instance, this has been achieved via the coregistration of separately acquired PET and CT images using fusion software, a technique which has especially proved to be effective for brain imaging. However, coregistration of images of other anatomical regions poses significant challenges [5]. Difficulties in the registration process primarily arise from varying patient positioning between the two image sets. Whilst such difficulties are minimal for brain scans, they may be significant in other anatomical regions wherein there may be substantial organ movement or deformation. In light of this, the development and subsequent commercialization of the PET/CT scanner in 2001 has generally addressed the issues affecting the coregistration of separately acquired PET and CT images [6].

The use of integrated PET/CT in oncologic imaging has since become a widespread field of research, particularly with the utilization of ^{18}F-FDG. This combined modality overcomes some of the drawbacks that are characteristic of standalone PET scanning, namely, the significant presence of noise in attenuation correction factors, the lengthy duration of scans, and the absence of anatomic markers [7]. While the vast majority of published research concerns standalone PET, PET/CT has begun to show great promise across a multitude of clinical settings [3]. Studies have shown significantly improved accuracy in the staging of nonsmall cell lung cancer with PET/CT over the separate performance of PET and CT [8–10]. Some centres perform a series of PET/CT scans on nonoperative head and neck cancer patients following either radiation therapy or chemotherapy due to its indispensable combined anatomical and functional information [3, 7, 11]. For patients with recurrent or metastasized thyroid cancer, the localization capabilities of ^{18}F-FDG-PET/CT can lead to improved diagnostic accuracy [12]. Finally, the incorporation of FDG-PET/CT data into the radiation treatment planning process has repeatedly shown to improve target delineation and enhance the therapeutic ratio via its increased cancer staging accuracy [13]. The ongoing clinical evaluation and further innovation of PET/CT, namely, via technological improvements and the establishment of new tracers, will continue to propel this technology into widespread use in oncology [3].

The significant interpatient variability in both tumour behaviour and normal tissue response to cancer therapy has prompted increasing demand for individualized treatment planning. Simulations of radiobiological processes in malignancies provide scope for the further optimization of individual treatment plans and to ultimately improve patient outcomes. The development of computer or *in silico* models of these radiobiological processes is particularly beneficial [14]. It is here in which PET and PET/CT, as noninvasive functional imaging modalities, can play a crucial role [15–17]. The aim of this work is to review the application and effectiveness of PET and PET/CT for the *in silico* modelling of tumours. In particular, it examines the various parameters that can be determined via PET and PET/CT imaging along with the assortment of radiotracers available for these purposes. The biophysical quantities used to quantify PET and PET/CT data, including SUV and compartmental models, are assessed. Finally, an overview of the existing *in silico* models that utilize PET or PET/CT data is provided.

TUMOUR MODEL PARAMETERS OBTAINABLE FROM PET IMAGING

With the objective of optimal, individualized treatment planning, many biological characteristics of a given tumour can be considered for computer simulation. In the case of radiation therapy, the proliferative potential of a tumour and its radiosensitivity are of particular interest. Parameters which describe tumour cell proliferation and repopulation characteristics are useful for *in silico* modelling of tumour growth and treatment response. Similarly, since it is well established that tumour's oxygenation level contributes significantly to its radioresistance, modelling parameters that characterize tumour hypoxia and angiogenesis are also common. Increasingly, *in silico* models have expanded to incorporate additional contributing factors to tumour radiosensitivity, including intracellular tumour pH, gene expression, and cell-cycle simulation. The complexity of any given model dictates the number of characteristics simulated, as well as the accuracy with which each tumour characteristic is simulated. Simpler models may use averaged, macroscopic measures of a given tumour characteristic, while more complex models may simulate a tumour and its behaviour at the microscopic, molecular, and even atomic levels. Indeed, some of the most accurate, contemporary models are "multiscale"; they simulate tumour behaviour across multiple biological scales, reconciling the macroscopic and microscopic levels [18]. Details of the various parameters utilized for tumour modelling obtainable from PET and PET/CT data are provided below.

Cell Proliferation

There are many approaches to modelling cell proliferation but it is useful to separately consider modelling at the macroscopic and microscopic scales. At the macroscopic scale, tumours are commonly simulated using continuum models. Using this approach, gross tumour morphology and behaviour are modelled under various environmental conditions and typically governed by a set of differential equations whose initial conditions serve as input parameters [18]. Input parameters include such quantities as cell density (both viable and necrotic), cell volume fractions, and gross tumour proliferation and metabolic rates [19]. Such parameters may be gathered from PET/CT data, where a corresponding radiotracer can be utilized [20]. Details of various radiotracers and the functional information they provide are given in Section 3.

In contrast to continuum tumour models, discrete tumour models developed at the microscopic scale simulate individual cell behaviour. Their input parameters correspondingly describe a set of biophysical rules applied to the modelled cells and the diversity of these rules varies across different models. Each cell is assigned an initial state and its status is subsequently tracked throughout the simulation. Consequently, discrete models directly simulate cell proliferation; a process which is mediated by input parameters such as proliferative potential, cell-cycle positions, and durations for different cell types, probability of cell division, cell loss rates, and maximum tumour radius or cell number [19]. Patient-specific parameters such as tumour radius and proliferating cells per voxel are directly obtainable from functional imaging data [15].

Hypoxia

The adverse effects of tumour hypoxia on patient outcomes following radiotherapy have been well established. Indeed, hypoxia is often attributed to poor tumour control probabilities in locally advanced head and neck cancers, for which low oxygenation levels are common [21]. Accordingly the simulation of oxygenation in tumours has been a prolific area of research in the tumour modelling community for more than 60 years [22]. The simulation of tumour hypoxia varies greatly across different models; the most complex models concurrently simulate oxygen diffusion, oxygen consumption by tissue, and the interdependence of oxygen with tumour proliferation and vasculature [23].

Oxygen information is generally modelled using an analytic approach. That is, sets of differential equations are used to describe oxygen diffusion and/or consumption rates both during tumour growth and in response to treatment. Even stochastic (discrete) tumour cell proliferation models that incorporate oxygen information typically employ a composite approach; oxygen and other substrate concentrations are generally governed by continuous fields. Since the oxygen distribution within a tumour is directly related to the extent and nature of its vasculature, many tumour hypoxia models incorporate blood vessel information into their simulations. The most sophisticated models simulate angiogenesis (see Section 2.3) and its effect on tumour hypoxia [16, 23, 24].

Simulated oxygen distributions are typically quantified using partial oxygen tension or pO_2 values. Thus, for the successful simulation of tumour hypoxia, realistic initial pO_2 conditions must be set. In order to categorize the oxygen status of tissue, binary approaches to oxygen modelling establish a defined threshold pO_2 value below which the region is considered to be hypoxic. More robust models establish a more detailed relationship between tumour growth or treatment response and oxygen information and may incorporate both oxygen diffusion and oxygen consumption by tissue [23, 25]. Additional parameters of interest may be included for oxygen effects, such as reoxygenation probability distributions and hypoxic thresholds for which cell quiescence is initiated. Multiple studies have demonstrated the utility of PET for obtaining pO_2 distributions for use in patient-specific tumour models [16, 23, 26].

Angiogenesis

The extent and nature of tumour vasculature significantly influence tumour growth and oxygen status. Consequently, the simulation of angiogenesis serves as a useful complement to models of tumour cell proliferation and hypoxia. Likecell proliferation, models of tumour-induced angiogenesis may be either continuous or discrete in nature. More advanced models utilize both approaches for simulating the various mechanisms involved in angiogenesis, such as capillary sprout formation, endothelial cell migration, blood flow, and vessel adaptation [19].

Input parameters related to vessel branching generally include probabilities of random endothelial cell migration, chemotaxis with tumour angiogenesis factor concentration (especially vascular endothelial growth factor or VEGF), and haptotaxis with fibronectin gradients [18]. The utility of PET for the imaging of specific angiogenic markers, such as $_{v}\,_{3}$ integrins and tumour expression of VEGF, has been demonstrated by multiple groups [28, 29]. To date, tumour models of angiogenesis have traditionally not incorporated angiogenesis-specific PET data. However, since cellular oxygenation and tumour vasculature are intimately related phenomena, hypoxia-specific PET data has been utilised in models of tumour vasculature [16, 23]. Specifically, the value of PET information for the conversion of oxygen maps into capillary density maps has been demonstrated. The potential of angiogenesis-specific PET-based imaging for input in modelling the temporal development of vasculature or angiogenesis is well understood and will develop with ongoing studies [16].

pH

Though the intracellular pH of solid tumours is maintained in a range similar to that of normal cells, the extracellular pH of solid tumours is commonly acidic. The increased glucose metabolism of solid tumours, assisted by characteristically poor perfusion, is the most probable cause of their low extracellular pH. This is because glucose catabolism results in net acid production and insufficient vasculature cannot remove excess acid from the extracellular environment [30].

Perhaps the most compelling value of including pH in tumour growth models arises from the acid-mediated tumour invasion hypothesis. This suggests that tumour cells develop phenotypic adaptations to the harmful effects of acidosis during carcinogenesis, traits that are not present in normal cells. Consequently, tumour cells are rendered relatively impervious to the decreased pH in the tumour microenvironment resulting from increased anaerobic glycolysis, which is otherwise toxic to normal tissue. Effectively the tumour provides for itself a selective growth advantage and a useful mechanism for invasion [31]. Investigation of this phenomenon is particularly suited for *in silico* tumour growth modelling and the role played by acid gradients in triggering tumour invasion has been evaluated this way [31]. Excess H^+ ion concentration may be simulated in a continuum model utilizing reaction-diffusion partial differential equations, where input parameters such as acid production rate, reabsorption rate, and H^+ ion diffusion coefficients may be obtained from measurement [32].

As with angiogenesis, *in silico* models incorporating tumour pH have not utilized noninvasive PET data for input. Though PET has been used for the measurement of pH since the 1970s and was the first noninvasive *in vivo* pH meter, it has historically been both an inaccurate and imprecise measurement tool [30]. However, the development of novel radiotracers that selectively target acidic tumours will enable the incorporation of pH related PET data into *in silico* models of tumour growth [33].

RADIOTRACERS USED FOR TUMOUR MODELLING

PET and PET/CT are able to image an increasing variety of physiological phenomena. This versatility arises from the ability to select a radiotracer that specifically targets a particular mechanism. Additionally, the diversity of PET tracers continues to expand with ongoing innovations in radiopharmaceutical production. Today, radiotracers exist for the imaging of metabolism, proliferation, perfusion, drug/receptor interactions, and gene expression. Despite this variety, the extent of PET data incorporated into *in silico* tumour models has so far been limited to radiotracers specific to glucose metabolism, cell proliferation, and hypoxia. There is significant potential for the use of alternative radiotracers to obtain additional functional information for *in silico* models using PET. A list of radiotracers whose information has been directly incorporated into *in silico* models as well as those that show significant promise for such applications is provided in Table 1.

Table 1: PET Radiotracers whose data has been/the potential to be incorporated into *in silico* models

PET radio-tracer	Functional characteristic	Corresponding tumour model parameters	Use in *in silico* models?
FDG	Glucose metabolism	(i) Intracellular Volume Fraction (ICVF)	Yes
		(ii) Acid production rates*	
FLT	DNA replication	Tumour cell proliferative rates (vector- or voxel-based)	Yes
FMISO	Hypoxia	(i) Partial oxygen tension (pO_2)	Yes
		(ii) Relative hypoxic fraction (RH)	
Cu-ATSM	Hypoxia	Partial oxygen tension (pO_2)	Yes
pHLIP	Acidosis	Extracellular pH (pH_e)	Potential
Galacto-RGD	Angiogenesis	(i) expression rate	Potential
FDOPA	Malignancy	(ii) L-DOPA activity	Potential
FES	Malignancy	(iii) Oestrogen overexpression	Potential

The specificity of FDG to glucose metabolism provides in indirect measure of acid production rates in tumour cells, since anaerobic glycolysis is net acid producing.

FDG

^{18}F-2-Fluoro-2-deoxy-glucose or FDG is by far the most commonly used and extensively researched PET radiotracer. Today, FDG-PET plays an important role in oncology. It has been recommended for use as an imaging tool additional to traditional radiological modalities in the appropriate clinical setting. In particular, it has demonstrated efficacy in the diagnosis, staging, unknown primary discovery, and the detection of cancer recurrence [1].

Increased glucose consumption is a typical characteristic of most cancers. In hypoxic regions, the Pasteur effect results in the upregulation of anaerobic glycolysis and the GLUT 1 glucose transporter in tumour cells. However, even if oxygen is plentiful, cancers undergo accelerated glycolysis. This observation, called the Warburg effect, is widely attributed to mutations in oncogenes and tumour suppressor genes [34]. Since FDG is a glucose analogue, it is a particularly suitable radiotracer to measure the increased glucose utilization typical of cancers. Along with increased glucose consumption, the upregulation of appropriate enzymatic activity further amplifies FDG uptake in tumour cells.

The primary drawback of FDG-PET for oncologic imaging is that FDG uptake is not specific to cancer. That is, FDG-PET exhibits a poor level of specificity for certain applications. FDG uptake may be intense in benign diseases as well as in areas of infectious disease and inflammatory tissue. That is, there are many potential causes of false-positive PET signals in oncologic imaging [1, 35, 36]. Conversely, some malignant diseases do not exhibit high glycolytic activity. Bronchioloalveolar carcinoma and carcinoid tumours are examples of cancers for which false-negative signals may occur for standalone FDG-PET imaging [35]. Combining FDG-PET with other imaging modalities has served to mediate this drawback somewhat; FDG-PET/CT has demonstrated superior performance than standalone FDG-PET in common cancers [37]. Additionally, the emergence of novel radiotracers that target biochemical processes that are more specific to cancer promises to overcome the relative nonspecificity of FDG-PET in oncologic imaging.

Commensurate with the predominance of FDG as the PET radiotracer of choice, metabolic information provided by FDG-PET is often utilized in *in silico* models of both tumour growth and treatment response [16, 20]. The specific information employed from FDG-PET varies across such models. Images may be used solely to identify existing cancerous tissue, particularly in simulations for the prediction of tumour response to therapy [16]. Alternatively, glucose metabolism data from FDG-PET can be quantitatively used to simulate metabolic processes in predictive models of tumour growth [20].

FLT

PET radiotracers specific to cell proliferation are an effective alternative to those specific to glucose metabolism, such as FDG. Of these tracers, ^{18}F-3-fluoro-3-deoxy-thymidine (FLT) is perhaps the most researched and the most utilized. FLT-PET is typically a less sensitive imaging modality than FDG-PET: the difference in FLT uptake between normal and malignant tissues is usually less pronounced than that for FDG [38]. However, FLT uptake correlates very well with Ki-67, an index of cell proliferation. Consequently, FLT-PET is useful for aiding in the grading of tumours. The combined FLT-PET/CT modality has demonstrated efficacy for the early prediction of treatment response [39–41] as well as the assessment of cancer aggressiveness [42]. The uptake of FLT in infectious or inflammatory tissue is less than that of FDG and FLT has lower background activity in the brain and thorax [43, 44]. Consequently, the specificity of FLT-PET/CT exceeds that of FDG-PET/CT in certain imaging applications [44].

The magnitude of FLT that is trapped in a tumour cell is proportional to the amount of DNA/RNA synthesis undertaken by the cell. Since the growth of malignant tissue is intricately related to DNA replication, the degree of FLT uptake is strongly correlated with proliferation rate. Accordingly, FLT-PET data is particularly useful for patient-specific tumour modelling, where proliferative rates are often of paramount interest. A group led by Benjamin Titz at the University of Wisconsin has correspondingly acquired cell proliferation information from FLT-PET data for *in silico* models of tumour growth and treatment response [15, 16].

FMISO

In an effort to develop accurate, noninvasive measurement techniques of tumour hypoxia, a number of PET radiotracers have been produced which irreversibly bind to cells in poorly oxygenated conditions. Of these, ^{18}F-fluoromisonidazole (FMISO) is the most extensively studied and clinically validated. FMISO uptake is inversely proportional to O_2 level and perfusion does not restrict its delivery to malignant tissue. Several studies have demonstrated FMISO-PET to be a viable prognostic indicator of tumour response to treatment [45].

FMISO-PET has not been universally adopted into routine clinical application because of a number of limitations inherent in the radiotracer. Since it relies on passive transport mechanisms, its uptake is relatively slow in hypoxic tumours, usually requiring 2–4 hours to be selectively retained in the target following injection [46]. FMISO-PET imaging also exhibits relatively low tumour-to-background ratios, since its binding to malignant, hypoxic cells is highly nonspecific. Finally, considerable levels of unwanted, radioactive metabolite products result from the nonoxygen dependent metabolism of FMISO [45]. Improvements in the quantification of hypoxia by modelling FMISO-PET dynamics, as opposed to using the standardized uptake value (SUV) in a binary manner, may aid in overcoming contrast limitations. Several studies have demonstrated reasonable success in the simulation of tracer transport and its application to tumour models (see Section 5) [25–27].

Cu-ATSM

Alternative hypoxia-specific PET radiotracers have been developed in order to overcome the various limitations of FMISO. Several other nitroimidazole compounds have been developed for this purpose [47]. In 1997, an alternative hypoxia PET tracer was proposed that does not suffer from the undesirable radioactive residues of nitroimidazoles. This tracer is Cu(II)-diacetyl-bis(N^4-methylthiosemicarbazone) or Cu-ATSM [48]. Cu-ATSM has evolved to become one of the most promising PET agents for hypoxia imaging. It has demonstrably high hypoxic tissue selectivity [45]. It is able to rapidly identify hypoxic tissue with high tumour-to-background ratios, due to a combination of small molecular weight, high cell membrane permeability, rapid blood clearance, and prompt retention in hypoxic tissues [45].

The effectiveness of Cu-ATSM for providing clinically relevant tumour oxygenation information has been confirmed in multiple studies and its predictive value of tumour behaviour and treatment response has been demonstrated [49–51]. Perhaps unsurprisingly, it is a preferred radiotracer for the determination of oxygenation information using PET for incorporation into *in silico* tumour models [15, 16, 23]. For example, Titz and Jeraj chose a sigmoidal relationship between the SUV of Cu-ATSM and local tissue oxygenation [15], following the findings of Lewis et al. [52]. Using pretreatment Cu-ATSM-PET spatial maps of tumour oxygenation, it was demonstrated that lower oxygen levels resulted in reduced treatment efficacy. However, the group did note that further investigation into the quantitative relationship between partial oxygen tension and Cu-ATSM uptake is warranted.

Other Radiotracers

Although *in silico* models are yet to incorporate PET or PET/CT information beyond glucose metabolism, cell proliferation, and tumour oxygenation, there is scope for the use of tracers that image additional processes. Tumour acidosis arising from amplified glycolysis is a common feature of cancers and is a likely trigger of invasion into surrounding tissue [53]. Consequently, several mathematical models, inclusive of tumour acidity, have been developed to study the glycolytic phenotype and the tumour-host interface [31, 54]. There is suggestive potential of PET and particularly PET/CT for directly obtaining parameters of interest for such models, including glucose metabolic rate and acid production rates. One promising, novel PET tracer that specifically targets acidosis is pH low insertion peptide (pHLIP) [33]. pHLIP binds to acidic cell membranes and has demonstrated ability to target areas of hypoxia and carbonic anhydrase IX (CAIX) overexpression, an acid-extruding protein [55].

As discussed in Section 2, the simulation of angiogenesis is of significant interest for *in silico* tumour modelling. Though parameters related to angiogenesis can be indirectly obtained using hypoxia-specific PET-imaging, alternative markers of angiogenesis can instead be targeted. For example, vascular integrins are targeted by PET radiotracers containing the tripeptide sequence arginine-glycine-aspartic acid (RGD) [28]. In particular, the$\alpha_v\beta_3$ integrin is a receptor related to cell adhesion and involved in tumour-induced angiogenesis that can be imaged using radiotracers such as ^{18}F-Galacto-RGD [29].

In models that simulate specific tumour types, PET information from alternative tracers might be useful. For example, neuroendocrine tumours are typically characterised by an increased L-DOPA decarboxylase activity [56]. The imaging of advanced neuroendocrine tumours has been validated with PET using ^{18}F-dihydroxyphenylalanine (FDOPA) [57] and may be of value in the modelling of these tumours. Similar arguments may be made for the simulation of breast cancers and the imaging of oestrogen receptor expression using ^{18}F labelled oestrogens such as ^{18}F-fluoroestradiol (FES) [58].

BIOPHYSICAL PARAMETERS USED IN PET AND PET/CT

The integration of reliable imaging-based information into *in silico* models of tumour growth and treatment response greatly relies on the accurate and precise quantification of imaging data. This is especially relevant for PET and PET/CT, for which radiotracer uptake is dependent on a host of factors. SUV is the most extensively used parameter clinically for the analysis of PET tracers, but its high degree of sensitivity to multiple variables can render the comparison of SUVs taken at different times or between different centres to be extremely difficult [59].

Details of the biophysical parameters used to quantify PET and PET/CT data are provided below. Their applications and limitations are discussed and compared, as well as the various methods that have been developed to overcome the potential pitfalls of a given measure. The present use of PET and PET/CT quantification measures in *in silico* models of tumour growth and treatment response is also discussed.

SUV

The standardized uptake value is the quintessential parameter employed to analyse and quantify PET radiotracer data. It is defined as follows:

$$\mathrm{SUV} = \frac{\text{radiotracer concentration in ROI}}{\text{total injected activity}/N}\ \mathrm{g/mL}, \quad (1)$$

where the concentration is as measured with PET in kBq/mL, ROI is the region or volume elements of interest, and N is a factor normalizing for body weight, body surface area, or lean body mass. The overall denominator has units kBq/g. The radiotracer is commonly computed by scanning the patient for a 5–15-minute interval after a predetermined period (e.g., 1 hour) after radiotracer injection.

In general, SUV depends on the time between injection and scanning as well as multiple image acquisition settings such as the reconstruction algorithm and scatter and attenuation corrections [59]. Its comparative value is hampered by methodology differences, such as choice of normalization factor and choice of *max*, *mean*, *peak*, or *total* SUV [60]. SUV may also be confounded by biological mechanisms, such as variations in plasma clearance before and after treatment and plasma glucose concentration (in the case of FDG) [59].

Despite its limitations, SUV poses advantages over alternative quantification techniques such as compartmental methods (see Section 4.2). It is the only method of PET quantitative analysis that can be realistically employed for routine clinical use, due to its sheer simplicity and the efficiency of the associated scan protocols. In addition, the effectiveness of SUV for the assessment of cancer therapy response by comparing values for scans taken before and after treatment has been extensively validated. This is particularly true for FDG-PET, whose efficacy has been confirmed for multiple cancer types [59]. Accordingly, in spite of its large variation in

SECTION 3:
NUCLEAR FORCES

6

Nuclear Forces in the Medium: Insight From the Equation of State

Francesca Sammarruca and Randy Millerson

Physics Department, University of Idaho, Moscow, ID, United States

In this review, we concentrate on recent efforts of our group aimed at investigating the nuclear equation of state of symmetric nuclear matter (equal concentrations of protons and neutrons) and the one of pure neutron matter. Although idealized, these systems are suitable "laboratories" to probe nuclear forces in the many-body system. The energy per particle as a function of density can reveal rich information about the nature of nuclear forces in the medium and how they impact observable properties. For instance, the pressure in neutron-rich matter has been found to have profound impact on very diverse systems, ranging from the thickness of the neutron skin in a heavy nucleus to the properties of compact stars. The

Citation: Sammarruca F and Millerson R (2019) "Nuclear Forces in the Medium: Insight From the Equation of State". Front. Phys. 7:213. https://doi.org/10.3389/fphy.2019.00213

current trend in nuclear physics is to build few-nucleon forces according to the prescription of chiral effective field theory. We open by reviewing in depth how we develop equations of state based on state-of-the-art chiral interactions. We then highlight some applications in neutron-rich nuclei and neutron stars.

Keywords: nuclear forces, chiral effective field theory, symmetry energy, neutron skin thickness, neutron stars

INTRODUCTION

Understanding the interaction of hadrons in nuclei is a most fundamental problem in nuclear physics. Our present knowledge of the nuclear force in vacuum is still incomplete, although decades of efforts have been devoted to this problem. The study of nuclear forces in many-body systems is, of course, much more challenging because additional aspects are involved beyond those which can be constrained by free-space nucleon-nucleon (*NN*) scattering. Predictive power with respect to the properties of nuclei is the true test for a successful microscopic theory.

The system known as "nuclear matter" is a suitable, although idealized, theoretical "test bench" for many-body theories. Nuclear matter is defined as an infinite system of nucleons interacting *via* strong forces in the absence of electromagnetic interactions. Nuclear matter›s "signature" is its energy per particle as a function of density and potentially additional "variables" (for instance, isospin polarization or temperature). The nuclear matter equation of state (EoS) is precisely the energy per particle as a function of density and other appropriate quantities. Naturally, the idealized nature of this system, which implies translational invariance, simplifies theoretical calculations. Furthermore, within the "local density approximation" (LDA), one can utilize the EoS directly in calculations of actual nuclei. (We recall that LDA amounts to the assumption that the properties at a point with density ρ in a nucleus are the same as they would be in infinite nuclear matter at the same density).

When the densities of protons and neutrons are equal, we speak of isospin-symmetric nuclear matter. The latter has been studied since the earlier works by Brueckner and others [1–4], who introduced what became known as the Brueckner-Hartree-Fock (BHF) theory. The BHF theory seeks to find the ground state energy of a many-body system [1–6] as a linked-cluster perturbation expansion. The main point was the realization that regrouping the linked-cluster diagrams by the number of hole lines allowed the series to converge.

Other approaches to the development of the EoS were also pursued, one of them being the variational method [7, 8]. The latter yielded predictions in close agreement with those from Brueckner theory if realistic *NN* potentials were adopted [9].

The BHF theory, or "conventional approach," was followed by the Dirac-Brueckner-Hartree-Fock (DBHF) approach [10–13], developed during the 1980's. The novel, and most striking feature of the DBHF theory was its ability to describe the saturation properties (both energy and density) of nuclear matter, a fundamental aspect which reflects the saturating nature of the nuclear force. The DBHF method contains important relativistic features through the description of the nuclear mean field in terms of a scalar and a vector components, strong and of opposite sign. In their combination, they provide an explanation for the binding of nucleons and the spin-orbit splitting in nuclear states. The reader is referred to Muether et al. [14] for a relatively recent review of the DBHF method and a variety of applications to both nuclear matter and nuclei.

Irrespective of the many-body framework, a quantitative *NN* potential must be part of its input. Presently, forces based on nuclear chiral effective field theory (χEFT) [15–18] are the most popular. Chiral effective field theory respects the symmetries of quantum chromodynamics (QCD) but, at the same time, makes use of the degrees of freedom typical of low-energy nuclear physics, nucleons, and pions. Furthermore, it provides a clear systematics to determine the few- and many-body diagrams which must be present at each order of the perturbation.

Deriving nuclear forces directly from QCD is problematic. For starters, each nucleon is, itself, a complicated many-body system consisting of quarks, quark-antiquark pairs, and gluons, thus rendering the two-nucleon problem an even more complex many-body problem. Second, the interaction among quarks, which is due to the exchange of gluons, is very strong at the low energies involved in nuclear physics processes. For this reason, it is difficult to find converging perturbative solutions. Therefore, the first attempts to incorporate QCD in nuclear physics consisted mostly of QCD-inspired quark models. On the positive side, these models sought to explain nucleon structure (which consists of three quarks) and nucleon-nucleon processes (involving six-quarks) in an internally consistent manner. Some global features of the two-nucleon force, like the "hard core," could be explained by these quark models. On the other hand, quark-based approaches are, in fact, models rather than a theory. From an entirely different point of view, one may confront the six-quark problem by putting this system on a four dimensional discrete lattice representing three dimensions in space and one in time. This method is known as lattice QCD. Although progress in lattice QCD goes on, such calculations are computationally very demanding and thus the approach is not (currently) feasible as a standard tool to make predictions of nuclear properties.

A new era for the theory of nuclear forces started when Steven Weinberg worked out an effective field theory (EFT) for low-energy QCD [16, 19]. He argued that all one needs to do is to write the most general Lagrangian consistent with all the properties of low-energy QCD, as this action would render the theory equivalent to low-energy QCD. A crucially important property for this discussion is $SU(2)_R \times SU(2)_L$ symmetry, or chiral symmetry, which is "spontaneously" broken, as briefly reviewed next. Massless spin- $\frac{1}{2}$ fermions have their spin and momentum either parallel to each other ("right-handed") or anti-parallel ("left-handed"), a property which is referred to as having definite chirality. Since nucleons are made of "up" and "down" quarks, which have nearly zero mass, chiral symmetry holds approximately. As a consequence of this symmetry, one might expect to find in nature mesons of the same mass but opposite parity. However, such parity "doublets" are not observed, which amounts to a "spontaneous" breaking of the symmetry. According to a theorem first proven by Goldstone, the spontaneous symmetry breaking implies the existence of a pseudoscalar meson, the pion. Thus, the pion plays an outstanding role in generating the nuclear force. Pions and nucleons interact weakly at low energies as compared to the gluons and quarks. Therefore, calculations of pion-nucleon processes pose no problems. Moreover, in EFT one makes use of expansions in powers of momentum over an appropriate "scale," which is the "chiral symmetry breaking scale," close to 1 GeV. In short, this is the essence chiral perturbation theory or ChPT, and the reason why it allows to calculate the various contributions to the potential systematically order by order, where each order refers to a particular power of the momentum. Furthermore, χEFT can generate not only the force between two nucleons, but also many-nucleon forces in a consistent manner [17]. The χEFT approach continues to gain popularity and is applied with great success in contemporary theoretical nuclear physics [18, 20–22].

However, it is important to keep in mind that a low-momentum expansion has a limited range of applicability. For that reason, interactions derived from chiral perturbation theory are not meant for applications to high energy processes or in dense matter, where high Fermi momenta are involved, as is the case in the interior of compact stars. In such situations, strategies to extend chiral predictions must be adopted, and we will discuss some instances where extensions become necessary.

Mean-field models, both relativistic and non-relativistic (see, for instance [23, 24]) are still a popular, although non-microscopic alternative to methods based on the in-medium reaction matrix. They continue to be utilized frequently in the development of the EoS and related predictions.

Describing the properties of (dense) systems from elementary forces and including all required contributions is an extremely challenging program, whose completion is not in sight. However, χEFT provides a path on which to proceed systematically toward that goal. We share the point of view that χEFT is currently the most fundamental approach due to its strong link with QCD. At the same time, the degrees of freedom of the theory make calculations of low-energy observables a manageable task.

Our main objective in this article is to provide a self-contained review of the recent work with isospin symmetric and asymmetric matter done systematically by our group and based mainly on chiral interactions, comparing with empirical constraints when available. We will place particular emphasis on neutron-rich matter, which is currently the focus of numerous empirical investigations both in terrestrial laboratories (especially through experiments aimed at constraining the thickness of neutron skins), or through astrophysical observations of neutron stars and related phenomena.

This article is organized as follows. After these opening remarks, in section 2 we describe in detail the calculations of the EoS, starting with the two-nucleon forces (2NF) and the three-nucleon forces (3NF) which we apply (see sections 2.2.1 and 2.2.2, respectively). In section 2.3, we review and discuss some of our results of the energy per particle in both symmetric and neutron matter [25]. In section 3, we focus specifically on the symmetry energy and the chief role its density dependence plays for neutron-rich systems. We then develop a discussion on the EoS in neutron stars (see section 4). We conclude with a summary and an outlook in section 5.

NUCLEAR MATTER AND THE EQUATION OF STATE

The G-Matrix and the Energy per Particle

In the previous section, we mentioned the linked-cluster perturbation series for the energy of a many-body system [1–6]. To facilitate convergence (otherwise problematic in view of the strong repulsive core of the *NN* force), the linked-cluster expansion for the energy per particle in nuclear matter [3] is written in terms of the reaction matrix or "*G*-matrix," which itself is solution of the Bethe-Goldstone equation. Schematically, the Bethe-Goldstone equation can be written as

$$G(E_0) = V + V\frac{Q}{E - E_0}G(E_0) \,, \tag{1}$$

where V is the *NN* potential, Q is the Pauli operator, and E_0 the starting energy of the two nucleons. The second term on the RHS of Equation (1) represents the infinite ladder sum which builds short-range correlations (SRC) into the wave function. The correlated (ψ) and the uncorrelated (φ) wave functions satisfy

$$G\phi = V\psi \,, \tag{2}$$

from which one can write

$$\psi = \phi + V\frac{Q}{E - E_0}G\phi \,. \tag{3}$$

At large distances, the correlated wave function is expected to approach the uncorrelated one (a behavior known as the "healing" property), whereas the two can be very different at short range. Hence, the difference between the correlated and the uncorrelated wave functions, or "defect function" $f = \psi - \varphi$, can be associated to the degree of SRC.

Usually, its momentum-dependent Bessel transform is considered instead, so as to bring out the dependence on specific partial waves. For each angular momentum state [5], we then have

$$f^{JST}_{LL'}(k) = \frac{k\,\bar{Q}(k_F, k, P^{c.m.}_{avg})G^{JST}_{LL'}(P^{c.m.}_{avg}, k, k_0)}{E_0 - E} \,, \tag{4}$$

3NF as an effective density dependent 2NF was first proposed in Baldo et al. [96] within the BHF theory.

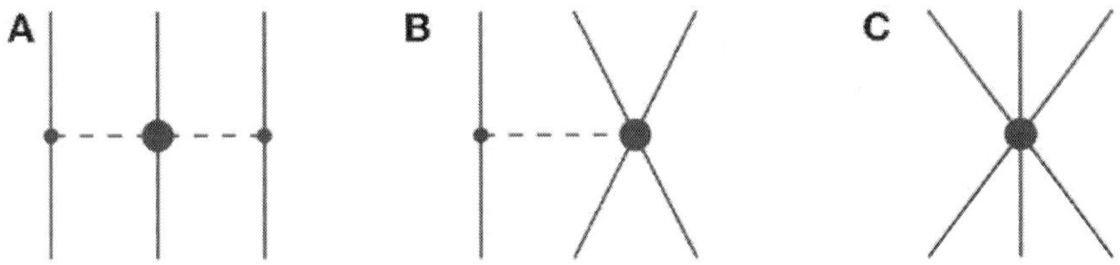

Figure 1: The 3NF at N^{2}LO with (A) the 2PE, (B) the 1PE, and (C) the contact terms.

The effective density-dependent two-nucleon interactions can be regrouped into six topologies involving one loop. Three of them originate from the 2PE graph of the chiral 3NF (Figure 1A), and depend on the LECs $c_{1,3,4}$, which already appear in the 2PE part of the *NN* force. Two one-loop topologies are derived from the 1PE diagram (Figure 1B), and contain the LEC c_D. Last, we have the one-loop topology related to the 3NF contact diagram (Figure 1C), associated with the LECs c_E. Note that, in pure neutron matter, the contributions proportional to the LECs c_4, c_D, and c_E vanish [56]. In recent nuclear matter calculations [63, 93], progress has been made toward including N^{3}LO three-body interactions in the two-body normal-ordering approximation as well as including the residual three-body normal-ordered force. Our group is in the process of including effective density-dependent 3NF at N^{3}LO as from Kaiser et al. [97, 98]. We have preliminary evidence that the contributions from the short-range terms [97] may be negligibly small.

The LECs c_D and c_E which we use are determined *via* the three-nucleon system. They are constrained to reproduce the A = 3 binding energies and the Gamow-Teller matrix element of tritium β-decay through the procedure described in Gardestig and Phillips [99], Gazit et al. [100], and Marcucci et al. [101]. The regulator function applied to the 3NF is

$$f(q) = \exp[(-q/\Lambda)^4] \,, \qquad (11)$$

as in Navrátil [102], with $q = |\vec{p}\,' - \vec{p}\,|$ the momentum transfer. Note that this choice makes the 3NF local in coordinate space, which, in turn, facilitates the construction of the $A = 3$ wave functions [103].

The complete 3NF at orders higher than the third (N^2LO) is very challenging, both in its development and applications, and, therefore, it is frequently excluded from nuclear structure studies. Note, though, that good progress is being made toward the inclusion of the subleading 3NF at N^3LO [63, 93, 97, 98, 104, 105]. However, in Krebs et al. [106] it was shown that the 2PE 3NF has nearly the same analytical structure at the third (N^2LO), fourth (N^3LO), and fifth (N^4LO) orders. Thus, one can parametrize the sum of all the three orders of 3NF contributions in terms of a set of effective LECs. Therefore, at least for this very important component of the 3NF, complete calculations up to N^4LO are possible.

In the N^4LO rows of Table 2 we give the effective LECs $c_{1,3,4}$ obtained in Krebs et al. [106]. Concerning the 2PE 3NF at N^3LO, Equation (2.8) of Bernard et al. [107] provides the corrections to the c_i. (Note, though, that there is an error in the values given below that equation. The correct values for δc_3 and δc_4 are $\delta c_3 = -\delta c_4 = 0.89$ GeV^{-1}.) With these corrections, we obtain the values given in the N^3LO rows of Table 2. Then, inserting the c_i of Table 2 in the expression for the N^2LO 3NF, we are able to include the 2PE parts of the 3NF up to N^3LO and up to N^4LO in a straightforward way, with the LECs c_D and c_E refitted. Their values, also listed in Table 2, are different from those listed in Table 1 but of the same order and with the same sign.

Table 1: Values of the LECs $c_{1,3,4}$, c_D, and c_E for different orders of the 2NF in the χEFT expansion, and the 3NF at N^2LO, and different values of the momentum-space cutoff Λ

	Λ (MeV)	*n*	c_1	c_3	c_4	c_D	c_E
N^2LO	450	2	−0.74	−3.61	2.44	0.935 (0.215)	0.12 (0.04)
	500	2	−0.74	−3.61	2.44	0.495 (0.195)	−0.07 (0.04)
N^3LO	450	2	−1.20	−4.43	2.67	0.670 (0.210)	0.41 (0.05)
	500	2	−1.20	−4.43	2.67	−0.750 (0.210)	−0.41 (0.04)
N^4LO	450	2	−0.73	−3.38	1.69	0.560 (0.220)	0.46 (0.05)
	500	2	−0.73	−3.38	1.69	−0.745 (0.225)	−0.15 (0.04)

*That is, at each order, the 2PE term of the 3NF is included summing up all contributions up to that order. (The N^2LO numbers are the same as in **Table 1**).*

We close this section by highlighting that, of all possible 3NF contributions, the 2PE 3NF is the first to have been calculated [108]. The prescriptions outlined above allow to include this very important 3NF up to the highest order we consider at this time.

Predictions for the Equation of State

Symmetric Matter Predictions

We begin with the symmetric nuclear matter (SNM) EoS. This is displayed in Figure 2, where, on the left, the momentum-space cutoff is fixed at 450 MeV but the chiral order of the two-body force is varied from leading to fifth order. The 3NFs are chosen with LECs in Table 2, which at N^3LO and N^4LO include the 2PE 3NF at fourth and fifth order, respectively. (We note that, in all that follows, when we refer to predictions obtained with 3NF at N^3LO or at N^4LO, we mean to say that the 2PE 3NF is included up to those orders). The dashed lines indicate results at N^2LO and above with no three-body forces present, while the solid lines include the 3NF when appropriate, that is, at N^2LO and up. Formally, we observe a good convergence pattern at the two-body level with this family of *NN* potentials, but naturally we do not expect realistic saturation behavior when soft two-body forces alone are included in the calculation of the EoS. We see that the inclusion of 3NFs is necessary beyond about half nuclear matter saturation density and that for this set of nuclear potentials the total 3NF contribution to the EoS decreases with the chiral order from N^2LO to N^4LO.

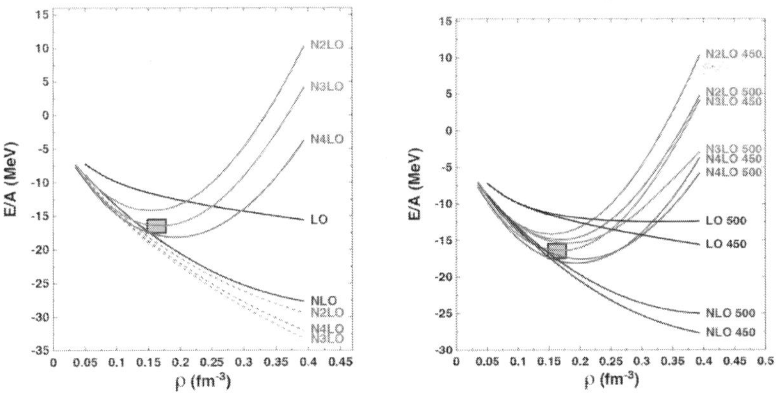

Figure 2: (Left) Ground state energy per particle of SNM as a function of density from the chiral two- and three-body forces with cutoff $\Lambda = 450$ MeV. The three dotted curves show predictions which include only two-body forces. For the 3NF contributions at N^2LO and above, the LECs of Table 2 are used. The shaded box locates the approximate empirical saturation energy and density. (Right) Ground state energy per particle of SNM as a function of density at the indicated orders and with varying cutoff parameters. Other details as on the left.

Table 2: Same as Table 1, but including the 2PE 3NF at N^3LO and N^4LO

	Λ (MeV)	*n*	c_1	c_3	c_4	c_D	c_E
N^2LO	450	2	−0.74	−3.61	2.44	0.935 (0.215)	0.12 (0.04)
	500	2	−0.74	−3.61	2.44	0.495 (0.195)	−0.07 (0.04)
N^3LO	450	2	−1.20	−4.43	2.67	0.670 (0.210)	0.41 (0.05)
	500	2	−1.20	−4.43	2.67	−0.750 (0.210)	−0.41 (0.04)
N^4LO	450	2	−0.73	−3.38	1.69	0.560 (0.220)	0.46 (0.05)
	500	2	−0.73	−3.38	1.69	−0.745 (0.225)	−0.15 (0.04)

That is, at each order, the 2PE term of the 3NF is included summing up all contributions up to that order. (The N^2LO numbers are the same as in ***Table 1****).*

We note that the uncertainty band obtained by varying the chiral order from N^2LO to N^4LO while keeping Λ fixed to 450 MeV encloses the empirical saturation point. The saturation energy varies in the range $-14\text{MeV} \lesssim E_0 \lesssim -18\text{MeV}$ while the saturation density varies between 0.155fm−3≲ρ0≲0.195fm−30.155fm-3≲ρ0≲0.195fm-3. We stress that, once the two- and three-nucleon forces are fixed by the *NN* data and the properties of the three-nucleon system, no parameters are readjusted, making the many-body calculation parameter-free. Since the predicted binding energies and charge radii of intermediate-mass nuclei are closely related to the corresponding saturation point in SNM, we see the possibility that the new class of chiral potentials constructed in Entem et al. [87] and used in this work may lead to more reliable predictions in ab initio calculations of finite nuclei. For densities larger than $\rho \gtrsim 0.20\text{fm}^{-3}$, the predictions shown on the LHS of Figure 2 display a trend that does not suggest satisfactory convergence, since the three (saturating) solid curves are about equally spaced. This is most likely due to the incompleteness of the 3NF at orders above N^2LO. It is natural to expect that such trend will be a recurrent theme in later results. As discussed in section 2.2.2, we believe that including the important 2PE contribution consistently across all orders is important and insightful. For instance, our results suggest that the missing 3NF contributions at orders higher than N^3LO can be expected to play a substantial role toward a successful convergence.

On the RHS of Figure 2 we show the dependence of the SNM EoS on the choice of momentum-space cutoff Λ in the two- and three-body forces as well as the order in the chiral expansion. In the present work we consider only the two cases Λ = 450, 500 MeV, see comments in section

2.2.1. At orders N²LO, N³LO, and N⁴LO, the cutoff dependence appears to be comparable but generally smaller than the truncation errors.

In Figure 2, we show the impact of choosing at fourth (N³LO) and fifth (N⁴LO) order in the chiral expansion either the N²LO 3NF coupling strengths shown in Table 1 (labeled "I" in the figure) or those obtained by including the 2PE 3NF contributions at higher order shown in Table 2 (labeled "II" in the figure). We only show results for potentials with momentum-space cutoff Λ = 450 MeV, but we expect similar results for the Λ = 500 MeV cutoff potentials due to the identical change in the important c_i LECs (i = 1, 3, 4). We see that at N⁴LO the impact is rather large and roughly of the same size as variations in the chiral order from N²LO to N⁴LO. However, the additional theoretical uncertainty resulting from the choice of LECs entering into the 2PE 3NF would extend the overall error band inferred from the RHS of Figure 2 only moderately and only at the largest densities considered. In other words, Figure 3 shows that the truncation error (compare N³LO II and N⁴LO II) can be much smaller than the uncertainty arising from using different sets of LECs (compare N⁴LO I and N⁴LO II), indicating the importance of completeness in the 3NF at all orders.

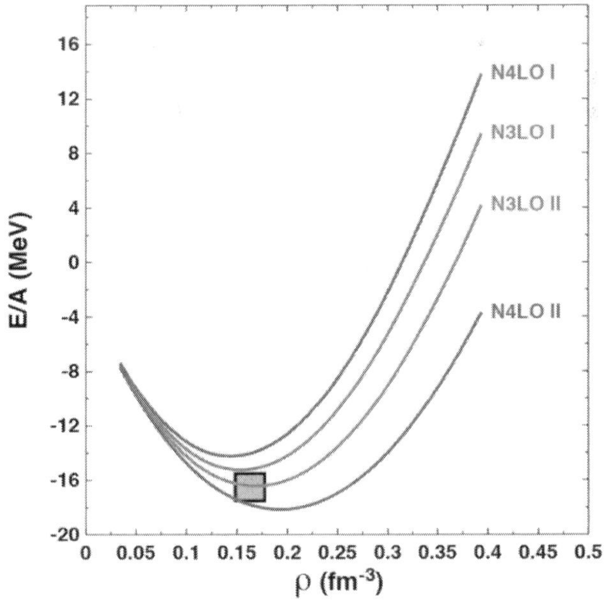

Figure 3: Energy per particle in SNM as a function of density at N³LO and N⁴LO with a cutoff of Λ = 450 MeV. For the 3NF contributions, the LECs of either Table 1 or Table 2 are applied as indicated by labels "I" or "II", respec-

tively. Case "II" is characterized by including the 2PE 3NF up to the given order. The approximate empirical saturation energy and density are indicated by the gray box.

Before closing this section, we summarize the saturation properties of SNM at the various orders and cutoff values. In Table 3, we show saturation density, saturation energy, and the isoscalar incompressibility. For the latter, constraints can be obtained from giant monopole resonance energies. In De et al. [109], the authors obtain a range of 211.9 ± 24.5 MeV. Our predictions at N^3LO are consistent with this range, whereas the larger values at N^4LO reflect the larger saturation density at that order.

Table 3: Saturation properties from N^2LO to N^3LO and two values of the cutoff

	Λ (MeV)	ρ_0 (fm^{-3})	$e_0(\rho_0)$ (MeV)	K_0 (MeV)
N^2LO	450	0.155 ± 0.015	-14.2 ± 2.3	217.4 ± 3.6
	500	0.170 ± 0.010	-14.9 ± 0.5	214.4 ± 8.0
N^3LO	450	0.170 ± 0.025	-16.4 ± 1.7	221.2 ± 51.5
	500	0.180 ± 0.023	-15.4 ± 2.2	206.4 ± 49.2
N^4LO	450	0.195 ± 0.012	-18.1 ± 0.8	272.7 ± 24.3
	500	0.203 ± 0.010	-17.6 ± 0.9	255.6 ± 21.2

Parameters which involve isospin asymmetry will be discussed later.

Neutron Matter Predictions

We next consider the ground state energy of neutron matter (NM) as a function of density, employing the same set of chiral potentials and many-body methods discussed previously in the case of symmetric nuclear matter. The EoS for both SNM and NM are crucial to determine the density-dependent nuclear symmetry energy and to better understand the properties of neutron-rich nuclei and neutron stars, aspects which will be addressed in later sections.

In Figure 4, on the LHS, we show the energy per particle of NM as a function of density starting from chiral two- and three-body forces with the same value of the momentum-space cutoff Λ = 450 MeV but at different orders in the chiral expansion. As in the case of symmetric nuclear matter, we observe good convergence at the level of 2NF alone. When 3NFs are included, we find somewhat smaller truncation errors compared to the case of SNM. This may be due in part to the absence of large, central isospin-0 partial waves in NM, which appear to be more sensitive to differences among interactions. Clearly, the 3NF plays an outstanding role in very neutron-rich systems at and beyond nuclear saturation density, where its contribution to the EoS grows more strongly with the density than the 2NF contributions.

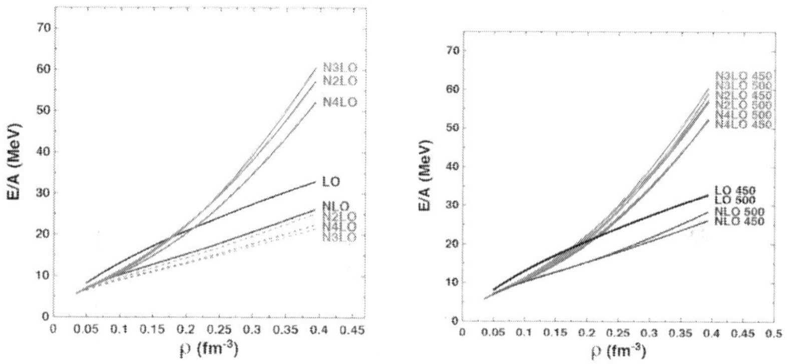

Figure 4: (Left) Ground state energy per particle of NM as a function of density at the indicated orders in the chiral expansion. The three dotted curves show predictions including only the 2NF. The cutoff parameter is fixed at $\Lambda = 450$ MeV and the 3NF LECs from Table 2 are used. (Right) Ground state energy per particle of NM as a function of density at the indicated chiral orders and with varying cutoff. The LECs of Table 2 are used.

On the RHS of Figure 4 we display the energy per particle of pure neutron matter as a function of density when varying both the order in the chiral expansion and the momentum-space cutoff Λ from 450 to 500 MeV. We see that, in comparison to the analogous study in symmetric nuclear matter, the pure neutron matter results display a much weaker cutoff dependence, which may again be due to the absence of strong isospin-0 partial waves. Interestingly, even in the case of the relatively large density $\rho = 0.4$ fm^{-3}, corresponding to a Fermi momentum of $k_F = 450$ MeV which lies at the effective breakdown scale of the expansion, there is relatively little cutoff dependence.

Once again, we observe that the order-by-order pattern is not satisfactory when moving from N^{3}LO to N^{4}LO. The impact of including the 2PE 3NF up to fourth (N^{3}LO) and fifth (N^{4}LO) order (consistent with the order of the 2NF), compared to including only the third-order (N^{2}LO) contributions, through the adoption of the LECs given in Table 2, is demonstrated in Figure 5. As in the case of symmetric nuclear matter, the effect at N^{4}LO is much larger than at N^3LO due to the larger change $\Delta c_3 = 2.16$ GeV^{-1} vs. $\Delta c_3 = 0.89$ GeV^{-1}, respectively, in the c_3 LEC at these two orders in the chiral expansion. Moreover, the choice of LECs entering into the 2PE 3NF contributions again results in a moderate systematic increase in the pure neutron matter energy per particle at the highest densities considered. As we mentioned earlier, the investigation of higher-order 3NF contributions is in progress.

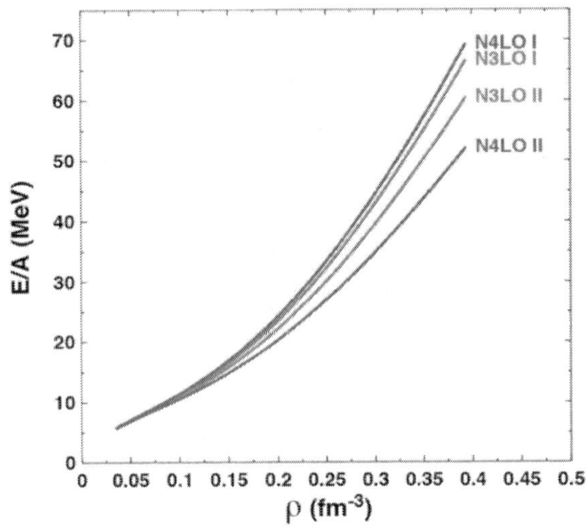

Figure 5: Ground state energy per particle of NM as a function of density at N^3LO and N^4LO with cutoff equal to 450 MeV. Similar to Figure 3, for the 3NF contributions the LECs of either Table 1 or Table 2 are applied as indicated by labels "I" or "II," respectively.

Before closing this section, we take the opportunity to comment on how our SNM and NM EoS compare with those from Drischler et al. [93]. There, chiral interactions including full 3NF and 4NF at N^3LO are applied to investigate nuclear saturation. Judging from the RHS of Figure 4 in Drischler et al. [93], where the EoS for both SNM and NM [63] are displayed, we conclude that our EoS at N^3LO are qualitatively comparable with them within the density range covered in Figure 4 in Drischler et al. [93], namely up to ρ=0.20 fm^{-3}, with ours revealing more attraction. We also point out that, in Drischler et al. [93], the 3NF couplings are fit to triton and to saturation properties, whereas we do not impose any constraints other than those from the two- and the three-nucleon systems. Even so, we find (confirming the conclusion from [93]) that realistic saturation properties are possible at N^3LO.

THE SYMMETRY ENERGY AND RELATED ASPECTS

Review of Some Basic Concepts and Definitions

The properties of isospin-polarized matter have relevance for a number of open questions in nuclear physics and nuclear astrophysics. For instance, the existance of the neutron drip lines, the thickness of neutron skins, and the properties of neutron stars all have in common a strong sensitivity to the EoS of neutron-rich matter. The symmetry energy determines to a good approximation the energy per particle in homogeneous nuclear matter with any degree of isospin asymmetry (cf. Equation 14 below). The symmetry energy and its density dependence are therefore a key focus of contemporary theoretical and experimental investigations, and much effort has been devoted to identifying nuclear observables which correlate with this important property of infinite matter [25, 45, 53, 56–64, 110–120].

The isospin asymmetry parameter is a measure of the relative densities of neutrons and protons and is defined as

$$\alpha = \frac{(\rho_n - \rho_p)}{(\rho_n + \rho_p)} , \tag{12}$$

where ρ_n and ρ_p are the neutron and proton densities. It is useful to write the energy per particle in isospin asymmetric matter at some density as an expansion with respect to α:

$$e(\rho,\alpha) = e(\rho,\alpha=0) + \frac{1}{2}\left(\frac{\partial^2 e(\rho,\alpha)}{\partial \alpha^2}\right)_{\alpha=0} \alpha^2 + \mathcal{O}(\alpha^4) . \tag{13}$$

Frequently, the expansion above is truncated at the term quadratic in α, resulting in the popular parabolic approximation:

$$e(\rho,\alpha) \approx e_0(\rho) + e_{sym}(\rho)\,\alpha^2 , \tag{14}$$

where $e_0(\rho) = e(\rho, \alpha = 0)$. [Equation (14) has been verified to be valid up to fairly high densities [119].] Within the assumption of Equation (14), the symmetry energy, e_{sym}, is the difference between the energy per particle in neutron matter and the one in symmetric matter. We can expand the symmetry energy about the saturation density, ρ_0,

$$e_{sym}(\rho) \approx e_{sym}(\rho_0) + \frac{L}{3}\frac{\rho - \rho_0}{\rho_0} + \frac{K_{sym}}{18}\left(\frac{\rho - \rho_0}{\rho_0}\right)^2 + \dots . \tag{15}$$

The slope parameter, L, is defined as

$$L = 3\rho_0\left(\frac{\partial e_{sym}(\rho)}{\partial \rho}\right)_{\rho_0}, \tag{16}$$

and therefore is a measure of the density dependence of the symmetry energy around saturation density. We recall that L is an important quantity because of its significance for the skin thickness in neutron-rich nuclei. Experiments which plan to measure the neutron radius of ^{208}Pb and ^{48}Ca using electroweak probes, such as PREX II [121] and CREX [122], respectively, are expected to provide accurate measurements of the neutron skin. As a consequence, one hopes for reliable constraints on the *symmetry pressure*, clearly related to the slope parameter (see section 3.3.1 below). Also, the radius of the average-mass neutron star is known to be sensitive to the pressure in neutron matter at normal density, P_0^{NM}, which is simply related to L (for fixed ρ_0) due to the vanishing of the density derivative of $e(\rho, \alpha = 0)$ at saturation. That is:

$$P_0^{NM} \approx L\,\rho_0/3 . \tag{17}$$

The reader is referred, for instance, to Sammarruca and Millerson [123] and the comprehensive list of citations therein.

The isovector incompressibility, K_{sym}, is associated with the next higher-order derivative, that is, it measures the curvature of the symmetry energy at saturation density. It is defined as:

$$K_{sym} = 9\rho_0^2\left(\frac{\partial^2 e_{sym}(\rho)}{\partial \rho^2}\right)_{\rho_0} . \tag{18}$$

Correlations between L and both K_{sym} and the symmetry energy at saturation, $e_{sym}(\rho_0)$ [124–126] have been examined. Predictions for the isovector incompressibility carry large uncertainty, as is the case for the isoscalar one. Attempts to constrain the second derivative of the symmetry energy (that is, its curvature) are discussed in Vidaña et al. [127], Ducoin et al. [128], and Santos et al. [129].

Predictions of Symmetry Energy and Related Properties

Figure 6 displays the energy per particle in isospin asymmetric matter as a function of density and for increasing degree of asymmetry, cf. Equation (14), for one selected order and cutoff [130].

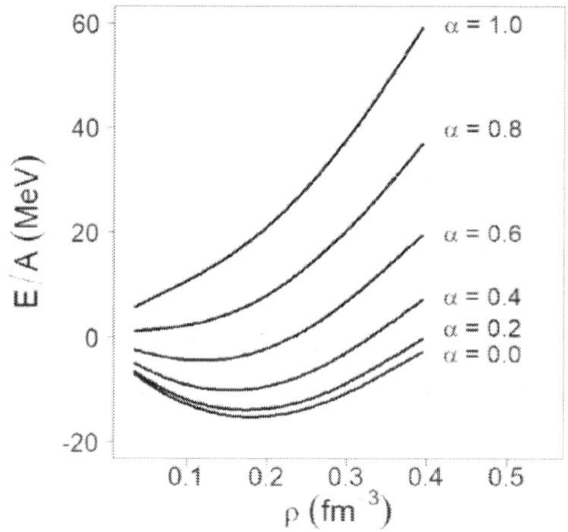

Figure 6: Energy per particle in isospin-asymmetric nuclear matter. In each case, the isospin asymmetry parameter is given. Calculations conducted at N^3LO of the 2NF (and the 2PE 3NF included up to N^3LO) with cutoff equal to 500 MeV.

As we already noted in conjunction with Figure 2, the saturation properties of the chiral interactions we are considering are different from one another, with the saturation density varying between about 0.16 and 0.20 fm^{-3}. Clearly, this will impact the expansion parameters contained in Equation (15), see definitions in Equations (16–18), differently than if the derivative were evaluated, in all cases, at some common, nominal saturation density ρ_0. On the other hand, analyses of correlations between the symmetry energy, its density slope, and the neutron skin thickness are typically done utilizing families of phenomenological models, such as large sets of Skyrme interactions or relativistic mean-field (RMF) models [131]. These models are constructed so as to have in common good saturation properties (usually by adjusting parameters to empirical properties of nuclei) while differing in the slope of the symmetry energy which, at saturation, is essentially a measure for the pressure in pure neutron matter (see Equation 17). Already almost two decades ago, Brown [132] considered a set of Skyrme interactions whose predictions of the density slope of the NM EoS around normal density differed dramatically and found a linear relation between such derivative and the neutron skin thickness in ^{208}Pb. Similar investigations have been and continue to be done with RMF models, with families of interactions constructed so as to span a large range of L values. For instance, RMF models such as NL3 [133] and IU-FSU [134] give values of L equal to 118.2 and 47.2 MeV, respectively. (Not surprisingly, these models span a large range of both neutron skin values and stellar radii). In Roca-Maza et al. [24], the authors utilize a large set of RMF models all of which describe accurately the nuclear binding energies and charge radii across the periodic table (which should constrain tightly the binding energy and saturation density of SNM). On the other hand, the same models predict very different neutron root-mean-squared (*r.m.s.*) radii, since the isovector channel is poorly constrained [24].

Our EoS are microscopic and parameter-free and we are not in the practice of constructing families of parameterized EoS models to establish phenomenological correlations. Nevertheless, for the purpose of demonstration, next we wish to perform a study meant to highlight the role of neutron matter pressure for the neutron skin thickness once the uncertainty associated with the saturation point in SNM, cf. Figure 2, is removed. To that end, we will construct "semi-microscopic" models of asymmetric matter as follows: for the symmetric part, we will use an established phenomenological EoS, such as the one from Alam et al. [135]. For the neutron matter part, currently our focal point, we will continue to use the chiral EoS presented in section 2.3.2. We then proceed treating these six cases (three chiral orders and two cutoffs) as six EoS models differing in their NM components. Figure 7 shows the phenomenological EoS of SNM in comparison with our chiral predictions with cutoff equal to 450 MeV. Figure 8 displays the symmetry energy obtained with our microscopic NM EoS combined with SNM EoS represented by the black curve in Figure 7. We also include in the figures the results of several analyses and constraints [136–139]. The predictions based on our microscopic NM EoS are considerably softer than those constraints above normal density. Table 4 contains values for the parameters defined previosly through Equation (15), for the six EoS models which we have constructed as described.

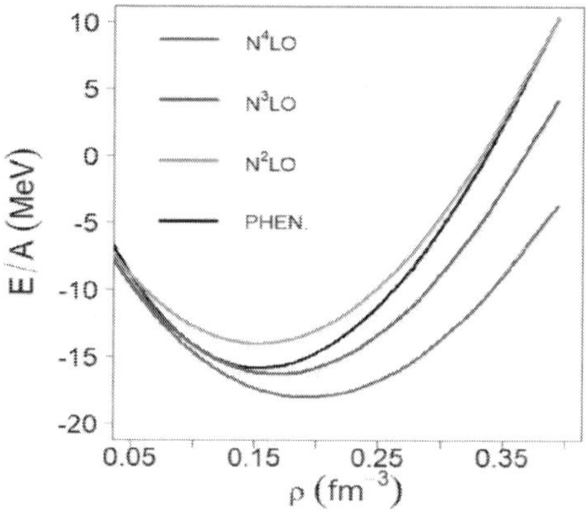

Figure 7: The EoS of SNM at the three chiral orders considered here (cutoff fixed at 450 MeV) compared with the phenomenological EoS of Alam et al. [135].

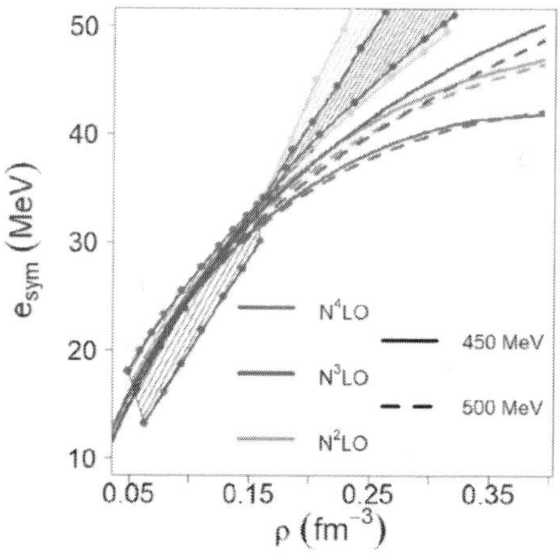

Figure 8: The symmetry energy vs. density. The curves are obtained from the various microscopic EoS for NM at the indicated chiral orders and cutoff values, combined with the phenomenological EoS for SNM [135] as explained in the text. The additional predictions and various constraints are from: Danielewicz and Lee [136], dark green; Tsang et al. [137], magenta contour; Russotto et al. [138, 139], yellow and brown shaded areas. (The data points were extracted from the graphs assuming $\rho_0 = 0.16$ fm^{-3} and using WebPlotDigitizer opensource software, https://automeris.io/WebPlotDigitizer).

Table 4: Predicted values of symmetry energy and related properties at three orders of chiral perturbation theory and two values of the cutoff parameter obtained as explained in the text

	Λ (MeV)	$e_{sym}(\rho_0)$ (MeV)	L (MeV)	K_{sym} (MeV)
N^2LO	450	32.8 ± 0.4	52.2 ± 0.8	-117.6 ± 36.1
	500	32.2 ± 0.5	50.2 ± 1.5	-106.3 ± 31.1
N^3LO	450	32.4 ± 1.0	53.0 ± 7.2	-81.5 ± 32.8
	500	31.7 ± 0.6	48.8 ± 4.6	-75.2 ± 32.8
N^4LO	450	31.4 ± 0.5	45.8 ± 3.1	-114.3 ± 14.3
	500	31.1 ± 0.2	44.2 ± 1.8	-108.0 ± 12.9

We now proceed to discuss the spread of our values for the symmetry energy, the L parameter, and the isovector incompressibility in the framework of chiral uncertainties of the NM EoS. We recall that one of the strengths of χEFT is the opportunity of order-by-order improvement of the predictions. Naturally, the truncation error at a given order should be a reasonable measure of the uncertainty which arises from omitting the next order contributions. If the value of the observable X has been calculated at order $n+1$, than the truncation error at order n can be estimated as the difference between the values at order $n+1$ and n:

$$\epsilon_n = |X_{n+1} - X_n| \,, \tag{19}$$

which is a reasonable way to estimate what one is missing by retaining only terms up to order n. On the other hand, if X_{n+1} is not known, then some alternative prescription must be used. We use the definition [87]

$$\epsilon_n \approx |X_n - X_{n-1}| \frac{Q}{\Lambda} \,, \tag{20}$$

where Q is a momentum characteristic for the observable under consideration and Λ is the cutoff. For the fifth (and highest) order, we use Equation (20) and we find it reasonable to define Q as the *r.m.s.* value of the relative momentum of two neutrons in neutron matter at the given density [see [123] and references therein].

We wish to express our final results for the symmetry energy, the slope parameter, and the isovector incompressibility at N^3LO. To that end, we average the predictions for the quantity X obtained with the two values of the cutoff separately at N^{3}LO and N^{4}LO, yielding $\bar{X}_4$ and $\bar{X}_5$, respectively. The truncation uncertainty at N^{3}LO can then be estimated as $\Delta_X = |\bar{X}_4 - \bar{X}_5|$. As an alternative, we choose to take the largest of the errors at the two cutoff values.

Applying that prescription, we obtain for the symmetry energy, the slope parameter, and the isovector incompressibility at N^{3}LO (all numbers in MeV):

$$e_{sym} = 32.1 \pm \Delta_{e_{sym}} \qquad \Delta_{e_{sym}} = 1.0 \,, \tag{21}$$

$$L = 50.9 \pm \Delta_L \qquad \Delta_L = 7.2 \,, \tag{22}$$

$$K_{sym} = -78.4 \pm \Delta_{K_{sym}} \qquad \Delta_{K_{sym}} = 32.8 \,. \tag{23}$$

We see that K_{sym} shows large variations, which reflect the extreme sensitivity of the second derivative to the details of the interactions for each of the curves in Figure 8. We emphasize that variations among those curves are due entirely to the NM predictions.

A phenomenological study of the EoS based on Skyrme density functionals [135] reports the slope parameter to be L = 65.4±13.5 MeV, whereas the isovector incompressibility is found to be within the range K_{sym} = −22.9±73.2 MeV. Lattimer and Lim [140] determined L to be between 40.5 and 61.9 MeV. For the isovector incompressibility, they suggest a linear relation between K_{sym} and L, that is, $K_{sym} \approx aL - b$, with a, b equal to 3.33 and 281 MeV, respecively [128], or 2.867 and 260 MeV [127]. More recent constraints obtained from tidal deformabilitie inferred from GW170817, report $30 < L < 86$ MeV and $-140 < K_{sym} < 16$ MeV or $40 < L < 62$ MeV and $-112 < K_{sym} < -52$ MeV [141].

Before closing this section, we take the opportunity to address the pressure in neutron matter at saturation density, which, for the EoS of SNM which we have chosen is equal to 0.155 fm^{-3}. Using Equation (17) and the uncertainty on L, we find (in MeV/fm^3):

$$P_0^{NM} = 2.66 \pm \Delta_P \qquad \Delta_P = 0.37 \,. \tag{24}$$

Symmetry Energy Slope and Neutron Skins

The neutron skin is defined as the difference between the r.m.s. radii of the neutron and proton density distributions:

$$S_n = R_n - R_p , \tag{25}$$

where

$$R_i = \sqrt{\frac{1}{T_i} \int_0^\infty \rho_i(r)\, r^2\, d^3 r} , \tag{26}$$

$i = n, p$ and T_n, $T_p = N, Z$ respectively.

As mentioned before, the neutron skin thickness, particularly for ^{208}Pb, is of great contemporary interest due to the possibility of constraining the slope of the symmetry energy through skin measurements [24, 142–147].

The Experimental State-of-the-Art

While electron scattering has been very successful in providing accurate information on the proton distributions within the nucleus, mapping neutron densities is a much more challenging task. In particular, measurements which make use of hadronic probes carry large uncertainties due to the model dependence of the nuclear interactions used in the analyses.

On the other hand, parity-violating electron scattering is in principle capable of providing accurate information on the weak charge distribution in the nucleus through the coupling of the neutron to the *Z*-boson. The typical parity-violating electron scattering experiment measures the difference between the cross sections for scattering of right-handed and left-handed electrons, that is

$$A_{PV} = \frac{\sigma_R - \sigma_L}{\sigma_R + \sigma_L} , \tag{27}$$

which is proportional to the ratio of the weak to the charge form factor of the nucleus [110] and thus can be related to coordinate space densities by Fourier transform. The challenging aspects of measuring observables related to parity violation is that they can be etremely small, in the case of A_{PV} between 10^{-4} to 10^{-7} [110].

The first PREX experiment [148, 149] provided a value of 0.33(+0.16, −0.18) fm for the skin of ^{208}Pb, which carries a large experimental error due to technical problems which resulted into poor statistics. However, the planned PREX-II and CREX experiments have a target uncertainty of ±0.06 fm and ±0.02 fm for the neutron skin of ^{208}Pb and ^{48}Ca, respectively [121, 122].

Furthermore, additional constraints are expected from the forthcoming MESA accelerator in Mainz [150], which promises to constraint the neutron skin of ^{208}Pb within ±0.03 fm and the one of ^{48}Ca within ±0.02 fm, same as the target uncertainty of CREX. Note that these two nuclei are both stable, doubly-magic, and with a relatively large neutron to proton asymmetry, which is part of the reasons why investigations have concentrated on them.

For an extensive review of correlation analyses based on a large set of relativistic and non-relativistic nuclear density functionals see [110].

Predictions

We now move to neutron skins, specifically for the ^{208}Pb and ^{48}Ca nuclei, as predicted by the EoS models based on the six chiral interactions in NM as described previously. Using the energy per particle in infinite isospin-asymmetric matter as given in Equation (14), we can establish a simple but direct connection with the energy per nucleon in a spherically symmetric nucleus through the semi-empirical mass formula:

$$E(Z,A) = \int d^3r\, \rho(r)\, e(\rho,\alpha) + \int d^3r\, f_0\, |\nabla\rho|^2 + E_{Coul}\,, \tag{28}$$

where the Coulomb contribution is written as:

$$E_{Coul} = \frac{e^2}{\epsilon_0} \int_0^\infty dr' \,[r' \rho_p(r') \int_0^{r'} d^3r\, \rho_p(r)]\,. \tag{29}$$

The parameter f_0 is a fitted constant for which we used a value of 65 MeV fm^5, consistent with the range determined in Oyamatsu et al. [151].

We use the two-parameter Thomas-Fermi distribution function to describe the nucleon density:

$$\rho(r) = \frac{\rho_a}{1 + e^{(r - r_b)/c}}\,. \tag{30}$$

The "radius" r_b and the "diffuseness" c are themselves evaluated through minimization of the energy per nucleon, while ρ_a is easily obtained from normalization.

Table 5 shows the values of the neutron skin thickness predictions along with the truncation error for ^{208}Pb and ^{48}Ca. Proceeding as described previously, and taking the largest of the errors at the two cutoff values, we state our final estimates for the neutron skins of ^{208}Pb and ^{48}Ca as

$$S_n(^{208}Pb) = 0.147 \pm \Delta_{S_n} \qquad \Delta_{S_n} = 0.009 \ , \tag{31}$$

$$S_n(^{48}Ca) = 0.145 \pm \Delta_{S_n} \qquad \Delta_{S_n} = 0.005 \ . \tag{32}$$

Note that the skin depends considerably on the constant f_0 appearing in Equation (28). We have not included that uncertainty in Equations (31–32) as we are focusing on chiral truncation errors. We report, however, that varying f_0 between 60 and 70 MeV fm^5 introduces an uncertainty of 0.01 fm, essentially independent of chiral order or cutoff.

Table 5: Predicted neutron skin of ^{208}Pb and ^{48}Ca

Nucleus	Order	Λ = 450 MeV (fm)	Λ = 500 MeV (fm)
^{208}Pb	N2LO	0.155 ± 0.005	0.147 ± 0.004
	N3LO	0.150 ± 0.009	0.143 ± 0.004
	N4LO	0.141 ± 0.004	0.139 ± 0.002
^{48}Ca	N2LO	0.150 ± 0.003	0.144 ± 0.002
	N3LO	0.147 ± 0.005	0.142 ± 0.002
	N4LO	0.141 ± 0.003	0.140 ± 0.001

We close this section by showing in Figure 9 a typical correlation between L and the thickness of the neutron skin in ^{208}Pb obtained with a large set of successful RMF models. As we discussed previously, the ranges we give in Equations (22) and (31) are relatively small, which is, of course, desirable, since they originate from chiral uncertainties in the NM rather than variations of phenomenological parameterizations. We note that our range of values seem to be located on the low end of the correlation in Figure 9, with L approximately between 44 and 58 MeV and the skin between approximately 0.14 and 0.16 fm.

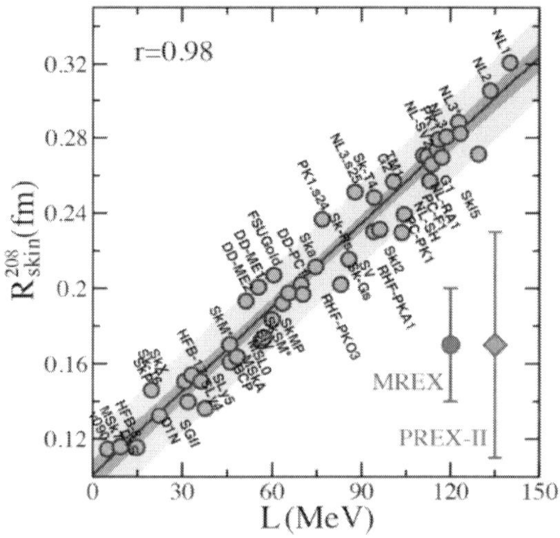

Figure 9: Correlation between the slope parameter (L) and the neutron skin thickness of ^{208}Pb obtained with RMF models. The error bars represent the target precision for the future PREX-II [121] and MREX [150] experiments. Reproduced from Physics Today **72**, 7, 30 (2019) (https://doi.org/10.1063/PT.3.4247) with the permission of the American Institute of Physics.

THE EQUATION OF STATE AND NEUTRON STARS

Some General Aspects

It is remarkable that the relation between the mass and the radius of neutron stars is uniquely determined by the EoS together with the star's self-gravity through the Tolman-Oppenheimer-Volkoff (TOV) equations of General Relativity [152]. In fact, although the detailed structure of a neutron star is complex and varies as a function of density, the part of its core mostly composed of a uniform liquid of neutrons, protons, and leptons in β-equilibrium accounts for almost all the mass and the volume. Therefore, these compact systems are intriguing testing grounds for both nuclear physics [153–156] and General Relativity. Extensive effort has been and continue to be devoted to constraining properties of compact stars from astrophysical observations see, for instance [156–160].

The largest possible value for the mass of a neutron star was estimated by Rhoades and Ruffini [161] based on the following assumptions: (1) General Relativity is the appropriate theory of gravitation; (2) the EoS obeys the Le Chatelier's principle ($\partial P/\partial \epsilon \geq 0$) and is consistent with causality, $\partial P/\partial \epsilon \leq c^2$; and (3) the EoS is reliably known below some density. Subject to these conditions, it was determined that the maximum mass of a neutron star cannot exceed 3.2 solar masses. Note that, releasing the causality constraint, the limit can be as high as 5 solar masses [162, 163] due to the increased stiffness of the EoS at supernuclear densities.

While the maximum mass is mostly determined by the stiffness of the EoS at densities greater than a few times saturation density, the star radius is impacted mainly by the slope of the symmetry energy. More precisely, it is closely connected to the internal pressure (that is, the energy gradient) of matter at densities between about $1.5\rho_0$ and $2\text{-}3\rho_0$ [157]. The mass and the radius of the neutron star are predicted by the TOV equation as we review next.

The equations for a perfect fluid in hydrostatic equilibrium allow to determine the pressure and the total mass-energy density as a function of the radial distance from the center of the star. These coupled equations are

$$\frac{dP(r)}{dr} = -\frac{G}{c^2}\frac{(P(r)+\epsilon(r))(M(r)+4\pi r^3 P(r)/c^2)}{r(r-2GM(r)/c^2)}, \tag{33}$$

with

$$\frac{dM(r)}{dr} = 4\pi r^2 \epsilon(r)/c^2, \tag{34}$$

where ϵ is the total mass-energy density. The star *gravitational* mass is

$$M(R) = \int_0^R 4\pi r^2 (\epsilon(r)/c^2) dr, \tag{35}$$

with R the value of r where the pressure vanishes. It›s worth recalling that no mass limit exists in Newtonian gravitation.

Recently, the LIGO/Virgo [164] detection of gravitational waves originating from two neutron stars spiraling inward and merging, the neutron star merger GW170817, has generated even more interest and excitement around these highly exotic systems.

The dimensionless *tidal deformability* is related to the neutron star response to the tidal field induced by the companion star and is defined as

$$\Lambda = \frac{2}{3}k_2\left(\frac{R}{M}\right)^5, \tag{36}$$

where the *Love number* k_2 reflects the quadrupole component of the gravitational potential induced by the companion star at the surface [165]. It depends on the neutron star *compactness*, M/R, and the energy density and pressure profile of the star. The tidal deformability can be obtained by solving the appropriate equations together with the TOV equations which yield the $M(R)$ relation [166]. Hence, the merger detection can provide constraints on the star radius based on the tidal deformabilities of the colliding system [167]. In fact, the August 2017 first direct detection of a binary neutron star merger helped establish new limits on the radius of a 1.4 $M_\odot$ neutron star. Additional references addressing the radius of a 1.4 $M_\odot$ neutron star include [166, 168–172].

The correlation between the neutron skin thickness (discussed in section 3.3) and the radius of a neutron star originates from the sensitivity of the star radius to the *pressure at normal density*. Note that such correlation weakens as the mass increases see, for instance [110], which is why the radii of lighter stars are good candidate to help constrain the neutron skin of ^{208}Pb and, in turn, the slope of the symmetry energy around saturation density. Based on these considerations, an upper limit of 0.25 fm was found for the neutron skin thickness of ^{208}Pb. Additional observations from the LIGO-Virgo collaboration scheduled for 2019 are likely to provide stronger constraints.

In the remainder of this section, after reviewing how the EoS of β-stable matter is obtained from conditions of charge neutrality and energy minimization (section 4.2), we will address (spherical) neutron star properties, with emphasis on the radius of a "typical" neutron star, namely one with a mass approximately equal to 1.4 $M_\odot$. The reasons for this choice have been given in the previous paragraph.

The EoS of β-Stable Matter

In this section, we review the basic equations which we use to obtain the EoS for stellar matter in β-equilibrium.

The total energy per particle, e_{tot}, related to the total energy density, ϵ_{tot}, by $e_{tot} = \epsilon_{tot}/\rho$, for neutrons and protons in β equilibrium with leptons (electrons and muons) is given by:

$$e_{tot} = e_0 + e_{sym}(Y_n - Y_p)^2 + \sum_{i=n,p} Y_i m_i + e_e + e_\mu \, , \tag{37}$$

where Y_i, $i = n(p)$, stands for the neutron(proton) fraction. On the right-hand side are the baryon contributions including their rest energies (first three terms), and the relativistic electron and the muon energies per baryon (last two terms). Note that, in the equation above, e_0 and e_{sym} are the EoS of symmetric nuclear matter and the symmetry energy, respectively. All terms are functions of density.

The relativistic energy density for particle species "i" having Fermi momentum $(k_F)_i$ is given by

$$\epsilon_i = \frac{\gamma}{2\pi^2} \int_0^{(k_F)_i} \sqrt{k^2 + m_i^2}\, k^2 \, dk \, , \tag{38}$$

where γ is an appropriate degeneracy factor. The partial densities are related to the respective Fermi momenta as

$$\rho_i = \frac{\gamma}{2\pi^2} \int_0^{(k_F)_i} k^2 \, dk \, , \tag{39}$$

which gives, for spin-$\frac{1}{2}$ fermions (γ = 2),

$$\rho_i = \frac{(k_F)_i^3}{3\pi^2} \, , \tag{40}$$

with the corresponding particle fractions given by

$$Y_i = \frac{\rho_i}{\rho} \, . \tag{41}$$

The chemical potential for species "i" is defined as

$$\mu_i = \frac{\partial \epsilon_i}{\partial \rho_i} = \frac{\partial \epsilon_i}{\partial (k_F)_i} \frac{\partial (k_F)_i}{\partial \rho_i} = \sqrt{(k_F)_i^2 + m_i^2} \, , \tag{42}$$

where we have used Equations (38–39) to perform the derivatives with respect to the upper integration limit.

The standard procedure is to minimize the total energy per particle with the constraints of fixed baryon density and global charge neutrality:

$$\rho_p + \rho_n = \rho \Rightarrow Y_p + Y_n = 1 \tag{43}$$

and

$$\rho_p = \rho_e + \rho_\mu \Rightarrow Y_p = Y_e + Y_\mu \, . \tag{44}$$

The resulting set of equations allow to solve for the various lepton fractions from which one can easily obtain the corresponding energy densities.

For the purpose of applying the Lagrange multipliers method, we define the functional

$$F = e_{tot} + \lambda_1 \eta_1 + \lambda_2 \eta_2 \, , \tag{45}$$

where

$$\eta_1 = 1 - Y_n - Y_p \tag{46}$$

and

$$\eta_2 = Y_p - Y_e - Y_\mu \, , \tag{47}$$

and set

$$\frac{\partial F}{\partial Y_i} = 0 \, . \tag{48}$$

Equations (45–48) then yield

$$\frac{\partial F}{\partial Y_p} = \frac{\partial e_{tot}}{\partial Y_p} - \lambda_1 + \lambda_2 = 0 \Rightarrow \mu_p = \lambda_1 - \lambda_2 \tag{49}$$

$$\frac{\partial F}{\partial Y_n} = \frac{\partial e_{tot}}{\partial Y_n} - \lambda_1 = 0 \Rightarrow \mu_n = \lambda_1 \tag{50}$$

$$\frac{\partial F}{\partial Y_e} = \frac{\partial e_{tot}}{\partial Y_e} - \lambda_2 = 0 \Rightarrow \mu_e = \lambda_2 \tag{51}$$

$$\frac{\partial F}{\partial Y_\mu} = \frac{\partial e}{\partial Y_\mu} - \lambda_2 = 0 \Rightarrow \mu_\mu = \lambda_2 \, . \tag{52}$$

Thus,

$$\mu_\mu = \mu_e \tag{53}$$

and

$$\mu_p = \mu_n - \mu_e \; . \tag{54}$$

The equations above allow to solve for the various lepton fractions and, through Equation (38), the corresponding energies are easily obtained. For electrons, we find the ultra-relativistic approximation to be appropriate and set their rest energy to zero in Equation (38).

In Figure 10, we show the predicted fractions for the various species (neutrons, protons, and leptons) at the three highest orders of χEFT which we consider. We note that the proton fraction goes up to just above 10% at the highest densities being considered. This is a rather low value, most likely related to the relatively soft nature of the symmetry energy displayed in Figure 8. It implies that neutron stars (with central densities up to those included in the figure) will not cool down *via* direct Urca processes.

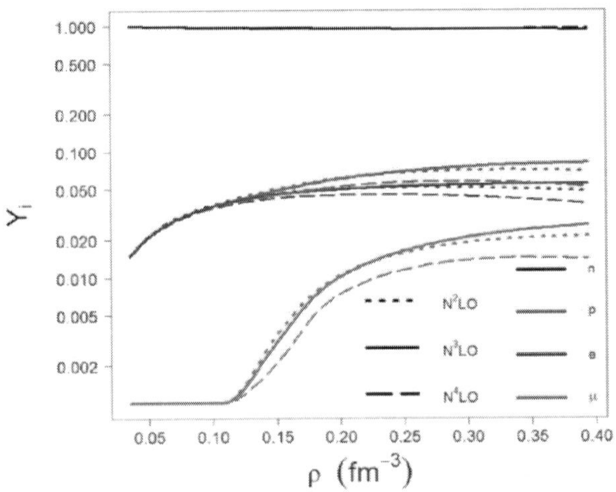

Figure 10: Fractions of neutrons, protons, electrons, and muons as a function of density at the indicated orders for Λ = 450 MeV.

One is now in the position to calculate the pressure in β-stable matter. The pressure is related to the energy density through

$$P(\rho) = \rho^2 \frac{d(e_{tot})}{d\rho} \; . \tag{55}$$

In order to continue the discussion started in section 3.2 and extend it to neutron star radii in a consistent manner, in this section we will use the same interactions constructed in section 3 in terms of an empirical SNM EoS and the microscopic NM chiral EoS. Note, from section 4.2, that the symmetry energy, and thus the EoS of both SNM and NM are needed to obtain the various particle fractions. As mentioned earlier, those fractions tend to be rather small (see Figure 10), and thus the results shown in this and the next sections are to a large extent determined by our chiral predictions in NM.

We close this section by showing in Figure 11 the calculated pressure in β-stable matter at the third, fourth, and fifth orders of the 2NF together with the 3NF constructed as described in section 2.2.2. Of course we are referring to the chiral 3NF appropriate for NM, where the LECs c_4, c_D, and c_E vanish [56].

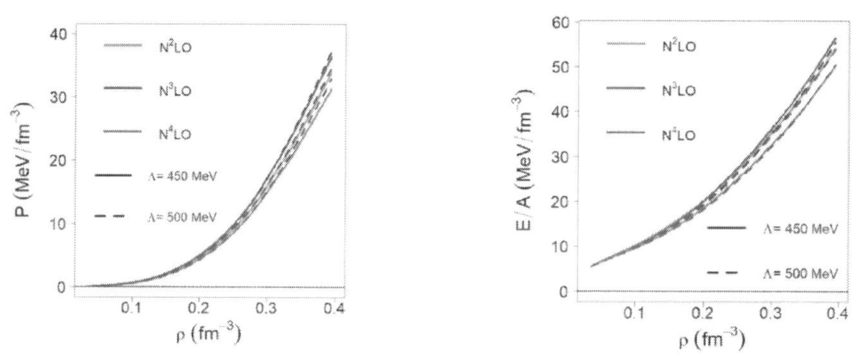

Figure 11: (Left) pressure in β-equilibrated matter *vs*. density at the given chiral orders (of the NM chiral EoS). Predictions with both cutoffs are shown, revealing that the two sets of curves are very similar. (Right) the corresponding energy per particle. These are the predictions which we retain up to $2\rho_0$.

The BHF approach to nuclear matter is appropriate for the description of homogeneous matter, such as a homogeneous fluid of nucleons. Below nuclear densities, the chiral EoS are joined with the crustal equations of state from Harrison and Wheeler [173] and Negele and Vautherin [174], performing a smooth interpolation between the two EoS. The crust has crystal-like composition, and contains light [173] or heavy [174] metals together with a gas of electrons.

Predicting Neutron Star Masses and Radii

We now proceed to discuss specifically neutron star predictions. As mentioned above, we will focus on the radius of a star with mass equal to 1.4 $M_{\odot}$. We note, in passing, that the increased population of neutron stars observed around the mass range of 1.4 $M_{\odot}$ may be related to the physics of white dwarfs, atomic stars supported by electron (rather than neutron) degeneracy pressure. Since the Chandrasekhar limit of white dwarfs is approximately 1.4 $M_{\odot}$ [175], their collapse is likely to generate neutron stars in that mass range.

As stated in section 1, χEFT is a low-energy theory and thus limitations to its domain of applicability must be carefully considered. To begin with, the chiral symmetry breaking scale, $\Lambda_{\chi} \approx 1$ GeV, imposes clear limitations on the momentum or energy ranges where pions and nucleons can be taken as suitable degrees of freedom [15, 17]. Furthermore, the cutoff parameter Λ appearing in the regulator function (cf. Equation 10), has the purpose to remove high momentum components. Naturally, the strength of the cutoff determines to which degree such high-momentum components are suppressed. On the other hand, central densities of compact astrophysical systems can reach as high as several times the density of normal saturated matter, resulting, of course, in the presence of Fermi momenta which are beyond the reach of χEFT. Therefore, if one wishes to make predictions based, to some extent, on χEFT, methods to extend those predictions must be devised.

It has been observed that the pressure as a function of baryon density (or mass density) for a very large number of existing EoS can be fitted by piecewise polytropes, namely functions of the form $P(\rho) = \alpha\rho^{\Gamma}$ [176]. (Note that, in our definitions, ρ denotes the baryon density). Guided by this observation, we find it reasonable to extrapolate the pressure predictions obtained from the EoS shown in Figure 11 using polytropes, as we have done in Sammarruca and Millerson [123]. More precisely, we employ our semi-microscopic predictions up to about $2\rho_0$, where ρ_0 is defined to be 0.16 fm^{-3}, approximately the density of saturated matter. The reason for choosing $2\rho_0$ as a matching density is as follows: since we are dealing with a perturbative expansion in the parameter Q/Λ, we base our arguments on the size of the expansion parameter for typical momenta of the system under consideration. The highest momentum for pure neutron matter around twice normal density is approximately 420 MeV, as obtained from the usual relation $\rho = \frac{k_{F,n}^3}{3\pi^2}$, with $k_{F,n}$ the neutron Fermi momentum. And of course, the highest momentum in β-stable matter is slightly lower due to the presence of a small proton fraction. In conclusion, we are still below (although getting close to) $\Lambda \sim 450 - 500$ MeV. Furthermore, the *r.m.s.* value of the relative momentum for two nucleons in infinite matter is lower than their maximum momentum, and in fact it can be estimated to be about 60% of the Fermi momentum [177]. Thus, on statistical grounds, we should be safe from "cutoff artifacts," even in the presence of smooth regulators.

We then proceed to match polytropes with diverse adiabatic indices, preserving continuity of the pressure. The range of the polytropic index was taken to be between 1.5 and 4.5 [123] (based on guidelines from the literature [176]), and these extensions are calculated up to about $3\rho_0$. At this density, every polytrope is again joined continuously with another set of polytropes spanning the same range in values of Γ. In this way, we are able to cover a large set of possible EoS continuations, simulating scenarios where the EoS displays different degrees of "softness" or "stiffness" in different density regions, and thus we can estimate a realistic uncertainty. We stress again that this procedure is a way to simulate the uncertainty arising from reasonable parameterizations of the EoS as determined by phenomenological studies in the literature, and is not to be understood as a replacement for a theoretical model. A demonstration of this procedure is shown in Figure 12 (top row) for the case of N^{3}LO with Λ = 450 MeV. (Note that cgs units are adopted in Figure 12 as those are popular in astronomy and may facilitate comparison by other authors).

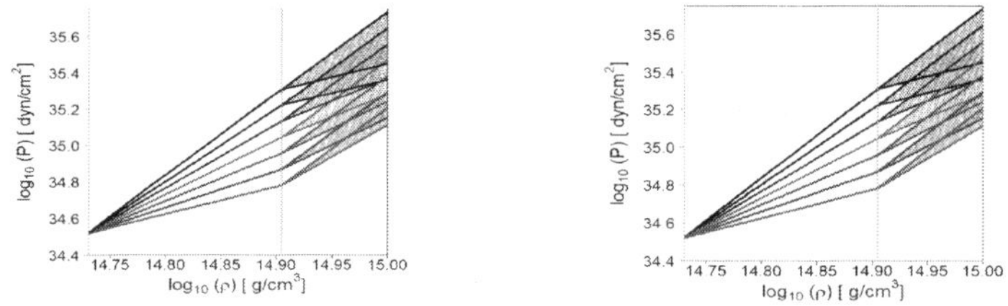

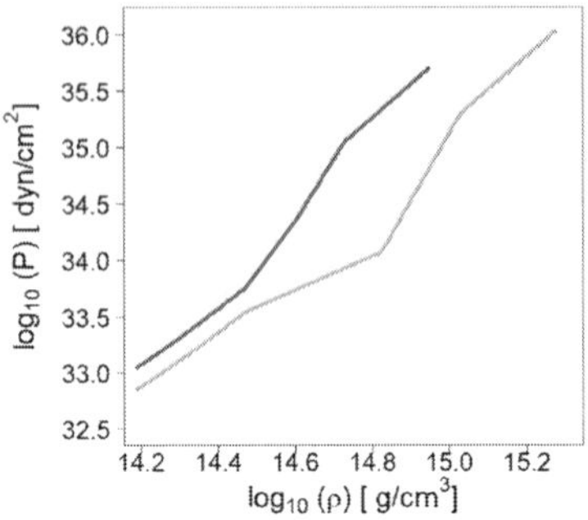

Figure 12: (Top) Pressure in β-equilibrated matter *vs*. density at N^3LO for Λ=450 MeV (left) and Λ=500 MeV (right) extended with polytropes as explained in the text. The vertical axis and the vertical yellow line mark the two matching densities (see text for details). (Bottom) Vaues of the pressure as a function of density taken from Table 5 of Hebeler et al. [178]. The lower (higher) values correspond to the "soft" ("stiff") predictions shown in that table.

In Hebeler et al. [178], the β-equilibrated EoS based on microscopic chiral interactions in neutron matter, is extended to high densities employing a general piecewise polytropic extrapolation, which leads to a very large number of EoS. Applying causality and the requirement that the EoS must be able to support a mass of 1.97 $M_{\odot}$, the author select a range of possible EoS, ranging from "soft" to "stiff." We show the resulting uncertainty band in the bottom row of Figure 12, noting that our uncertainty band from Figure 12 is consistent with it.

Having built up the EoS at all needed densities, we are now in the position to solve the TOV [152] star structure equations and obtain the mass as a function of the radius for a sequence of stars differing in their central densities, up to several times normal density. The *M*(*R*) relations we obtain are shown on the bottom row of Figure 13. Note that only combinations of polytropic indices which can support a maximum mass of at least 1.97 $M_{\odot}$ have been retained for the purpose of Figure 13, to account for the observation of a pulsar with a mass of 2.01 ± 0.04 $M_{\odot}$ [179]. It is appropriate to point out here that most recent observations [180] are consistent with the even higher value of 2.14 $M_{\odot}$. In future work, we will apply this new constraint, which will result in a more limited set of acceptable EoS.

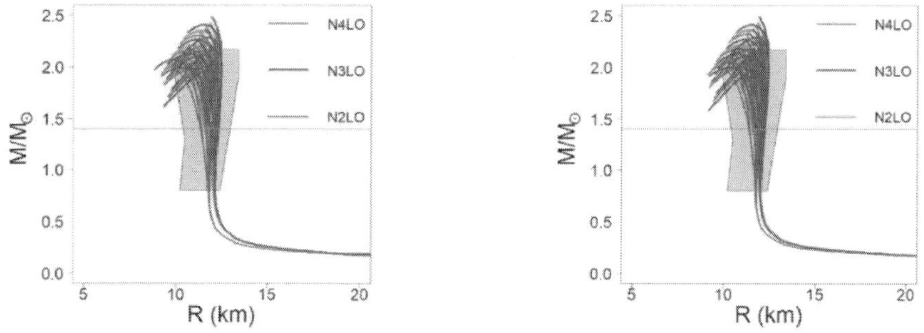

Figure 13: The mass vs. radius relation at the given chiral order. (Left) Λ=450 MeV; (Right) Λ=500 MeV. The purple curves are the result of extending the predictions at N^{4}LO, while the red and the green curves are obtained extrapolating the predictions at N^{3}LO and at N^{2}LO, respectively. The horizontal yellow lines marks the value of 1.4 $M_\odot$. The shaded area in the background is the constraint taken from Steiner et al. [159].

The causality constraint imposes limitations and those are applied in Figure 13. That is, one must require that the speed of sound in stellar matter is less than the speed of light, a condition which can be expressed as $\frac{dP}{d\epsilon} < 1$, where ϵ is the total energy density.

Table 6 shows the the radius and the central density of the 1.4 $M_\odot$ neutron star when the pressure curves at the fourth and fifth orders from Figure 11 are extrapolated *via* piecewise polytropes with adiabatic indices Γ_1 and Γ_2 as shown. The speed of sound at central density is also included. Confirming what we found in Sammarruca and Millerson [123], Table 6 demonstrates quite clearly that the radius is practically insensitive to how the continuation is done. In particular, no changes are observed due to variations of the polytrope attached at $3\rho_0$, and changes by less that one kilometer occurr in response to varying the first polytropic index. Note that the central densities we predict for the canonical-mass star are typically in the order of, and can exceed $3\rho_0$. These densities are at or above those where the spreading of the pressure can be quite large (see Figure 12). Evidently, the radius of a star with this kind of mass responds to pressures at much lower than central densities, in line with earlier observations (see, for instance [178, 181]), where the insensitivity of the radius to the higher densities was pointed out. Tables similar to Table 6 with Λ=500 MeV and at N^{4}LO with changing value of Λ are not included but do lead to very similar observations.

Table 6: Adiabatic indices, Γ_1 and Γ_2, of the polytropes attached at the two matching densities, followed by the radius and the central density of the 1.4 $M_\odot$ neutron star

Γ_1	Γ_2	R (km)	ρ (fm^{-3})	$v_s(c)$
1.5	3.5	11.19	0.653	0.718
1.5	4.0	11.28	0.615	0.767
1.5	4.5	11.34	0.589	0.813
2.0	3.0	11.49	0.620	0.642
2.0	3.5	11.56	0.585	0.695
2.0	4.0	11.60	0.564	0.741
2.0	4.5	11.62	0.548	0.786
2.5	3.0	11.83	0.543	0.621
2.5	3.5	11.84	0.529	0.677
2.5	4.0	11.85	0.521	0.734
2.5	4.5	11.86	0.514	0.789
3.0	2.5	12.05	0.488	0.555
3.0	3.0	12.06	0.485	0.615
3.0	3.5	12.06	0.482	0.671
3.0	4.0	12.05	0.479	0.721
3.0	4.5	12.06	0.478	0.766
3.5	1.5	12.19	0.445	0.650
3.5	2.0	12.19	0.445	0.653
3.5	2.5	12.18	0.444	0.660
3.5	3.0	12.18	0.444	0.673
3.5	3.5	12.18	0.444	0.665
3.5	4.0	12.18	0.444	0.671
3.5	4.5	12.18	0.444	0.676
4.0	1.5	12.27	0.420	0.759
4.0	2.0	12.27	0.420	0.744
4.0	2.5	12.27	0.420	0.734
4.0	3.0	12.27	0.419	0.731
4.0	3.5	12.27	0.420	0.726
4.0	4.0	12.27	0.420	0.710
4.0	4.5	12.27	0.420	0.699
4.5	1.5	12.32	0.404	0.792
4.5	2.0	12.32	0.404	0.781
4.5	2.5	12.32	0.403	0.775
4.5	3.0	12.32	0.403	0.772
4.5	3.5	12.32	0.403	0.772
4.5	4.0	12.32	0.403	0.766

REFERENCES

1. Brueckner KA, Levinson CA, Mahmoud HM. Two-body forces and nuclear saturation. I. Central forces. *Phys Rev.* (1954) **95**:217–28. doi: 10.1103/PhysRev.95.217
2. Bethe HA. Nuclear many-body problem. *Phys Rev.* (1956) **103**:1353–90. doi: 10.1063/1.2915762
3. Goldstone J. Derivation of the Brueckner many-body theory. *Proc R Soc.* (1957) A239:267–79. doi: 10.1098/rspa.1957.0037
4. Bethe HA. Theory of nuclear matter. *Annu Rev Nucl Sci.* (1971) **21**:93–214. doi: 10.1146/annurev.ns.21.120171.000521
5. Haftel MI, Tabakin F. Nuclear saturation and the smoothness of nucleon-nucleon potentials. *Nucl Phys.* (1970) A158:1–42. doi: 10.1016/0375-9474(70)90047-3
6. Sprung DWL. Nuclear matter calculations. *Adv Nucl Phys.* (1972) **5**:225–343. doi: 10.1007/978-1-4615-8231-1_2
7. Pandharipande VR, Wiringa RB. Variations on a theme of nuclear matter. *Rev Mod Phys.* (1979) **51**:821–59. doi: 10.1103/RevModPhys.51.821
8. Lagaris IE, Pandharipande VR. Variational calculations of realistic models of nuclear matter. *Nucl Phys.* (1981) A359:349–64. doi: 10.1016/0375-9474(81)90241-4
9. Day BD, Wiringa RB. Brueckner-Bethe and variational calculations of nuclear matter. *Phys Rev C.* (1985) **32**:1057–62. doi: 10.1103/PhysRevC.32.1057
10. Anastasio MR, Celenza LS, Pong WS, Shakin CM. Relativistic nuclear structure physics. *Phys Rep.* (1983) **100**:327–92. doi: 10.1016/0370-1573(83)90060-1
11. Horowitz CJ, Serot BD. Two-nucleon correlations in a relativistic theory of nuclear matter. *Phys Lett.* (1984) 137B:287–93. The relativistic two-nucleon problem in nuclear matter, Nucl. Phys. **A464**, 613-699 (1987).
12. Brockmann R, Machleidt R. Nuclear saturation in a relativistic Bruckner-hartree-fock Approach. *Phys Lett.* (1984) 149B:283–7.
13. Brockmann R, Machleidt R. Relativistic nuclear structure. I: Nuclear matter. *Phys Rev C.* (1990) **42**:1965.
14. Muether H, Sammarruca F, Ma Z. Relativistic effects and three-nucleon forces in nuclear matter and nuclei. *Int J Mod Phys*

E. (2017) **26**:1730001. doi: 10.1142/S0218301317300016

15. Weinberg S. Nuclear forces from chiral Lagrangians. *Phys Lett B.* (1990) **251**:288–92. doi: 10.1016/0370-2693(90)90938-3
16. Weinberg S. Effective chiral lagrangians for nucleon-pion interactions and nuclear forces. *Nucl Phys B.* (1991) **363**:3–18. doi: 10.1016/0550-3213(91)90231-L
17. Weinberg S. Three-body interactions among nucleons and pions. *Phys Lett B.* (1992) **295**:114–21. doi: 10.1016/0370-2693(92)90099-P
18. Machleidt R, Entem DR. Chiral effective field theory and nuclear force. *Phys Rep.* (2011) **503**:1–75. doi: 10.1016/j.physrep.2011.02.001
19. Weinberg S. Phenomenological lagrangians. *Physica.* (1979) 96A:327–40.
20. Epelbaum E. Hammer HW, Meissner UG. Modern theory of nuclear forces. *Rev Mod Phys.* (2009) **81**:1773–825. doi: 10.1103/RevModPhys.81.1773
21. Meißner UG. The long and winding road from chiral effective Lagrangians to nuclear structure. *Phys Scr.* (2016) **91**:033005. doi: 10.1088/0031-8949/91/3/033005
22. Machleidt R, Sammarruca F. Chiral EFT based nuclear forces: achievements and challenges. *Phys Scr.* (2016) **91**:083007. doi: 10.1088/0031-8949/91/8/083007
23. Fattoyev FJ, Piekarewicz J. Accurate calibration of relativistic mean-field models: correlating observables and providing meaningful theoretical uncertainties. *Phys Rev C.* (2011) **84**:064302. doi: 10.1103/PhysRevC.84.064302
24. Roca-Maza X, Centelles M, Viñas X, Warda M. Neutron skin of ^{208}Pb, nuclear symmetry energy, and the parity radius experiment. *Phys Rev Lett.* (2011) **106**:252501. doi: 10.1103/PhysRevLett.106.252501
25. Sammarruca F, Marcucci LE, Coraggio L, Holt JW, Itaco N, Machleidt R. Nuclear and neutron matter equations of state from high-quality potentials up to fifth order of the chiral expansion. *arXiv:1807.06640* (2018).
26. Sammarruca F. Short-range correlations in the deuteron: chiral effective field theory, meson-exchange, and phenomenology. *Phys Rev C.* (2015) **92**:044003. doi: 10.1103/PhysRevC.92.044003
27. Hen O, Miller GA, Piasetzky E, Weinstein LB. Nucleon-nucleon correlations, short-lived excitations, and the quarks within. *Rev Mod Phys.* (2017) **89**:045002. doi: 10.1103/RevModPhys.89.045002

28. Marcucci LE, Sammarruca F, Viviani M, Machleidt R. Momentum distributions and short-range correlations in the deuteron and ^{3}He with modern chiral potentials. *Phys Rev C.* (2019) **99**:034003. doi: 10.1103/PhysRevC.99.034003

29. Machleidt R. High-precision, charge-dependent Bonn nucleon-nucleon potential. *Phys Rev C.* (2001) **63**:024001. doi: 10.1103/PhysRevC.63.024001

30. Stoks VGJ, Klomp RAM, Terheggen CPF, de Swart JJ. Construction of high-quality NN potential models. *Phys Rev C.* (1994) **49**:2950–63. doi: 10.1103/PhysRevC.49.2950

31. Wiringa RB, Stoks VGJ, Schiavilla R. Accurate nucleon-nucleon potential with charge-independence breaking. *Phys Rev C.* (1995) **51**:38–51. doi: 10.1103/PhysRevC.51.38

32. Machleidt R. Historical perspective and future prospects for nuclear interactions. *Int J Mod Phys E.* (2017) **26**:1730005. doi: 10.1142/S0218301317300053

33. Epelbaum E, Nogga A, Gl'ockle W, Kamada H, Meißner UG, Witala H. Three nucleon forces from chiral effective field theory. *Phys Rev C.* (2002) **66**:064001. doi: 10.1103/PhysRevC.66.064001

34. Navratil P, Roth R, Quaglioni S. *Ab initio* many-body calculations of nucleon scattering on ^{4}He, ^{7}Li, ^{7}Be, ^{12}C and ^{16}O. *Phys Rev C.* (2010) **82**:034609. doi: 10.1103/PhysRevC.82.034609

35. Viviani M, Girlanda L, Kievsky A, Marcucci LE. Effect of three-nucleon interaction in p-3He elastic scattering. *Phys Rev Lett.* (2013) **111**:172302. doi: 10.1103/PhysRevLett.111.172302

36. Golak J, Skibiński R, Topolnicki K, Witala H, Epelbaum E, Krebs H, et al. Low-energy neutron-deuteron reactions with N^{3}LO chiral forces. *Eur Phys J A.* (2014) **50**:177–89. doi: 10.1140/epja/i2014-14177-7

37. Kalantar-Nayestanaki N, Epelbaum E, Messchendorp JG, Nogga A. Signatures of three-nucleon interactions in few-nucleon systems. *Rept Prog Phys.* (2012) **75**:016301. doi: 10.1088/0034-4885/75/1/016301

38. Navratil P, Quaglioni S, Hupin G, Romero-Redondo C, Calci A. Unified ab initio approaches to nuclear structure and reactions. *Phys Scr.* (2016) **91**:053002. doi: 10.1088/0031-8949/91/5/053002

39. Coraggio L, Covello A, Gargano A, Itaco N, Kuo TTS, Entem DR, et al. Low-momentum nucleon-nucleon interactions and shell-

model calculations. *Phys Rev C.* (2007) **75**:024311. doi: 10.1103/PhysRevC.75.024311

40. Coraggio L, Covello A, Gargano A, Itaco N. Shell-model calculations for neutron-rich carbon isotopes with a chiral nucleon-nucleon potential. *Phys Rev C.* (2010) **81**:064303. doi: 10.1103/PhysRevC.81.064303
41. Coraggio L, Covello A, Gargano A, Itaco N, Kuo TTS. Effective shell-model hamiltonians from realistic nucleon-nucleon potentials within a perturbative approach. *Ann Phys.* (2012) **327**:2125–51. doi: 10.1016/j.aop.2012.04.013
42. Hagen H, Hjorth-Jensen M, Jansen GR, Machleidt R, Papenbrock T. Continuum effects and three-nucleon forces in neutron-rich oxygen isotopes. *Phys Rev Lett.* (2012) **108**:242501. doi: 10.1103/PhysRevLett.108.242501
43. Hagen H, Hjorth-Jensen M, Jansen GR, Machleidt R, Papenbrock T. Evolution of shell structure in neutron-rich calcium isotopes. *Phys Rev Lett.* (2012) **109**:032502. doi: 10.1103/PhysRevLett.109.032502
44. Barrett BR, Navratil P, Vary JP. Ab initio no core shell model. *Prog Part Nucl Phys.* (2013) **69**:131–81. doi: 10.1016/j.ppnp.2012.10.003
45. Gezerlis A, Tews I, Epelbaum E, Gandolfi S, Hebeler K, Nogga A, et al. Quantum Monte Carlo calculations with chiral effective field theory interactions. *Phys Rev Lett.* (2013) **111**:032501. doi: 10.1103/PhysRevLett.111.032501
46. Hergert H, Bogner SK, Binder S, Calci A, Langhammer J, Roth R, Schwenk A. In-medium similarity renormalization group with chiral two- plus three-nucleon interactions. *Phys Rev C.* (2013) **87**:034307. doi: 10.1103/PhysRevC.87.034307
47. Hagen G, Papenbrock T, Hjorth-Jensen M, Dean DJ. Coupled-cluster computations of atomic nuclei. *Rept Prog Phys.* (2014) **77**:096302. doi: 10.1088/0034-4885/77/9/096302
48. Somà V, Cipollone A, Barbieri C, Navratil P, Duget T. Chiral two- and three-nucleon forces along medium-mass isotope chains. *Phys Rev C.* (2014) **89**:061301(R). doi: 10.1103/PhysRevC.89.061301
49. Hebeler K, Holt JD, Menéndez J, Schwenk A. Nuclear forces and their impact on neutron-rich nuclei and neutron-rich matter. *Annu Rev Nucl Part Sci.* (2015) **65**:457–84. doi: 10.1146/annurev-nucl-102313-025446
50. Hagen G, Ekström A, Forssén C, Jansen GR, Nazarewicz W, Papenbrock

T, et al. Neutron and weak-charge distributions of the ^{48}Ca nucleus. *Nat Phys.* (2015) **12**:186–90. doi: 10.1038/nphys3529

51. Carlson J, Gandolfi S, Pederiva F, Pieper SC, Schiavilla R, Schmidt KE, Wiringa RB. Quantum Monte Carlo methods for nuclear physics. *Rev Mod Phys.* (2015) **87**:1067–118. doi: 10.1103/RevModPhys.87.1067
52. Hergert H, Bogner SK, Morris TD, Schwenk A, Tsukiyama K. The in-medium similarity renormalization group: a novel *ab initio* method for nuclei. *Phys Rep.* (2016) **621**:165–222. doi: 10.1016/j.physrep.2015.12.007
53. Holt J W, Kaiser N. Equation of state of nuclear and neutron matter at third-order in perturbation theory from chiral effective field theory. *Phys Rev C.* (2017) **95**:034326. doi: 10.1103/PhysRevC.95.034326
54. Simonis J, Stroberg SR, Hebeler K, Holt JD, Schwenk A. Saturation with chiral interactions and consequences for finite nuclei. *Phys Rev C.* (2017) **96**:014303. doi: 10.1103/PhysRevC.96.014303
55. Morris TD, Simonis J, Stroberg SR, Stumpf C, Hagen G, Holt JD, et al. Structure of the lightest Tin isotopes. *arXiv:1709.02786 [nucl-th]* (2017). doi: 10.1103/PhysRevLett.120.152503
56. Hebeler K, Schwenk A. Chiral three-nucleon forces and neutron matter. *Phys Rev C.* (2010) **82**:014314. doi: 10.1103/PhysRevC.82.014314
57. Hebeler K, Bogner SK, Furnstahl RJ, Nogga A, Schwenk A. Improved nuclear matter calculations from chiral low-momentum interactions. *Phys Rev C.* (2011) **83**:031301(R). doi: 10.1103/PhysRevC.83.031301
58. Baardsen G, Ekström A, Hagen G, Hjorth-Jensen M. Coupled-cluster studies of infinite nuclear matter. *Phys Rev C.* (2013) **88**:054312. doi: 10.1103/PhysRevC.88.054312
59. Hagen G, Papenbrock T, Ekström A, Wendt KA, Baardsen G, Gandolfi S, et al. Coupled-cluster calculations of nucleonic matter. *Phys Rev C.* (2014) **89**:014319. doi: 10.1103/PhysRevC.89.014319
60. Coraggio L, Holt JW, Itaco N, Machleidt R, Sammarruca F. Reduced regulator dependence of neutron-matter predictions with perturbative chiral interactions. *Phys Rev C.* (2013) **87**:014322. doi: 10.1103/PhysRevC.87.014322
61. Coraggio L, Holt JW, Itaco N, Machleidt R, Marcucci LE, Sammarruca F. Nuclear-matter equation of state with consistent two- and three-body

perturbative chiral interactions. *Phys Rev C.* (2014) **89**:044321. doi: 10.1103/PhysRevC.89.044321

62. Sammarruca F, Coraggio L, Holt JW, Itaco N, Machleidt R, Marcucci LE. Toward order-by-order calculations of the nuclear and neutron matter equations of state in chiral effective field theory. *Phys Rev C.* (2015) **91**:054311. doi: 10.1103/PhysRevC.91.054311
63. Drischler C, Carbone A, Hebeler K, Schwenk A. Neutron matter from chiral two- and three-nucleon calculations up to N^3LO. *Phys Rev C.* (2016) **94**:054307. doi: 10.1103/PhysRevC.94.054307
64. Tews I, Gandolfi S, Gezerlis A, Schwenk A. Quantum Monte Carlo calculations of neutron matter with chiral three-body forces. *Phys Rev C.* (2016) **93**:024305. doi: 10.1103/PhysRevC.93.024305
65. Wellenhofer C, Holt JW, Kaiser N, Weise W. Nuclear thermodynamics from chiral low-momentum interactions. *Phys Rev C.* (2014) **89**:064009. doi: 10.1103/PhysRevC.89.064009
66. Wellenhofer C, Holt JW, Kaiser N. Thermodynamics of isospin-asymmetric nuclear matter from chiral effective field theory. *Phys Rev C.* (2015) **92**:015801. doi: 10.1103/PhysRevC.92.015801
67. Bacca S, Hally K, Pethick CJ, Schwenk A. Chiral effective field theory calculations of neutrino processes in dense matter. *Phys Rev C.* (2009) **80**:032802. doi: 10.1103/PhysRevC.80.032802
68. Bartl A, Pethick CJ, Schwenk A. Supernova matter at subnuclear densities as a resonant fermi gas: enhancement of neutrino rates. *Phys Rev Lett.* (2014) **113**:081101. doi: 10.1103/PhysRevLett.113.081101
69. Rapaj E, Holt JW, Bartl A, Reddy S, Schwenk A. Charged-current reactions in the supernova neutrino-sphere. *Phys Rev C.* (2015) **91**:035806. doi: 10.1103/PhysRevC.91.035806
70. Buraczynski M, Gezerlis A. Static response of neutron matter. *Phys Rev Lett.* (2016) **116**:152501. doi: 10.1103/PhysRevLett.116.152501
71. Holt JW, Kaiser N, Miller GA. Microscopic optical potential for exotic isotopes from chiral effective field theory. *Phys Rev C.* (2016) **93**:064603. doi: 10.1103/PhysRevC.93.064603
72. Birkhan J, Miorelli M, Bacca S, Bassauer S, Bertulani CA, Hagen C, et al. Electric dipole polarizability of ^{48}Ca and implications for the neutron skin. *Phys Rev Lett.* (2017) **118**:252501. doi: 10.1103/PhysRevLett.118.252501
73. Rotureau J, Danielewicz P, Hagen G, Nunes F, Papenbrock T. Optical

potential from first principles. *Phys Rev C.* (2017) **95**:024315. doi: 10.1103/PhysRevC.95.024315

74. Lapoux V, Somà V, Barbieri C, Hergert H, Holt JD, Stroberg SR. Radii and binding energies in oxygen isotopes: a challenge for nuclear forces. *Phys Rev Lett.* (2016) **117**:052501. doi: 10.1103/PhysRevLett.117.052501
75. Binder S, Langhammer J, Calci A, Roth R. *Ab initio* path to heavy nuclei. *Phys Lett B.* (2014) **736**:119–23. doi: 10.1016/j.physletb.2014.07.010
76. Ekström A, Jansen GR, Wendt KA, Hagen G, Papenbrock T, Carlsson BD, et al. Accurate nuclear radii and binding energies from a chiral interaction. *Phys Rev C.* (2015) **91**:051301. doi: 10.1103/PhysRevC.91.051301
77. Alonso D, Sammarruca F. A Microscopic calculation of asymmetric nuclear matter properties. *Phys Rev C.* (2003) **67**:054301. doi: 10.1103/PhysRevC.67.054301
78. Sammarruca F. The '*Ab initio*' approach to the nuclear equation of state: review and discussion. *arXiv:0807.0263 [nucl-th]* (2008).
79. Sammarruca F, Chen B, Coraggio L, Itaco N, Machleidt R. Dirac-Brueckner-Hartree-Fock versus chiral effective field theory. *Phys Rev C.* (2012) **86**:054317. doi: 10.1103/PhysRevC.86.054317
80. Pudliner BS, Pandharipande VR, Carlson J, Wiringa RB. Quantum Monte Carlo calculations of $A \leq 6$ nuclei. *Phys Rev Lett.* (1995) **74**:4396–9. doi: 10.1103/PhysRevLett.74.4396
81. Akmal A, Pandharipande VR, Ravenhall DG. The Equation of state of nucleon matter and neutron star structure. *Phys Rev C.* (1998) **58**:1804–28. doi: 10.1103/PhysRevC.58.1804
82. Lonardoni D, Lovato A, Pieper SC, Wiringa RB. Variational calculation of the ground state of closed-shell nuclei up to $A = 40$. *Phys Rev C.* (2017) **96**:024326. doi: 10.1103/PhysRevC.96.024326
83. Pieper SC. The Illinois extension to the Fujita-Miyazawa three-nucleon force. *AIP Conf Proc.* (2008) **1011**:143–52. doi: 10.1063/1.2932280
84. Zuo W, Lejeune A, Lombardo U, Mathiot JF. Microscopic three-body force for asymmetric nuclear matter. *Eur Phys J A.* (2002) **14**:469–75. doi: 10.1140/epja/i2002-10031-y
85. Li ZH, Lombardo U, Schulze HJ, Zuo W. Consistent nucleon-nucleon potentials and three-body forces. *Phys Rev C.* (2008) **77**:034316. doi:

10.1103/PhysRevC.77.034316

86. Machleidt R. The meson theory of nuclear forces and nuclear structure. *Adv Nucl Phys.* (1989) **19**:189–376. doi: 10.1007/978-1-4613-9907-0_2
87. Entem DR, Machleidt R, Nosyk Y. High-quality two-nucleon potentials up to fifth order of the chiral expansion. *Phys Rev C.* (2017) **96**:024004. doi: 10.1103/PhysRevC.96.024004
88. Entem DR, Machleidt R. Accurate charge-dependent nucleon-nucleon potential at fourth order of chiral perturbation theory. *Phys Rev C.* (2003) **68**:041001. doi: 10.1103/PhysRevC.68.041001
89. Marji E, Canul A, MacPherson Q, Winzer R, Zeoli C, Entem DR, et al. Nonperturbative renormalization of the chiral nucleon-nucleon interaction up to next-to-next-to-leading order. *Phys Rev C.* (2013) **88**:054002. doi: 10.1103/PhysRevC.88.054002
90. Hoferichter M, de Elvira RJ, Kubis B, Meissner UG. Matching pion-nucleon roy-steiner equations to chiral perturbation theory. *Phys Rev Lett.* (2015) **115**:192301. doi: 10.1103/PhysRevLett.115.192301
91. Hoferichter M, de Elvira RJ, Kubis B, Meissner UG. Roy-Steiner-equation analysis of pion-nucleon scattering. *Phys Rep.* (2016) **625**:1–88. doi: 10.1016/j.physrep.2016.02.002
92. Hoppe J, Drischler C, Furnstahl RJ, Hebeler K, Schwenk A. Weinberg eigenvalues for chiral nucleon-nucleon interactions. *Phys Rev C.* (2017) **96**:054002. doi: 10.1103/PhysRevC.96.054002
93. Drischler C, Hebeler K, Schwenk A. Chiral interactions up to N^3LO and nuclear saturation. *Phys Rev Lett.* (2019) **122**:042501. doi: 10.1103/PhysRevLett.122.042501
94. Holt JW, Kaiser N, Weise W. Chiral three-nucleon interaction and the ^{14}C-dating β decay. *Phys Rev C.* (2009) **79**:054331. doi: 10.1103/PhysRevC.79.054331
95. Holt JW, Kaiser N, Weise W. Density-dependent effective nucleon-nucleon interaction from chiral three-nucleon forces. *Phys Rev C.* (2010) **81**:024002. doi: 10.1103/PhysRevC.81.024002
96. Baldo M, Bombaci I, Burgio GF. Microscopic nuclear equation of state with three-body forces and neutron star structure. *Astron Astrophys.* (1997) **328**:274–84.
97. Kaiser N, Niessner V. Density-dependent NN interaction from subleading chiral 3N forces: short-range terms and relativistic corrections. *Phys*

Rev C. (2018) **98**:054002. doi: 10.1103/PhysRevC.98.054002

98. Kaiser N, Singh B. Density-dependent NN interaction from subleading chiral three-nucleon forces: long-range terms. *Phys Rev C.* (2019) **100**:014002. doi: 10.1103/PhysRevC.100.014002
99. Gardestig A, Phillips DR. How low-energy weak reactions can constrain three-nucleon forces and the neutron-neutron scattering length. *Phys Rev Lett.* (2006) **96**:232301. doi: 10.1103/PhysRevLett.96.232301
100. Gazit D, Quaglioni S, Navrátil P. Three-nucleon low-energy constants from the consistency of interactions and currents in chiral effective field theory. *Phys Rev Lett.* (2009) **103**:102502. doi: 10.1103/PhysRevLett.103.102502
101. Marcucci LE, Kievsky A, Rosati S, Schiavilla R, Viviani M. Chiral effective field theory predictions for muon Capture on Deuteron and ^{3}He, *Phys Rev Lett.* (2012) **108**:052502; Erratum: Chiral effective field theory predictions for muon capture on deuteron and ^{3}He. (2018) **121**:049901(E). doi: 10.1103/PhysRevLett.108.052502
102. Navrátil P. Local three-nucleon interaction from chiral effective field theory. *Few-Body Syst.* (2007) **41**:117–40. doi: 10.1007/s00601-007-0193-3
103. Kievsky A, Rosati S, Marcucci LE, Girlanda L. A high-precision variational approach to three- and four-nucleon bound and zero-energy scattering states. *J Phys G.* (2008) **35**:063101. doi: 10.1088/0954-3899/35/6/063101
104. Tews I, Krüger T, Hebeler K, Schwenk A. Neutron matter at next-to-next-to-next-to-leading order in chiral effective field theory. *Phys Rev Lett.* (2013) **110**:032504. doi: 10.1103/PhysRevLett.110.032504
105. Hebeler K, Krebs H, Epelbaum E, Golak J, Skibinski R. Efficient calculation of chiral three-nucleon forces up to N^{3}LO for ab initio studies. *Phys Rev C.* (2015) **91**:044001. doi: 10.1103/PhysRevC.91.044001
106. Krebs H, Gasparyan A, Epelbaum E. Chiral three-nucleon force at N^{4}LO: longest-range contributions. *Phys Rev C.* (2012) **85**:054006. doi: 10.1103/PhysRevC.85.054006
107. Bernard V, Epelbaum E, Krebs H, Meißner UG. Subleading contributions to the chiral three-nucleon force: long-range terms. *Phys Rev C.* (2008) **77**:064004. doi: 10.1103/PhysRevC.77.064004
108. Fujita JI, Miyazawa H. Pion theory of three-body forces. *Prog Theor*

Phys. (1957) **17**:360–5. doi: 10.1143/PTP.17.360

109. De JN, Samadar SK, Agrawal BK. Reassessing nuclear matter incompressibility and its density dependence. *Phys Rev C.* (2015) **92**:014304. doi: 10.1103/PhysRevC.92.014304
110. Thiel M, Sfienti C, Piekarewicz J, Horowitz CJ, Vanderhaeghen M. Neutron skins of atomic nuclei: per aspera ad astra. *arXiv:1904.12269* (2019). doi: 10.1088/1361-6471/ab2c6d
111. Bombaci I, Logoteta D. Equation of state of dense nuclear matter and neutron star structure from nuclear chiral interactions. *Astron Astrophys.* (2018) **609**:A128. doi: 10.1051/0004-6361/201731604
112. Oertel M, Hempel M, Klahn T, Typel S. Equations of state for supernovae and compact stars. *Rev Mod Phys.* (2017) **89**:015007. doi: 10.1103/RevModPhys.89.015007
113. Baldo M, Burgio GF. The nuclear symmetry energy. *Prog Part Nucl Phys.* (2016) **91**:101016. doi: 10.1016/j.ppnp.2016.06.006
114. Le Fevre A, Leifels Y, Reisdorf W, Aichelin J, Hartnack Ch. Constraining the nuclear matter equation of state around twice saturation density. *Nucl Phys A.* (2016) **945**:112–3. doi: 10.1016/j.nuclphysa.2015.09.015
115. Horowitz CJ, Brown EF, Kim Y, Lynch WG, Michaels R, Ono A, et al. A way forward in the study of the symmetry energy: experiment, theory, and observation. *J Phys G.* (2014) **41**:093001. doi: 10.1088/0954-3899/41/9/093001
116. Verde G. Probing the nuclear equation-of-state and the symmetry energy with heavy-ion collisions. *EPJ Web Conf.* (2014) **66**:01018. doi: 10.1051/epjconf/20146601018
117. Trautmann W, Wolter HH. Elliptic flow and the symmetry energy at supra-saturation density. *Int J Mod Phys E.* (2012) **21**:1230003. doi: 10.1142/S0218301312300032
118. Steiner AW, Prakash M, Lattimer JM, Ellis PJ. Isospin asymmetry in nuclei and neutron stars. *Phys Rep.* (2005) **411**:325–75. doi: 10.1016/j.physrep.2005.02.004
119. Bombaci I, Lombardo U. Asymmetric nuclear matter equation of state. *Phys Rev C.* (1991) **44**:1892–900. doi: 10.1103/PhysRevC.44.1892
120. Sammarruca F. Recent advances in microscopic approaches to nuclear matter and symmetry energy. *Symmetry.* (2014) **6**:851–79. doi: 10.3390/sym6040851

121. Paschke K, Kumar K, Michael R, Souder PA, Urciuoli GM. *PREX-II: Precision Parity-Violating Measurement of the Neutron Skin of Lead.* Available online at: http://hallaweb.jlab.org/parity/prex/prexII.pdf

122. Mammei J, McNulty D, Michaels R, Paschke K, Riordan S, Souder P. C-REX: parity-violating measurement of the weak charge of 48Ca to an accuracy of 0.02 fm. Available online at: http://hallaweb.jlab.org/parity/prex/c-rex2013_v7.pdf

123. Sammarruca F, Millerson R. The radius of the canonical-mass neutron star and chiral effective field theory. *J Phys G.* (2019) **46**:024001. doi: 10.1088/1361-6471/aafbcd

124. Santos BM, Dutra M, Lourenco O, Delfino A. Correlations between bulk parameters in relativistic and nonrelativistic hadronic mean-field models. *Phys Rev C.* (2015) **92**:015210. doi: 10.1103/PhysRevC.92.0152

125. Tong H, Ren XL, Ring P, Shen SH, Wang SB, Meng J. Relativistic Brueckner-Hartree-Fock theory in nuclear matter without the average momentum approximation. *Phys Rev C.* (2018) **98**:054302. doi: 10.1103/PhysRevC.98.054302

126. Holt JW, Lim Y. Universal correlations in the nuclear symmetry energy, slope parameter, and curvature. *Phys Lett.* (2018) B784:77–81. doi: 10.1016/j.physletb.2018.07.038

127. Vidaña I, Providencia C, Polls A, Rios A. Density dependence of the nuclear symmetry energy: a microscopic perspective. *Phys Rev C.* (2009) **80**:045806. doi: 10.1103/PhysRevC.80.045806

128. Ducoin C, Margueron J, Providencia CMC, Vidaña I. Core-crust transition in neutron stars: predictivity of density developments. *Phys Rev C.* (2011) **83**:045810. doi: 10.1103/PhysRevC.83.045810

129. Santos BM, Dutra M, Lourenco O, Delfino A. Correlations between the nuclear matter symmetry energy, its slope, and curvature from a nonrelativistic solvable approach and beyond. *Phys Rev C.* (2014) **90**:035203. doi: 10.1103/PhysRevC.90.035203

130. Millerson R, Sammarruca F. Properties of isospin asymmetric matter derived from chiral effective field theory. *arXiv:1906.02905 [nucl-th]* (2019).

131. Piekarewicz J, Fattoyev FJ. Neutron rich matter in heaven and on Earth. *arXiv:1907.02561 [nucl-th].*

132. Brown BA. Neutron radii in nuclei and the neutron equation of state. *Phys*

Rev Lett. (2000) **85**:5296–9. doi: 10.1103/PhysRevLett.85.5296

133. Lalazissis GA, Konig J, Ring P. New parametrization for the Lagrangian density of relativistic mean field theory. *Phys Rev C.* (1997) **55**:540–3. doi: 10.1103/PhysRevC.55.540
134. Fattoyev FJ, Horowitz CJ, Piekarewicz J, Shen G. Relativistic effective interaction for nuclei, giant resonances, and neutron stars. *Phys Rev C.* (2010) **82**:055803. doi: 10.1103/PhysRevC.82.055803
135. Alam N, Agrawal BK, De JN, Samaddar SK, Colo G. Equation of state of nuclear matter from empirical constraints. *Phys Rev C.* (2014) **90**:054317. doi: 10.1103/PhysRevC.90.054317
136. Danielewicz P, Lee J. Symmetry energy II: isobaric analog states. *Nucl Phys A.* (2014) **922**:1–70. doi: 10.1016/j.nuclphysa.2013.11.005
137. Tsang MB, Zhang Y, Danielewicz P, Famiano M, Li Z, Lynch WG, et al. Constraints on the density dependence of the symmetry energy. *Phys Rev Lett.* (2009) **102**:122701. doi: 10.1103/PhysRevLett.102.122701
138. Russotto P, Gannon S, Kupny S, Lasko P, Acosta L, Adamczyk M, et al. Results of the ASY-EOS experiment at GSI: the symmetry energy at suprasaturation density. *Phys Rev C.* (2016) **94**:034608. doi: 10.1103/PhysRevC.94.034608
139. Russotto P, Wu PZ, Zoric M, Chartier M, Leifels Y, Lemmon RC, et al. Symmetry energy from elliptic flow in ^{197}Au + ^{197}Au. *Phys Lett B.* (2011) **697**:471–6. doi: 10.1016/j.physletb.2011.02.033
140. Lattimer JM, Lim Y. Constraining the symmetry parameters of the nuclear interaction. *Astrophys J.* (2013) **771**:51–65. doi: 10.1088/0004-637X/771/1/51
141. Malik T, Alam N, Fortin M, Providencia C, Agrawal BK, Jha TK, et al. GW170817: constraining the nuclear matter equation of state from the neutron star tidal deformability. *Phys Rev C.* (2018) **98**:035804. doi: 10.1103/PhysRevC.98.035804
142. Sammarruca F. Proton skins, neutron skins, and proton radii of mirror nuclei. *Front Phys.* (2018) **6**:90–8. doi: 10.3389/fphy.2018.00090
143. Sammarruca F. Investigation of constraints on few-neutron forces in neutron matter by empirical information on the neutron skin of ^{4}8Ca and ^{2}08Pb. *Phys Rev C.* (2016) **94**:054317. doi: 10.1103/PhysRevC.94.054317
144. Inakura T, Nakada N. Constraining the slope parameter of the symmetry energy from nuclear structure. *Phys Rev C.* (2015) **92**:064302. doi:

10.1103/PhysRevC.92.064302

145. Warda M., Viñas X, Roca-Maza X, Centelles M. Neutron skin thickness in the droplet model with surface width dependence: indications of softness of the nuclear symmetry energy. *Phys Rev C.* (2009) **80**:024316. doi: 10.1103/PhysRevC.80.024316
146. Xu C, Li B, Chen L, Ko CM. Analytical relations between nuclear symmetry energy and single-nucleon potentials in isospin asymmetric nuclear matter. *Nucl Phys A.* (2011) **865**:1–16. doi: 10.1016/j.nuclphysa.2011.06.027
147. Roca-Maza X, Brenna M, Agrawal BK, Bortignon PF, Colo G, Cao LG, et al. Giant quadrupole resonances in ^{2}08Pb, the nuclear symmetry energy, and the neutron skin thickness. *Phys Rev C.* (2013) **87**:034301. doi: 10.1103/PhysRevC.87.034301
148. Abrahamyan S, Ahmed Z, Albataineh H, Aniol K, Armstrong DS, Armstrong W, et al. Measurement of the neutron radius of ^{208}Pb through parity violation in electron scattering. *Phys Rev Lett.* (2012) **108**:112502. doi: 10.1103/PhysRevLett.108.112502
149. Mammei J, (for the PREX Collaboration). The Pb radius experiment (PREX). *AIP Conf Proc.* (2013) **1560**:620–3. doi: 10.1063/1.4826855
150. Becker D, Bucoveanu R, Grzesik C, Imai K, Kempf R, Molitor M, et al. The P2 experiment. *Eur Phys J.* (2018) A54:208–69. doi: 10.1140/epja/i2018-12611-6
151. Oyamatsu K, Iida K, Koura H. Neutron drip line and the equation of state of nuclear matter. *Phys Rev C.* (2010) **82**:027301. doi: 10.1103/PhysRevC.82.027301
152. Oppenheimer JR, Volkoff GM. On massive neutron cores. *Phys Rev C.* (1939) **55**:374–80. doi: 10.1103/PhysRev.55.374
153. Weber F. *Pulsar as Astrophysical Laboratories for Nuclear and Particle Physics*. Bristol; Philadelphia, PA: IOP (1999).
154. Glendenning NK. *Compact Stars*. 2nd Edn. New York, NY: Springer (2000).
155. Özel F, Freire P. Masses, radii, and the equation of state of neutron stars. *Annu Rev Astron Astrophys.* (2016) **54**:401–40. doi: 10.1146/annurev-astro-081915-023322
156. Lattimer JM, Prakash M. The physics of neutron stars. *Science.* (2004) **304**:536–42. doi: 10.1126/science.1090720
157. Lattimer JM, Prakash M. Neutron star observations: prognosis for

equation of state constraints. *Phys Rep.* (2007) **442**:109–65. doi: 10.1016/j.physrep.2007.02.003

158. Lattimer JM, Prakash M. The equation of state of hot, dense matter and neutron stars. *Phys Rep.* (2016) **621**:127–64. doi: 10.1016/j.physrep.2015.12.005
159. Steiner AW, Lattimer JM, Brown E.F. The neutron star mass-radius relation and the equation of state of dense matter. *Astrophys J.* (2013) **765**:L5. doi: 10.1088/2041-8205/765/1/L5
160. Özel F. Surface emission from neutron stars and implications for the physics of their interiors. *Rep Prog Phys.* (2013) **76**:016901. doi: 10.1088/0034-4885/76/1/016901
161. Rhoades CD, Ruffini R. Maximum mass of a neutron star. *Phys Rev Lett.* (1974) **32**:324–7. doi: 10.1103/PhysRevLett.32.324
162. Sabbadini AG, Hartle JB. Bounds on the moment of inertia of nonrotating neutron stars. *Ann Phys NY.* (1977) **104**:95–113. doi: 10.1016/0003-4916(77)90047-1
163. Hartle JB. Bounds on the mass and moment of inertia of non-rotating neutron stars. *Phys Rep.* (1978) **46**:201–47. doi: 10.1016/0370-1573(78)90140-0
164. Abbott BP, [LIGO Scientific and Virgo Collaborations]. GW170817: Observation of gravitational waves from a binary neutron star inspiral. *Phys Rev Lett.* (2017) **119**:161101. doi: 10.1103/PhysRevLett.119.161101
165. Hinderer T, Lackey BD, Lang RN, Read JS. Tidal deformability of neutron stars with realistic equations of state and their gravitational wave signatures in binary inspiral. *Phys Rev D.* (2010) **81**:123016. doi: 10.1103/PhysRevD.81.123016
166. Lim Y, Holt JW. Neutron star tidal deformabilities constrained by nuclear theory and experiment. *Phys Rev Lett.* (2018) **121**:062701. doi: 10.1103/PhysRevLett.121.062701
167. Annala E, Gorda T, Kurkela A, Vuorinen A. Gravitational-wave constraints on the neutron-star-matter equation of state. *Phys Rev Lett.* (2018) **120**:172703. doi: 10.1103/PhysRevLett.120.172703
168. Fattoyev FJ, Piekarewicz J, Horowitz CJ. Neutron skins and neutron stars in the multimessenger era. *Phys Rev Lett.* (2018) **120**:172702. doi: 10.1103/PhysRevLett.120.172702

169. Most ER, Weih LR, Rezzolla L, Schffner-Bielich J. New constraints on radii and tidal deformabilities of neutron stars from GW170817. *Phys Rev Lett.* (2018) **120**:261103. doi: 10.1103/PhysRevLett.120.261103
170. Raithel CA, Ozel F, Psaltis D. Tidal Deformability from GW170817 as a direct probe of the neutron star radius. *Astrophys J.* (2018) **857**:L23. doi: 10.3847/2041-8213/aabcbf
171. Nandi R, Char P, Pal S. Constraining the relativistic mean-field model equations of state with gravitational wave observations. *Phys Rev C.* (2019) **99**:052802. doi: 10.1103/PhysRevC.99.052802
172. Burgio GF, Drago A, Pagliara G, Schulze HJ, Wei JB. Are small radii of compact stars ruled out by GW170817/AT2017gfo?. *Astrophys J.* (2018) **860**:139-144. doi: 10.3847/1538-4357/aac6ee
173. Harrison BK, Wheeler JA. *Gravitation Theory and Gravitational Collapse*. Chicago, IL: University of Chicago Press (1965).
174. Negele JW, Vautherin D. Neutron star matter at sub-nuclear densities. *Nucl Phys.* (1973) A207:298–320. doi: 10.1016/0375-9474(73)90349-7
175. Liebert J. White dwarf stars. *Annu Rev Astron Astrophys.* (1980) **18**:363–98. doi: 10.1146/annurev.aa.18.090180.002051
176. Read JS, Lackey BD, Owen BJ, Friedman JL. Constraints on a phenomenologically parametrized neutron-star equation of state. *Phys Rev D.* (2009) **79**:124032. doi: 10.1103/PhysRevD.79.124032
177. Sammarruca F. Short-range correlations in isospin symmetric and asymmetric nuclear matter: a microscopic perspective. *Phys Rev C.* (2014) **90**:064312. doi: 10.1103/PhysRevC.90.064312
178. Hebeler K, Lattimer JM, Pethick CJ, Schwenk A. Equation of state and neutron star properties constrained by nuclear physics and observation. *ApJ.* (2013) **773**:11–25. doi: 10.1088/0004-637X/773/1/11
179. Antoniadis J, Freire PCC, Wex N, Tauris TM, Lynch RS, van Kerkwijk MH, et al. A massive pulsar in a compact relativistic binary. *Science.* (2013) **340**:6131. doi: 10.1126/science.1233232
180. Cromartie HT, Fonseca E, Ransom SM, Demorest PB, Arzoumanian Z, Blumer H, et al. Relativistic Shapiro delay measurements of an extremely massive millisecond pulsar. *Nat Astron.* (2019). doi: 10.1038/s41550-019-0880-2
181. Lattimer JM, Prakash M. Neutron star structure and the equation of state. *ApJ.* (2001) **550**:426–42. doi: 10.1086/319702

182. Miyatsu T, Saito K, Cheoun MK. Equation of state for neutron stars with hyperons and quarks in the relativistic hartree - fock approximation. *Astrophys J.* (2015) **813**:2–15. doi: 10.1088/0004-637X/813/2/135

183. Mishra RN, Sahoo HS, Panda PK, Barik N, Frederico T. Hyperon stars in a modified quark meson coupling model. *Phys Rev C.* (2016) **94**:035805. doi: 10.1103/PhysRevC.94.035805

184. Bombaci I. The hyperon puzzle in neutron stars. *arXiv:1601.05339 [nucl-th].*

185. Vidana I. Hyperons in neutron stars. *J Phys Conf Ser.* (2016) **668**:012031. doi: 10.1088/1742-6596/668/1/012031

186. Rapaj E, Roggero A, Holt JW. Microscopically constrained mean-field models from chiral nuclear thermodynamics. *Phys Rev C.* (2016) **93**:065801. doi: 10.1103/PhysRevC.93.065801

187. Endrizzi A, Logoteta D, Giacomazzo B, Bombaci I, Kastaun W, Ciolfi R. Effects of chiral effective field theory equation of state on binary neutron star mergers. *arXiv:1806.09832 [astro-ph.HE]* (2018). doi: 10.1103/PhysRevD.98.043015

188. van Paradijs J. Possible observation constraints on the Mass-Radius relation of neutron stars. *Astrophys J.* (1979) **234**:609–11. doi: 10.1086/157535

7

The Falsification of Chiral Nuclear Forces

E. Ruiz Arriola[1], J. E. Amaro[1] , and R. Navarro Perez[2]

[1]Departamento de Física Atómica, Molecular y Nuclear and Instituto Carlos I de Física Teórica y Computacional, Universidad de Granada E-18071 Granada, Spain
[2]Nuclear and Chemical Science Division, Lawrence Livermore National Laboratory Livermore, California 94551, USA

ABSTRACT

Predictive power in theoretical nuclear physics has been a major concern in the study of nuclear structure and reactions. The Effective Field Theory (EFT) based on chiral expansions provides a model independent hierarchy for many body forces at long distances but their predictive power may be undermined by the regularization scheme dependence induced by the counterterms and encoding the short distances dynamics which seem to dominate the uncertainties. We analyze several examples including zero

Citation: E. Ruiz Arriola, J. E. Amaro and R. Navarro Perez. "The falsification of Chiral Nuclear Forces" EPJ Web Conf., 137 (2017) 09006 DOI: https://doi.org/10.1051/epjconf/201713709006

energy NN scattering or perturbative counterterm-free peripheral scattering where one would expect these methods to work best and unveil relevant systematic discrepancies when a fair comparison to the Granada-2013 NN-database and partial wave analysis (PWA) is undertaken.

INTRODUCTION

Nuclear Physics has always been characterized by the fact that experiment is much more precise than theory. For nuclear masses one has $\Delta M(Z,N)^{\mathrm{exp}} < 1\mathrm{KeV} \ll \Delta M(Z,N)^{\mathrm{th}}$ but it is unclear what the theoretical uncertainty is. Traditionally, the theoretical and reductionist predictive power flow is expected to be from light to heavy nuclei form a Hamiltonian with multinucleon forces

$$H(A) = T + V_{2N} + V_{3N} + V_{4N} + \dots \quad \rightarrow \quad H(A)\Psi_n = E_n(A)\Psi_n \,. \tag{1}$$

In the absence of ab initio determinations, phenomenological V_{nN} interactions are adjusted to NN scattering and light nuclei binding energies. The chiral approach, originally suggested by Weinberg in 1990 [1] (see e.g. [2–4] for reviews) to nuclear forces provides a power counting in terms of the pion weak decay constant f_π, with the appealing feature of systematically providing a hierarchy

$$V_{2N}^{\chi} \gg V_{3N}^{\chi} \gg V_{4N}^{\chi} \gg \dots \tag{2}$$

Because the pion mass is so small, chiral approaches are unambiguous at long distances via 1π,2π,3π,... exchanges for relative distances above a short distance cut-off r_c, $V^{n\pi}(r_c) \sim e^{-n r_c m_\pi}$. For instance, NN chiral potentials, constructed in perturbation theory, are universal and contain chiral constants c_1, c_3, c_4,... which can be related to πN scattering [2–4]. At long distances we have

$$V_{NN}^{\chi}(r) = V_{NN}^{\pi}(r) + V_{NN}^{2\pi}(r) + V_{NN}^{3\pi}(r) + \dots \qquad r \gg r_c \,, \tag{3}$$

whereas they become singular at short distances

$$V_{NN}^{\chi}(r) = \frac{a_1}{f_\pi^2 r^3} + \frac{a_2}{f_\pi^4 r^5} + \frac{a_3}{f_\pi^6 r^7} + \dots \qquad r \ll r_c \,, \tag{4}$$

and some regularization must be introduced in any practical calculation. Thus, they trade the "old" model dependence for the "new" regulator dependence. What is the best theoretical accuracy we can get within "reasonable" cut-offs

? What is a reasonable cut-off ? Can the short distance piece be organized as a power counting compatible with the chiral expansion of the long distance piece ?

The huge effort which has been carried out over the last 25 years ellapsed since the seminal work of Weinberg, harvesting over 1000 citations, proves the computational feasibility of the chiral nuclear agenda requiring large scale calculations and many CPU computing hours. Here, we depart from the main streamline and wonder if chiral nuclear forces can be falsified or validated and, if yes, if they are useful for nuclear structure applications from the point of view of the predictive power.

Of course, all this has to do with proper assessment and evaluation of uncertainties of any sort and in particular in the NN interaction. Our original and simple estimates [5, 6] of ΔB th/A ~ 0.5MeV has been upgraded in Ref. [7] to be enlarged to ~ 2MeV. These crude estimates are not far from the recent uncertainty analysis and order-by-order optimization of chiral nuclear interactions [8] including threebody forces where it is found ΔB th(^{16}O)/16 ~ 4MeV. Most of the uncertainty stems from the cut-off variation within a "reasonable" range, and is much worse than the ancient Weiszacker semiempirical mass formula, where $\Delta B_{sem}/A \sim 0.1$MeV. If confirmed, it would be a rather pesimistic scenario for the chiral approach to nuclear structure pioneered by Weinberg. Motivated by this alarming possibility we have paid dedicated attention in the last five years to the issue of NN uncertainties [7, 9]. Here we focus on $V_{NN}^{2\pi}$, corresponding to chiral 2π exchange (χTPE).

VALIDATION AND FALSIFICATION: FREQUENTIST VS BAYESIAN

From our point of view, making first a fair statistical treatment is a prerequisite to credibly aim at any precision goal in low energy nuclear physics where information is extracted by fits. We remind the fact that least squares χ^2 -fitting any (good or bad) model to some set of data is always possible and corresponds to just minimizing a distance between the predictions of the theory and the experimental measurements. How can we disentangle between true and false models?.

The well-known statistical approach, to which we stick, provides one probabilistic answer and depends on the number of data, N_{Dat}, the number of parameters determined from this data, N_{Par}, and the nature of experimental uncertainties. The natural question is: What is the probability that given the data the theory is correct ? This corresponds to the Bayesian approach which requires some a priori expectations on the goodness of the theory regardless of the data and is dealt with often by augmenting the experimental $\chi^2_{\rm exp}$ with an additive theoretical contribution $\chi^2_{\rm th}$. However, it can be proven that when $N_{\rm Dat} >> N_{\rm Par}$ one can ignore these a priori expectations since $\chi^2_{\rm exp} \sim N_{\rm Dat} \gg \chi^2_{\rm th} \sim N_{\rm Par}$ and proceed with the frequentist approach where just the opposite question is posed: what is the probability of data given the model ?. 1 In our analysis below, where we have $N_{\rm Dat} \sim 8000$ and $N_{\rm Par} \sim 40$, we expect no fundamental differences. We thus simply ask: what is the probability q that the the model is false ?. The p-value is $p = 1-q$ and if p is smaller than a predetermined confidence level we will not trust the model and we will declare it to be false. Note that 1) we can never be sure that the model is true and 2) any experiment can be right if errors are sufficiently large and the theory cannot be falsified. This said, $p = 0.68$ when $\chi^2/\nu = 1 \pm \sqrt{2/\nu}$ with $\nu = N_{\rm Dat} - N_{\rm Par}$.

In general we expect discrepancies between theory and data and, ideally, if our theory is an approximation to the true theory we expect the optimal accuracy of the truncation to be comparable with the given experimental accuracy and both to be compatible within their corresponding uncertainties (see [12] for a Bayesian viewpoint). If this is or is not the case we validate or falsify the approximated theory against experiment and declare theory and experiment to be compatible or incompatible respectively. Optimal accuracy, while desirable, is not really needed to validate the theory. In the end largest errors dominate regardless of their origin; the approximated theory may be valid but inaccurate.

How should the discrepancies or residuals be interpreted ? Statistics has the obvious advantage that if we have no good reasons to suspect the theory we can test if residuals behave as, often gaussian, fluctuations and determine a confidence interval for fitting parameters within these fluctuations.

FITTING AND SELECTING DATA FORM COARSE GRAINED POTENTIALS

The NN scattering amplitude has 5 independent complex components for any given energy, which must and can be determined from a complete set of measurements involving differential cross sections and polarization observables. From this point of view it is worth reminding that phase shifts obtained in PWA are not data by themselves unless a complete set of 10 fixed energy and angle dependent measurements have been carried out, a rare case among the bunch of existing 8000 np+pp scattering data below 350MeV LAB energy and which corresponds to a maximal CM momentum of $p_{\rm CM}^{\rm max} = 2{\rm fm}^{-1}$. In order to intertwine all available, often incomplete and partially self-contradictory, information some energy dependence interpolation is needed. We assume a potential approach inspired by a Wilsonian point of view where we take a grid of equidistant radial "thick" points in coordinate space separated by the finite resolution given by the shortest de Broglie wavelength, $\Delta r = \hbar / p_{\rm CM}^{\rm max} \sim 0.6{\rm fm}$ up to the radius r_c = 3 fm, above which charge dependent 1π exchange gives the entire strong contribution. The counting of parameters [13] yields about 40 "thick" points, which can be represented by delta-shells (DS) [14] as originally proposed by Avilés [15]. The whole procedure needs long distance electromagnetic and relativistic contributions such as Coulomb, vacuum polarization and magnetic moments interactions. This approach allows to select the largest self-consistent existing NN database with a total of 6713 NN scattering data driven by the coarse grained potential [16, 17] with the rewarding consequence that statistical uncertainties can confidently be propagated. 2 Precise determinations of chiral coefficients, c_1, c_3, c_4 [18, 19], the isospin breaking pion-nucleon [20, 21], and the pionnucleon-delta [7] coupling constants have been made.

CHIRAL FITS, PERIPHERAL WAVES

The questions on the cut-of r_c raised above were answered by separating the potential as follows [19]

$$V(r) = V_{\rm short}(r)\theta(r_c - r) + V_{\rm long}^{\chi}(r)\theta(r - r_c), \qquad V_{\rm short} = \frac{1}{2\mu}\sum_n \lambda_n \delta(r - r_n), \tag{5}$$

with r_n = nΔr. Several fits varying r_c and $E_{\rm LAB}^{\rm max}$ were performed. The results were checked to be statistically consistent and are summarized in Table 1. It is striking that D-waves, nominaly N3LO and forbidden by Weinberg chiral counting at N2LO, are indispensable !.

Table 1: Fits of chiral TPE potentials depending on the cutoff radius and the maximum fitting energy [19]

Max T_{LAB} MeV	r_c fm	c_1 GeV^{-1}	c_3 GeV^{-1}	c_4 GeV^{-1}	Highest counterterm	χ^2/ν
350	1.8	−0.4(11)	−4.7(6)	4.3(2)	*F*	1.08
350	1.2	−9.8(2)	0.3(1)	2.84(5)	*F*	1.26
125	1.8	−0.3(29)	−5.8(16)	4.2(7)	*D*	1.03
125	1.2	−0.92	−3.89	4.31	*P*	1.70
125	1.2	−14.9(6)	2.7(2)	3.51(9)	*P*	1.05

Furthermore, data and N2LO do not support $r_c < 1.8$fm, while several χ-potentials [22, 23] take $r_c = 0.9-1.1$fm as "reasonable".

An alternative way of checking the failure of the power counting is provided by a deconstruction argument [19]. This corresponds to determine under what conditions are the short distance phases δ_{short}, .i.e. the phase shifts stemming solely from V_{short} compatible with zero within uncertainties, i.e. $|V_{short}| < \Delta V$?. This corresponds to check what partial waves fullfill $|\delta_{short}| \leq \Delta\delta_{stat}$ when $r_c = 1.8$ fm. Unfortunately, this does not work for D-waves, supporting the previous conclusions.

The long distance character of χTPE makes peripheral phases (large angular momentum) to be suitable for a perturbative comparison without counterterms [24–26]. However, one should take into account that 1) peripheral phases can only be obtained from a complete phase shift analyses and 2) their uncertainties are tiny [16]. The analysis of [26] just makes an eyeball comparison which looks reasonable but the agreement was not quantified. 3 We find that peripheral waves predicted by 5thorder chiral perturbation theory are not consistent with the Granada-2013 self-consistent NN database

$$|\delta^{\mathrm{Ch,N4LO}} - \delta^{\mathrm{PWA}}| > \Delta\delta^{\mathrm{PWA,stat}} . \qquad (6)$$

Sometimes we get even 3σ discrepancies. More details on this peripheral analysis will be presented elsewhere. Of course, one may thing that 125 MeV is too large an energy. We find that when we go down to 40 MeV, the χTPE potential becomes invisible being compatible with zero [13, 27].

The chiral potential (including Δ-degrees of freedom) of Ref. [23] explicitly violates Weinberg's counting since it has N2LO long distance and N3LO short distance pieces, and residuals are not gaussian. More recently, the local short distance components of this potential have been fitted up to 125 MeV LAB energy [28] improving the goodness of the fit, similarly to [19] (see also table 1).

CHIRAL INTERACTIONS AND ZERO ENERGY RENORMALIZATION

The low energy threshold parameters allow to probe the structure of chiral potentials against the NN interaction. The current approach to chiral interactions is to incorporate the χTPE tail and include short range counterterms fitted to pp and np phase-shifts or scattering data [29, 30]. 4 However, these approaches are subject to strong systematic uncertainties since a fit to phase-shifts may be subjected to off-shell ambiguities and so far low energy chiral potentials fitted to data have not achieved gaussian residuals [30] or even have huge [22] or moderate [23] χ^2/ν values. To avoid these shortcomings we use χTPE [13, 32] with a simpler short range structure inferred from low energy threshold parameters [7] with their uncertainties inherited from the 2013-Granada fit [16]. This corresponds to zero energy renormalization condition of the counterterms.

Table 2: Delta-Shell parameters located at r_1 = 0.9fm and r_2 = 1.8fm reproducing low energy threshold parameters for the indicated waves in the DS-χTPE potential (see main text)

	$^1S_0$	$^3P_0$	$^1P_1$	$^3P_1$	$^3S_1$	ε_1	$^3D_1$	$^3P_2$
λ_1	-0.572(7)	–	–	–	-0.368(9)	-0.706(7)	-4.15(1)	–
λ_2	-0.201(3)	-0.033(3)	0.103(7)	0.221(2)	-0.246(4)	-0.386(7)	0.35(1)	-0.125(1)

One could naively expect to be able to set any number of short range counterterms to reproduce the same number of low energy threshold parameters. Actually, in order to have as the 9 counterterms dictated by Weinberg to N2LO as in [29] we need to fix α_0 and r_0 for both $^1S_0$ and $^3S_1$ waves, the mixing αε and α_1 for the $^3P_0$, $^3P_1$, $^3P_2$, $^1P_1$ [7]. In practice this turned out to be unfeasible in particular for the J = 1 coupled channel where one has matrices a and r_0. If instead one includes two counterterms in each partial wave in the J = 1 coupled channel it is then possible to reproduce the coupled channel a and r_0 matrices. With this structure we have a total of 12 short range parameters set to reproduce 12 low energy threshold parameters from [7], and not the 9 expected from N2LO [29]. Statistical uncertainties can be propagated by making fits to each of the 1020 sets of threshold parameters that were calculated from the bootstrap generated DS potentials

[33]; this directly takes into account any statistical correlation between low energy parameters. Table 2 lists the resulting 12-λ_i parameters. In Figure 1 we show the phase-shifts corresponding to the DS-χTPE potential with the parameters of Table 2 and compare them to the DS-OPE potential [16, 34]. We observe a good agreement between both representations up to a laboratory energy of 20 MeV.

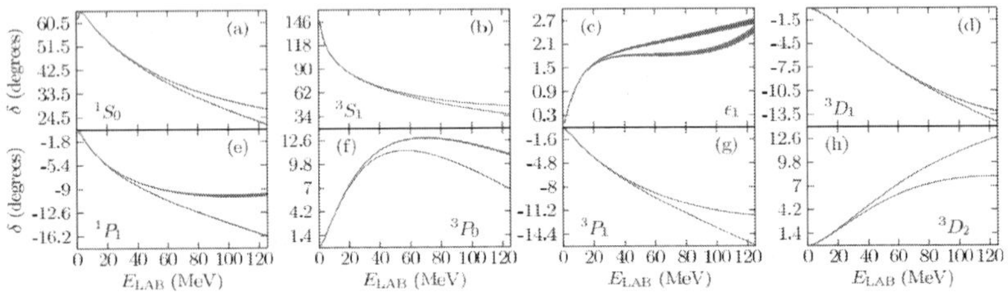

Figure 1: χTPE zero energy renormalized np phase-shifts fixing the low energy threshold parameters (see main text) [7] compared with the phases obtained from the fit to the 2013-Granada database [16].

CONCLUSIONS

Chiral nuclear forces have been massively implemented in Nuclear Physics in the last 25 years with the legitimate hope of providing a unified description of nuclear phenomena more rooted in QCD and less model dependent than most of the phenomenological approaches. This huge effort proves that they are not only calculable but also that they can be used in light nuclei studies, but their indispensability remains to be established. Their systematic uncertainties may be large and they might not be necessarily more predictive than the usual phenomenological and non-chiral approaches. Within the EFT approach there is a residual model dependence regarding the finite cut-off regularization scheme, which seems to dominate the uncertainties. Therefore, efforts should be placed on reducing this largest source of error. We stress that none of these results invalidates χTPE above r_c = 1.8fm, but it does question the status of Weinberg's power counting encoding short distance ignorance.

REFERENCES

1. S. Weinberg, Phys. Lett. B251, 288 (1990)
2. P.F. Bedaque, U. van Kolck, Ann. Rev. Nucl. Part. Sci. 52, 339 (2002), nucl-th/0203055
3. E. Epelbaum, H.W. Hammer, U.G. Meissner, Rev. Mod. Phys. 81, 1773 (2009), 0811.1338
4. R. Machleidt, D. Entem, Phys.Rept. 503, 1 (2011), 1105.2919
5. R. Navarro Pérez, J.E. Amaro, E. Ruiz Arriola (2012), 1202.6624
6. R. Navarro Pérez, J.E. Amaro, E. Ruiz Arriola, PoS QNP2012, 145 (2012), 1206.3508
7. R.N. Perez, J.E. Amaro, E. Ruiz Arriola, J. Phys. G43, 114001 (2016), 1410.8097
8. B.D. Carlsson, A. Ekström, C. Forssén, D.F. Strömberg, G.R. Jansen, O. Lilja, M. Lindby, B.A. Mattsson, K.A. Wendt, Phys. Rev. X6, 011019 (2016), 1506.02466
9. R. Navarro Pérez, J.E. Amaro, E. Ruiz Arriola, J. Phys. G42, 034013 (2015), 1406.0625
10. T. Ledwig, J. Nieves, A. Pich, E. Ruiz Arriola, J. Ruiz de Elvira, Phys. Rev. D90, 114020 (2014)
11. R.N. Perez, J.E. Amaro, E. Ruiz Arriola, Int. J. Mod. Phys. E25, 1641009 (2016), 1601.08220
12. S. Wesolowski, N. Klco, R.J. Furnstahl, D.R. Phillips, A. Thapaliya, J. Phys. G43,074001(2016)
13. R. Navarro Pérez, J.E. Amaro, E. Ruiz Arriola, Few-Body Systems pp. 1–5 (2014), 1310.8167
14. R. Navarro Pérez, J.E. Amaro, E. Ruiz Arriola, Phys.Lett. B724, 138 (2013), 1202.2689
15. J. Aviles, Phys.Rev. C6, 1467 (1972)
16. R. Navarro Pérez, J.E. Amaro, E. Ruiz Arriola, Phys.Rev. C88, 064002 (2013), 1310.2536
17. R. Navarro Pérez, J.E. Amaro, E. Ruiz Arriola, Phys.Rev. C89, 064006 (2014), 1404.0314
18. R. Navarro Pérez, J.E. Amaro, E. Ruiz Arriola, Phys. Rev. C91, 054002 (2015), 1411.1212
19. R. Navarro Pérez, J.E. Amaro, E. Ruiz Arriola, PoS CD12, 104 (2013),

1301.6949

20. R.N. Perez, J.E. Amaro, E. Ruiz Arriola (2016), 1606.00592
21. E. Ruiz Arriola, J.E. Amaro, R. Navarro Pérez, Mod. Phys. Lett. A31, 1630027 (2016)
22. A. Gezerlis, I. Tews, E. Epelbaum, M. Freunek, S. Gandolfi, K. Hebeler, A. Nogga, A. Schwenk, Phys. Rev. C90, 054323 (2014), 1406.0454
23. M. Piarulli, L. Girlanda, R. Schiavilla, R. Navarro Pérez, J.E. Amaro, E. Ruiz Arriola, Phys. Rev. C91, 024003 (2015), 1412.6446
24. N. Kaiser, R. Brockmann, W. Weise, Nucl. Phys. A625, 758 (1997), nucl-th/9706045
25. N. Kaiser, S. Gerstendorfer, W. Weise, Nucl. Phys. A637, 395 (1998), nucl-th/9802071
26. D.R. Entem, N. Kaiser, R. Machleidt, Y. Nosyk, Phys. Rev. C91, 014002 (2015), 1411.5335
27. J.E. Amaro, R. Navarro Pérez, E. Ruiz Arriola, Few-Body Systems pp. 1–5 (2013), 1310.7456
28. M. Piarulli, L. Girlanda, R. Schiavilla, A. Kievsky, A. Lovato, L.E. Marcucci, S.C. Pieper, M. Viviani, R.B. Wiringa (2016), 1606.06335
29. A. Ekström et al., Phys. Rev. Lett. 110, 192502 (2013), 1303.4674
30. A. Ekström, B.D. Carlsson, K.A. Wendt, C. Forssén, M. Hjorth-Jensen, R. Machleidt, S.M. Wild, J. Phys. G42, 034003 (2015)
31. E. Ruiz Arriola, Symmetry 8, 42 (2016)
32. R. Navarro Pérez, J.E. Amaro, E. Ruiz Arriola, Phys.Rev. C89, 024004 (2014), 1310.6972
33. R. Navarro Perez, J. Amaro, E. Ruiz Arriola, Phys.Lett. B738, 155 (2014), 1407.3937
34. R. Navarro Pérez, J.E. Amaro, E. Ruiz Arriola, Phys.Rev. C88, 024002 (2013), 1304.0895

SECTION 4:
α, β e γ Decay

8

Alpha Decay of the New Isotope ^{215}U

H.B. Yang[1,2,3], Z.Y. Zhang[1], J.G. Wang[1], Z.G. Gan[1],a, L. Ma[1,2,3], L. Yu[1,2], J. Jiang[1,2], Y.L. Tian[1], B. Ding[1], S. Guo[1], Y.S. Wang[1], T.H. Huang[1], M.D. Sun[1,2,3], K.L. Wang[1,2], S.G. Zhou[4], Z.Z. Ren[5], X.H. Zhou[1], H.S. Xu[1], and G.Q. Xiao[1]

[1] Institute of Modern Physics, Chinese Academy of Sciences, 730000 Lanzhou, China
[2] University of Chinese Academy of Sciences, 100049 Beijing, China
[3] School of Nuclear Science and Technology, Lanzhou University, 730000 Lanzhou, China
[4] Institute of Theoretical Physics, Chinese Academy of Sciences, 102413 Beijing, China
[5] Department of Physics, Nanjing University, 210093 Nanjing, China

ABSTRACT

The new neutron-deficient isotope ^{215}U was produced in the complete-fusion reaction ^{180}W (^{40}Ar, 5n) ^{215}U. Evaporation residues recoiled from the target were separated in flight from the primary beam by the gas-filled recoil separator SHANS and subsequently identified on the basis of energy-

Citation: Yang, H.B., Zhang, Z.Y., Wang, J.G. et al. "Alpha decay of the new isotope ". Eur. Phys. J. A 51, 88 (2015). https://doi.org/10.1140/epja/i2015-15088-9

position-time correlation measurement. The αα-particle energy and half-life of ^{215}U were determined to be 8.428(30)8.428(30) MeV and $0.73^{+1.33}_{-0.29}$ ms, respectively.

INTRODUCTION

It has been reported that the α-decay energies of ground state to ground state transitions for neutron-deficient isotopes (Z = 84–91) located beyond the N = 126 follow the trend of increase with the decrease of neutron number [1–3]. In order to see if the α-decay properties of very neutron-deficient uranium isotopes follow this trend, we performed experiments with the aim to synthesize ^{216}U (N = 124) and ^{215}U (N = 123). If their decay properties follow this trend, the α-particle energies and halflives of ground state to ground state transition for ^{215}U and ^{216}U will be similar to each other in analogy with the α-decay properties of other known 123-neutron isotopes and 124-neutron isotopes, such as ^{214}Pa and ^{215}Pa [4], ^{213}Th and ^{214}Th [5], ^{212}Ac and ^{213}Ac [4], ^{211}Ra and ^{212}Ra [6]. Recently we have observed the new isotope ^{216}U by the complete-fusion reaction $^{180}W(^{40}Ar, 4n)^{216}U$ [7]. The α-particle energies and half-lives were determined to be 8.384(30)MeV and $4.72^{+4.72}_{-1.57}$ ms for the ground state of ^{216}U and 10.582(30)MeV and $0.74^{+1.34}_{-0.29}$ ms for an isomeric state. In this paper, we will report the result of ^{215}U, synthesized by the $^{180}W(^{40}Ar, 5n)^{215}U$ reaction.

EXPERIMENT

The experiment was performed at the gas-filled recoil separator SHANS [8] and the ^{40}Ar beam was delivered by the sector focusing cyclotron of the Heavy Ion Research Facility in Lanzhou (HIRFL), China. The typical beam intensities were about 150–250pnA. The enriched 180W targets (91.4% ^{180}W and 8.5% ^{182}W) with an average thickness of 481 μg/cm^2 were evaporated on carbon foils of 43 μg/cm^2 and covered with carbon layers of 14 μg/cm^2. The targets were mounted on a fixed frame. The experiment can be separated into two runs depended on the selected two beam energies. In the first run, a beam energy of 204.5MeV was chosen on the basis of the expected maximum cross section for ^{215}U calculated by the HIVAP code [9]. When one correlated decay chain like the candidate of either ^{215}U or ^{216}U was observed after a irradiation time of 89hours, we found it was difficult to determine which of the two isotopes should be assigned to because of their very close production cross sections and the very similar α-decay properties of the nuclei in the $^{215}U \rightarrow {}^{211}Th \rightarrow {}^{207}Ra$ and $^{216}U \rightarrow {}^{212}Th \rightarrow {}^{208}Ra$ chains [2,10,11]. In the second run, the beam energy was increased to 207.6MeV. At this beam energy, the cross section of ^{215}U was estimated to be much higher than that of ^{216}U. Therefore, even though the decay properties of the two nuclei are much similar, we are able to distinguish them on the basis of their quite different cross sections. The estimated beam energies at the center of the target and the relevant irradiation times (in parentheses) in the two runs are 202.4MeV (89hours) and 205.5MeV (157hours), respectively. In the following, all beam energies mentioned in the text refer to the beam energies at the center of the target.

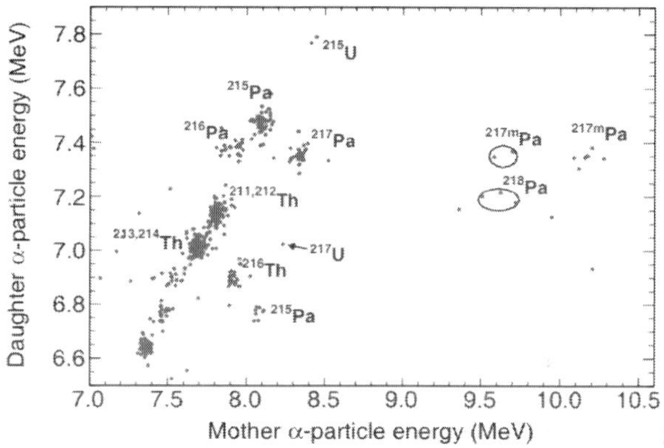

Figure 1: (Color online) Mother and daughter α-particle energies for all chains of the type ER $\rightarrow \alpha_1 \rightarrow \alpha_2$ observed in the 205.5 MeV ^{40}Ar + ^{180}W irradiation. Maximum search times were 100 ms for the ER$\rightarrow\alpha_1$ pair and 5 s for the $\alpha_1\rightarrow\alpha_2$ pair.

After filtering out by the separator, the evaporation residues (ERs) were implanted into three 300 μm thick position-sensitive strip detectors (PSSDs) installed side by side at the focal plane of the separator. Each PSSD, with an active area of 50 × 50mm2, was divided into 16 vertical strips in the front face. The energy resolution with all strips summed was about 70keV (FWHM) for 5–9MeV α particles and the vertical position resolution was about 2.5mm (FWHM). Upstream the PSSDs, eight non–position-sensitive silicon detectors (side detectors) of similar type were mounted. The PSSDs and the side detectors formed a Si-box detector. The efficiency of the Sibox detector for detecting an α particle with full energy was about 80% of 4π. In order to distinguish the α-decay events from the implantation events, two multi-wire proportional counters (MWPCs) were mounted 15cm and 25cm upstream from the PSSDs. α-particle energy calibrations were performed using a three-peak external α source as well as the peaks from nuclides produced in the test reaction $^{40}Ar + ^{175}Lu$.

RESULTS

The identification of the products was performed using the method of energy-position-time correlated α-decay chains. A two-dimensional plot showing the correlation between parent and daughter α-particle energies of type ER → α_1 → α_2 is shown in fig. 1. The analysis of the correlation was made with a vertical position window of ±2.5mm and the time windows of 100ms for ER → α_1 pair and 5s for α_1 → α_2 pair. In this plot different isotopes $^{211-214}Th$, ^{216}Th, $^{215-218}Pa$ and ^{217}U can be identified. However, the presence of ^{216}Th, $^{216,217}Pa$ and ^{217}U was explained by the admixture of ^{182}W isotope in the ^{180}W target. In addition a group of two correlated chains assigned to ^{215}U was also observed. These events are listed in table 1, together with the beam energies, vertical positions, measured energies, and also the time intervals between the subsequent decays.

Table 1: Beam energies (Elab), vertical positions, measured energies and time intervals (ΔT) of correlated chains assigned to the decays of ^{215}U

E_{lab} (MeV)	Nuclei	Type	Position (mm)	Energy (MeV)	ΔT
205.5	^{215}U	ER	−15.7	13.59	0
		α_1	−14.9	8.414	0.776 ms
		α_2	−15.9	7.767	59.7 ms
		α_3	−13.6	7.289^(a)^	0.73 s
205.5	^{215}U	ER	11.9	12.61	0
		α_1	11.4	8.442	1.33 ms
		α_2	13.4	7.792	0.43 ms
		α_3	11.7	7.127	0.23 s

(a) Escaped α-particle energy reconstructed by summing up the two energies deposited in the PSSD and side detector.

The two correlated chains were observed at the beam energy of 205.5MeV. Based on the two decay chains, the α-particle energy and half-life of the daughter nucleus were determined to be 7.780(30)MeV and

$20.8^{+37.9}_{-8.2}$ ms, which are consistent with the α decays from ^{211}Th (Eα = 7.792(14)MeV, $T_{1/2}$ = 37^{+28}_{-11} ms) [2] and ^{212}Th (Eα = 7.809(5)MeV, $T_{1/2}$ = 31.7(13)ms) [10]. The observed granddaughter decays are consistent with the α decays from ^{207}Ra and ^{208}Ra, whose α-particle energies and halflives are 7.131(5)MeV and 1.3(2)s for ^{207}Ra [11] and 7.131(5)MeV and 1.2(2)s for ^{208}Ra [11], respectively. The α_3 in the first decay chain escaped from the PSSD and its energy was recovered by summing up the energies deposited in the PSSD (2.740MeV) and the side detector (4.549MeV). The recovered value 7.289MeV is also in agreement with the literature value 7.131(5)MeV within the error limit of 200keV. The probability of random correlation [12], calculated on the basis of observed average counting rates in the detectors, was estimated to be less than 2.04×10^{-11} and the expected number of random correlation was about 1.4×10^{-4}. So these two decay chains were considered to be real correlation. Due to the similarities in the α decays of ^{215}U and ^{216}U, it is not possible for us to make an unambiguous identification of the two observed decay chains based on decay properties alone. However we prefer to assign these two decay chains to ^{215}U on the basis of other two important facts.

Firstly, we can be sure that these two isotopes have a mass number less than 216 because they are formed at higher excitation energy in the ^{40}Ar + ^{180}W system than that required to produce the ^{216}U. In the experiment to study ^{216}U [7], a beam of 40Ar at an energy of 187.2MeV was used. At this beam energy (E∗ = 45.7MeV) 4 nucleons evaporation channels are dominant, which can be confirmed by the observation of ^{216}U (4n channel), ^{216}Pa (p3n channel) and ^{218}U produced in the ^{182}W(40Ar, 4n)^{218}U reaction. However, at the beam energy of 205.5MeV in this experiment, the excitation energy of the compound nucleus is about 60.3MeV, which is 14.6MeV higher than that required to produce ^{216}U. With this beam energy, 5nucleons evaporation channels should be dominant in this experiment. This can be confirmed by the observation of ^{215}Pa (p4n channel) and ^{217}U produced in the ^{182}W(^{40}Ar, 5n)^{217}U reaction. All above discussions are accord with excitation functions calculated with HIVAP code [9] (shown in fig. 2). The calculations were carried out using its standard parameter set except that the fission barrier scaling factor Cf was adjusted into 0.63 according to the analysis in [13]. As shown in fig. 2, the cross section of 5n channel is much higher than that of 4n channel at the beam energy of 205.5MeV.

Secondly, the two decay chains are unlikely to be ^{216}U produced in 6n evaporation channel from ^{182}W component in the target, although some products such as $^{216,217}Pa$ and ^{217}U observed in this experiment could be more likely produced from ^{182}W component. Because at the beam energy of 205.5MeV the cross sections of ^{216}U produced in the $^{182}W(40Ar, 6n)^{216}U$ reaction was estimated to be about 9pb, which is much smaller than that of ^{216}Pa (15nb), ^{217}Pa (19nb) and ^{217}U (0.35nb). This can be partly seen in fig. 3. According to the proportions of these cross sections, it is unlikely to observe two ^{216}U when there are not many ^{216}Pa (25 chains), ^{217}Pa (40 chains) and ^{217}U (1 chain) to be produced.

Based on the two facts mentioned above, we concluded that the observed two correlated decay chains correspond to the α decays of the new isotope ^{215}U. The obtained α-particle energy and half-life for ^{215}U are 8.428(30)MeV and $0.73^{+1.33}_{-0.29}$ ms, respectively. In the present work the determination of the half-lives was made by the method reported in [14]. Using a transmission efficiency of 14% [8], the cross section for ^{215}U at the beam energy of 205.5MeV was determined to be 50^{+70}_{-30} pb.

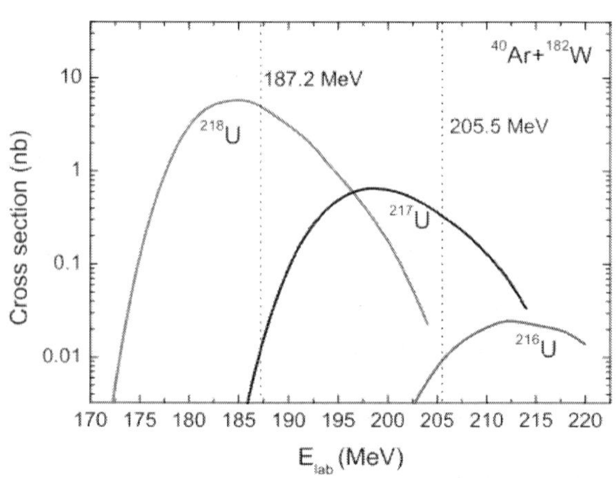

Figure 3: (Color online) Excitation functions of the 4–6n evaporation channels of the reactions $^{40}Ar + ^{182}W$ calculated with HIVAP code [9] using its standard parameter set except that the fission barrier scaling factor $C_f = 0.63$ was used.

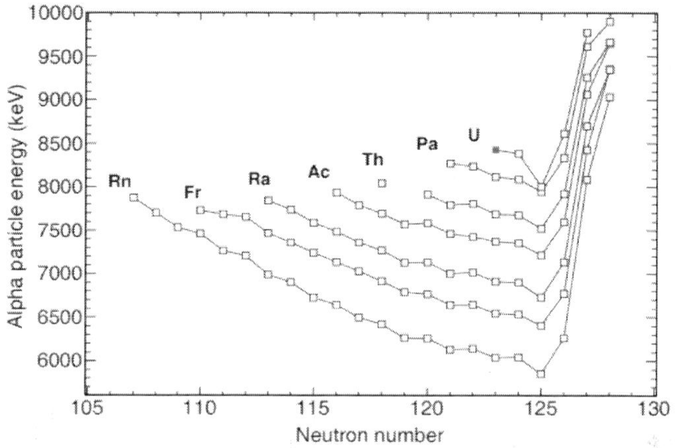

Figure 4: (Color online) α-particle energies of ground state to ground state transitions for neutron-deficient Rn, Fr, Ra, Ac, Th, Pa and U isotopes (except ^{209}Th). Open squares refer to literature values taken from refs. [2,4–7,10,11,15–18]. The red soild square refers to the value of ^{215}U measured in this work. Although one α-particle energy of 8.005(20) MeV for the α decay of ^{217}U has been reported in [15] and this value has been plotted in this figure, it may not be the α-particle energy of the ground state to ground state transition for ^{217}U, due to its significant deviation from the systematic regularities. However, this assumption needs to be confirmed by further experiments.

The production cross section for ^{216}U has been measured to be 300^{+170}_{-120} pb at the beam energy of 187.2MeV. The error bars of the measured cross sections only represent statistical errors determined by the method described in [14]. These measured values are consistent with the HIVAP calculation very well (see fig. 2).

The measured α-particle energy of ^{215}U fits well into the systematics of neutron-deficient U isotopes (shown in fig. 4) and it is also consistent with the systematic regularities observed in the α decay of neutron-deficient Rn isotopes through Pa isotopes. The reduced width δ^2 for the α decay of ^{215}U was calculated to be 173^{+316}_{-70} keV using the Rasmussen approach for $\Delta L = 0$ transitions [19]. These values, when compared to the values obtained for the closest even-even neighbors for the ground state to ground state decays (20–70keV), show that this decay is an unhindered α decay. In order to reproduce the halflive, empirical formulae ($\Delta L = 0$) given in ref. [20] relating half-life and Qα were used. And the calculated half-life for ^{215}U is 2.36ms, which is very close to our measured value $0.73^{+1.33}_{-0.29}$ ms.

The authors would like to thank the accelerator crew for delivering stable beams of high intensity. This work is supported by the National Basic Research Program of China under Grant No. 2013CB834403 and National Natural Science Foundation of China under Grant Nos. 11475226, 11405225, 11305230, 11120101005, 11405243 and 11205208.

REFERENCES

1. K. Valli, E.K. Hyde, Phys. Rev. 176, 1377 (1968).
2. J. Uusitalo et al., Phys. Rev. C 52, 113 (1995).
3. Z.Y. Zhang et al., Phys. Rev. C 89, 014308 (2014).
4. F.P. Heßberger et al., Eur. Phys. J. A 8, 521 (2000).
5. D. Vermeulen et al., Z. Phys. A 294, 149 (1980).
6. NNDC national Nuclear Data Center, Chart of nuclides, http://www.nndc.bnl.gov/chart.
7. L. Ma et al., Phys. Rev. C 91, 051302 (2015).
8. Z.Y. Zhang et al., Nucl. Instrum. Methods B 317, 315 (2013).
9. W. Reisdorf, M. Sch¨adel, Z. Phys. A 343, 47 (1992).
10. J.A. Heredia et al., Eur. Phys. J. A 46, 337 (2010).
11. K. Valli, W. Treytl, E.K. Hyde, Phys. Rev. 161, 1284 (1968).
12. C.M. Folden, PHD Thesis, University of California, 94-101 (2004).
13. A.N. Andreyev et al., Nucl. Phys. A 626, 857 (1997).
14. K.H. Schmidt, C.C. Sahm, K. Pielenz, H.G. Clerc, Z. Phys. A 316, 19 (1984). 1
15. 5. O.N. Malyshev et al., Eur. Phys. J. A 8, 295 (2000).
16. A.P. Lepp¨anen et al., Eur. Phys. J. A 25, s01, 183 (2005) DOI: 10.1140/epjad/i2005-06-116-y.
17. A.P. Lepp¨anen et al., Phys. Rev. C 75, 054307 (2007).
18. A.N. Andreyev et al., Z. Phys. A 345, 247 (1993).
19. J.O. Rasmussen, Phys.Rev. 113, 1593 (1959).
20. Y.B. Qian, Z.Z. Ren, D.D. Ni, Phys. Rev. C 83, 044317 (2011).

9

Alpha-decay and Spontaneous Fission Half-lives of Super-heavy Nuclei around the Doubly Magic Nucleus ^{270}Hs

I. Silisteanu[1] and C. I. Anghel[1,2]

[1]Horia Hulubei National Institute of Physics and Nuclear Engineering, Bucharest-Magurele, RO-077125, Romania
[2]University of Bucharest, Faculty of Physics, Bucharest-Magurele, RO-077125, Romania

ABSTRACT

The α-decay and spontaneous fission half-lives of nuclei around ^{270}Hs are calculated with formulas derived from an in-depth analysis of the available experimental data and of results of theoretical models. The parameters of these formulas result from the fit of half-lives with respect to the reaction energies, the height of the SF barrier and the fissionability. The calculated partial and total half-lives are compared with the experimental data and results of other approximations. Half-life predictions are made for many unknown nuclei.

Citation: I. Silisteanu and C. I. Anghel. "Alpha-decay and spontaneous fission half-lives of super-heavy nuclei around the doubly magic nucleus" EPJ Web of Conferences, 107 (2016) 07004 DOI: https://doi.org/10.1051/epjconf/201610707004

INTRODUCTION

The quantity and quality of experimental data on the production and decay of the SHN have increased considerably in the last two decades [1–13]. Systematic studies of the existing data are revealing new reaction mechanisms, a rich diversity of nuclear structures involving weakly bound states coupled to an environment of scattering states and new competing decay modes that are challenging our understanding of the nucleus at its limits of stability. Such studies have provided not only the access to the basic properties of SHN: masses, lifetimes, reaction energies, emission rates, but also the possibility to predict decay properties and nuclear structure of not yet produced SHN.

Up to now, new significant results are obtained: new magic numbers associated with shell gaps at nonzero deformations, the coexistence of structures and shapes at different deformations at low excitations, the existence of isomeric states, and the possible pairing states at near the ground state.

In the present work, our main aim is to correlate on solid theoretical basis the decay data (energy levels, reactions energies, and half-lives) with the current information on the microscopic nuclear structure and the reaction decay mechanism.

FORMALISM

The approach we used here for studying the essential features of α-decay of SHN is presented in Refs. [14, 15]. The first step is to match smoothly the four shell model wave functions of individual nucleons $(I_n^k(r)[S\,M])$, which describe the shell model formation amplitude (SMFA) of the outgoing α-particle (in channel n from the resonance state k) with a general solution of the system of differential equations

$$\left[\frac{\hbar^2}{2M}\left(\frac{d^2}{dr^2}-\frac{l(l+1)}{r^2}\right)-V_{nn}(r)+Q_n\right]u_n^{0(k)}(r)+$$
$$+\sum_{m\neq n}V_{nm}(r)u_m^{0(k)}(r)=\left\{\begin{array}{c}0\\ I_n^{k[S\,M]}\end{array}\right\}. \qquad (1)$$

The solutions $u_n^{0(k)}(r)$ describe the radial motion of the fragments at large and small separations, respectively, in terms of the reduced mass of the system M, the reaction energy Q_n, which includes screening energy, the angular momentum l, the formation amplitude (FA) $(I_n^k(r)[S\,M])$, and the matrix elements of interaction potential $V_{nm}(r)$.

$$I_n^k(r)[S\,M] = r\langle \Psi_k^{S\,M}(r_i) \mid \mathcal{A}\left[\Phi_D^{S\,M}(\eta_1)\Phi_p(\eta_2)Y_{lm}(\hat{r})\right]_n\rangle, \qquad (2)$$

where $\Phi_D^{S\,M}(\eta_1)$ and $\Phi_p(\eta_2)$ are the internal (space-spin) wave functions of the daughter nucleus and of the particle, $Y_{lm}(\hat{r})$ are the wave functions of the angular motion, r connects the centers of mass of the fragments, and the symbol < | > means integration over the internal coordinates and angular coordinates of relative motion. It should be pointed out that the spatial correlations imposed by Pauli principle on the nucleons in a simple shell model configuration are sufficient to determine the essential features of nuclear motion in the preformation stage. From the spectrum of single particle energies and wave functions we obtain the FA from active valence nucleons. The shell model overlap integral is estimated for the harmonic oscillator single particle wave functions of the parent and daughter nuclei. For SHN we use single particle proton states $1i_{13/2}$, $2\,f_{5/2}$, $3p_{3/2}$, and $3p_{1/2}$ for Z=102-120 and single particle neutron states $2g_{7/2}$, $3d_{5/2}$, $3d_{3/2}$ and $4s_{1/2}$, for N=150- 178.

The matrix elements $V_{nm}(r)$ include nuclear and Coulomb components defined with the quadrupole (β_2) and hexadecapole (β_4) deformation parameters of the daughter nucleus. To avoid the usual ambiguities encountered in formulating the potential for the resonance tunneling of the barrier we iterate directly the nuclear potential in the equations of motion.

Solving Eq. (1) and Eq. (2) with minimal computational effort we get the reaction amplitude and the α-decay width

$$\Gamma_n^{k[S\,M]} = 2\pi \left| \frac{\int_{r_{min}}^{r_{max}} I_n^{k[S\,M]}(r)u_n^0(r)dr}{\int_{r_{\min}}^{r_{\max}} I_n^{k[S\,M]}(r)u_n^k(r)dr} \right|^2, \qquad (3)$$

where the lower limit in the integrals is an arbitrary small radius $r_{min} > 0$ while the upper limit r_{max} is close to the first exterior node of $u_n^0(r)$.

Finally, from the systematics of calculated shell-model half-lives corrected by even-odd terms h_{e-o} (even-e, oddo), a practical formula for the α-decay half-lives has resulted [15]

$$\log T_\alpha(\mathrm{s}) = 10.591(Z_d^{0.6} Q_\alpha^{-1/2}) - 56.618, \quad rms = 0.078$$

(Z=e, N=e), (4)

$$\log T_\alpha(\mathrm{s}) = 10.148(Z_d^{0.6} Q_\alpha^{-1/2}) - 53.386, \quad rms = 0.161$$

(Z=e, N=o), (5)

$$\log T_\alpha(\mathrm{s}) = 10.148(Z_d^{0.6} Q_\alpha^{-1/2}) - 53.386, \quad rms = 0.161$$

(Z=o, N=e), (6)

$$\log T_\alpha(\mathrm{s}) = 10.225(Z_d^{0.6} Q_\alpha^{-1/2}) - 53.797, \quad rms = 0.047$$

(Z=o, N=o), (7)

where Q_α is the reaction energy in MeV units and Z_d is the atomic number of daughter nucleus. It was proven that the systematics of calculated SM half-lives well reproduces the systematics of available experimental data.

The SF-decay width is given by [16]

$$\log T_{SF}(s) = 1146.44 - 75.3153Z^2/A +$$
$$+1.63792(Z^2/A)^2 - 0.0119827(Z^2/A)^3 +$$
$$+B_f(7.23613 - 0.0947022Z^2/A) + H_{e-o}, \quad (8)$$

where B_f is the SF barrier height (in MeV) [17] and He−o are new even-odd corrections [18]: H_{e-o}=0 for (Z = e, N = e); 2.007 for (Z = e, N = o); 2.822 for (Z = o, N = e); and 3.357 for (Z = o, N = o).

RESULTS

The crucial quantity for a qualitative prediction of α-decay half-life is the Q_α-value that strongly affects the half-life calculation due to the exponential decay law.

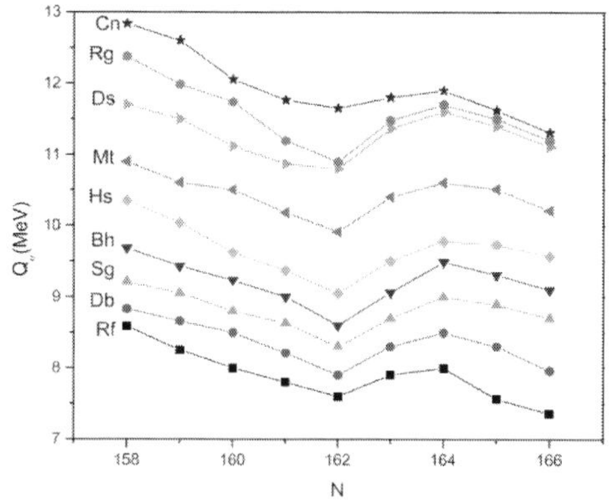

Figure 1: Q_α energy as a function of N for the isotopic series of Ra-Cn, Z=104-112 and N=158-166.

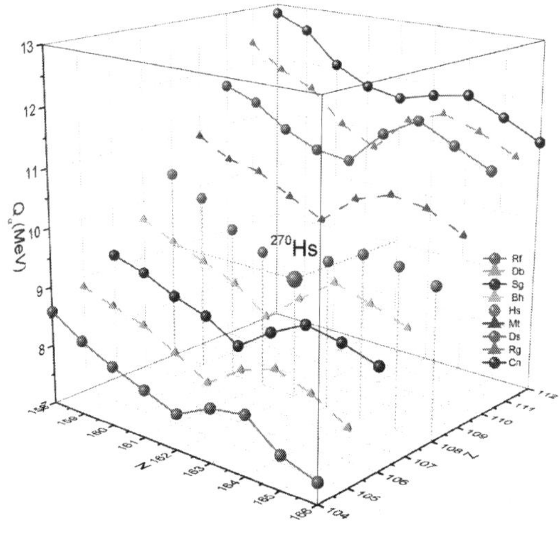

Figure 2: Q_α energy as a function of Z and N for the isotopic and isotonic series with Z=104-112 and N=158-166.

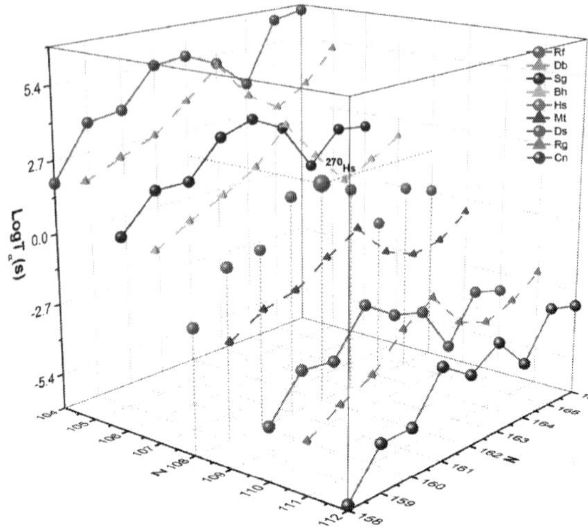

Figure 3: Calculated α-half-lives as a function of Z and N. Minimal α-half-lives correspond to isotones N=164 and N=160 (only Z=even). Maximal α-half-lives correspond to isotones N=161, 163, 165.

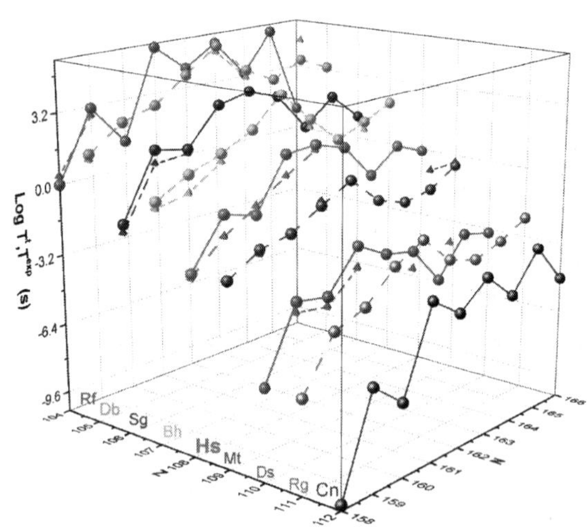

Figure 4: Experimental (triangles) and calculated (spheres) total half-lives of nuclei with Z=104-112 and N=158-166. Minimal values of calculated total half-lives are observed at N=160 isotones with Z=even, where SF is the main decay channel, and at N=164 isotones, where the dominant process is α-decay.

To estimate the unknown Q_α-values from the measured ones we employ the method [19], which takes into the account the effects of Coulomb energy and symmetry energy and validates not only the experimental measurements, but also predicts the Q_α-values with a high accuracy. Once the parent or daughter nuclei have neutron or proton magic numbers (Fig. 1 and Fig. 2) the behavior of Q_α-value is dramatically changed. Thus, the irregular behavior of Q_α-value confirms the existence of the magic shells. The Q_α-values in Fig. 1 and Fig. 2 present a linear decreasing trend for N=158-162 and N=164-166 and an increasing trend for N=162-164. Such a complicated trend was attributed [19] to the large symmetry energy that lowers the Q_α-value at the magic shells (Fig. 1 and Fig. 2). In Fig. 2, we can see an increasing trend of Q_α-value with increasing Z. Precise measurements and predictions of reaction energies are of importance for understanding of fine structure and transitions involving excited and isomeric states.

The α-decay half-life (Fig. 3) varies approximately with $Q^{-1/2}$ and shows an irregular trend at the magic shells Z=108 and N=162. In Fig. 3, Fig. 5, and Fig. 6, one can see that the values Tα or the even-even nuclei are always shorter than those corresponding to the even-odd, oddeven, and odd-odd neighbors. Notice that by successive αdecays the nuclear system becomes more and more stable. An example is the α-decay chain $^{274}Ds \rightarrow ^{270}Hs \rightarrow ^{266}Sg$ (see Fig. 3), where the corresponding Tα values increase: $T_\alpha(^{274}Ds) < T_\alpha(^{270}Hs) <$ Tα(^{266}Sg). In Fig. 3, we can observe a weak α-channel in isotonic sequence N=160 (Z=even), and a well defined α-chain in N=164 sequence. Figure 4 shows our calculated total half-lives and some experimental results known up to date. Figure 4 reveals a main SF channel in N=160 (Z=even) sequence, and a main α-decay channel in N=164 sequence. Comparing the results from Fig. 3 and Fig. 4 we may conclude that the SHN from unstable configurations undergo spontaneous α-decay or SF until the nuclear stability is reached. Our results for the decay properties of SHN around ^{270}Ds are in good agreement with existing data [1–13]. In general, the calculated α-decay half-lives are greater than the estimates based on previous predicted Q_α-values of models developed in Refs. [14, 20–22]. Now, there are new improved parametrization of the unified model for α-decay and αcapture [23, 24] that may describe very well the available data. The agreement to the results of Refs. [25–30] for partial and total half-lives seems to be also good.

CONCLUSIONS

In this work we give a review of the experimental and calculated nuclear decay properties of nuclei around ^{270}Hs with Z=104-112 and N=158-166. New half-life predictions are made for many unknown nuclei from this region of nuclei. The main conclusions of the present study are summarized as follows:

- Nuclei from this region are typically unstable with respect to both α-decay and SF.
- At ^{270}Hs the nuclear binding energy and the halflife increase strongly, while various reaction energies markedly decrease, these give rise to an increased stability.
- α-decay dominates in nuclei over closed shells $Z > 108$, $N > 162$ (^{274}Ds=^{270}Hs+α), while SF dominates in nuclei below the closed shells (^{266}Sg=^{270}Hs-α).
- Strong competition α-SF is observed in Z, N=even nuclei.
- T_{SF} increases considerably due to the effect of unpaired nucleons and T_α is much less sensitive to the unpaired nucleons.
- The calculated half-lives are in good agreement with existing data.

We proved that the main nuclear decay properties can be well described with simple models based on the doubly magic core and the valence nucleons.

The experimental and theoretical studies of nuclear decay properties are providing a deep insight into the structure of SHN and stringent new tests of our theoretical understanding of decay processes at the limits of nuclear stability.

In brief, by studying the decay channels it is possible to inter-relate most nuclear species on the energy-time content basis and to determine the regions of greater stability. We showed that the decay rates can be predicted with a fair degree of confidence and this may help in synthesis and identification of new heaviest nuclei.

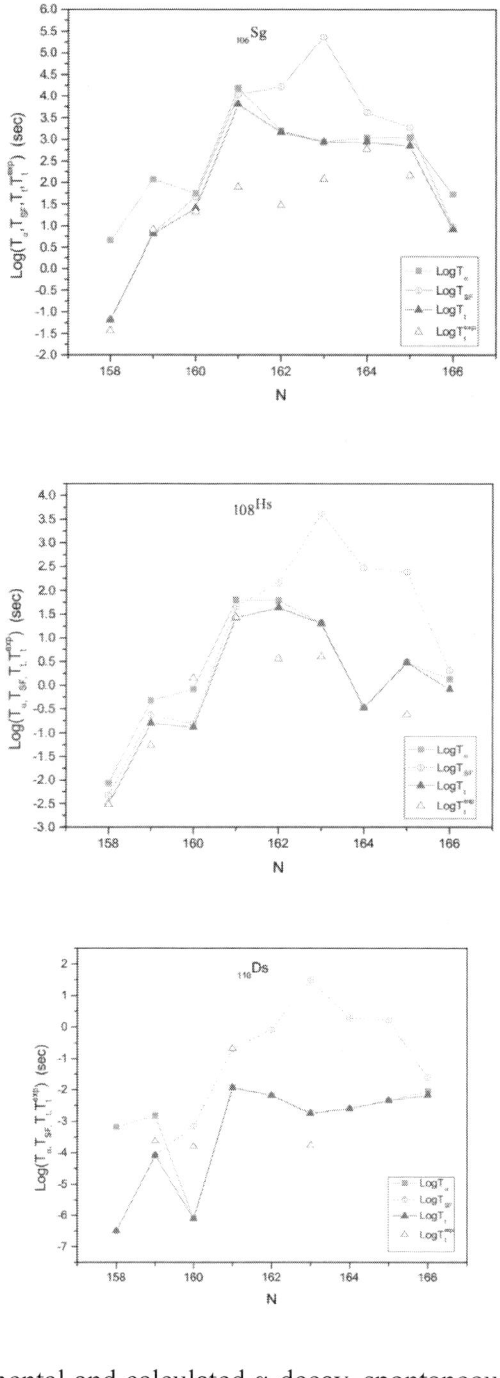

Figure 5: Experimental and calculated α-decay, spontaneous fission, and total half-lives for isotopes of the elements Sg, Hs and Ds.

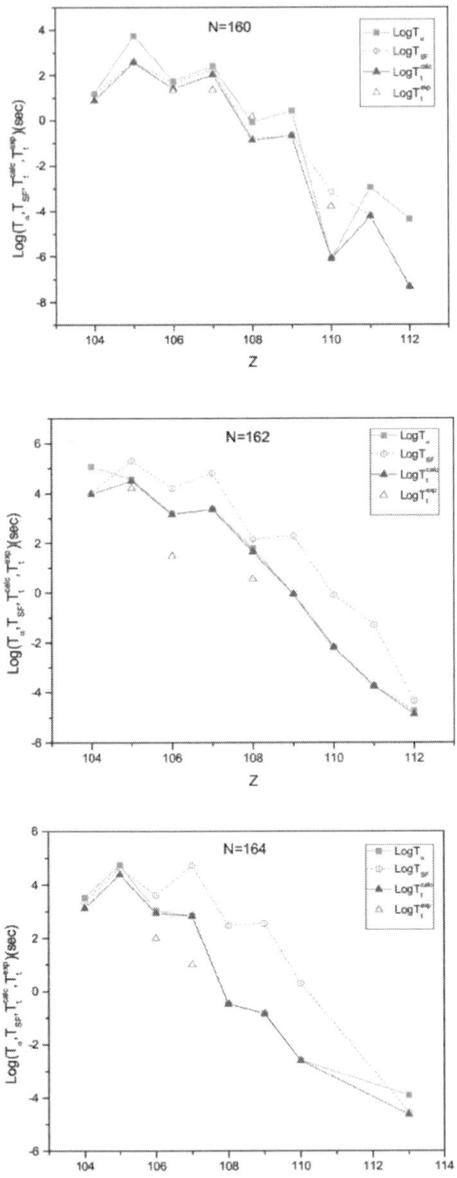

Figure 6: Experimental and calculated α-decay, spontaneous fission, and total half-lives for isotones with N=160, 162, 164.

REFERENCES

1. S. Hofmann and G. Münzenberg, Rev. Mod. Phys. 72, 733 (2000)
2. K. Morita et al., J. Phys. Soc. Jap. 73, 2593 (2004)
3. Yu.Ts. Oganessian et al., Phys. Rev. C 74, 044602 (2006)
4. Yu.Ts. Oganessian et al., Phys. Rev. C 87, 054621 (2013)
5. S. Hofmann et al., Eur. Phys. J. A 10, 5 (2001)
6. I. Muntian, Z. Patyk, and A. Sobieczewski, Phys. Lett. B 500, 241 (2001)
7. J. Dvorak et al., Phys. Rev. Lett. 97, 242501 (2006)
8. J. Dvorak et al., Phys. Rev. Lett. 100, 132503 (2008)
9. P.A. Ellison et al., Phys. Rev. Lett. 105, 182701 (2010)
10. R. Graeger et al. Phys. Rev. C 81, 061601(R) (2010)
11. A. Tuerler, Radiochimica Acta 100, 75 (2012)
12. Yu.Ts. Oganessian et al., Phys. Rev. C 87, 034605 (2013)
13. Chart of Nuclides, NNDC Brookhaven National Laboratory, http://www.nndc.bnl.gov/chart/
14. A.I. Budaca and I. Silisteanu, Phys. Rev. C 88, 044618 (2013)
15. I. Silisteanu and A.I. Budaca, At. Data Nucl. Data Tables 98, 1096 (2012)
16. I.V. Karpov et al., Int. J. Mod. Phys. E 21, 1250013 (2012)
17. P. Möller et al., Phys. Rev. C 79, 064304 (2009)
18. I. Silisteanu and C.I. Anghel, Rom. J. Phys. 60, 444 (2015)
19. J. Dong, W. Zuo, and W. Scheid, Phys. Rev. Lett. 107, 012501 (2011)
20. A.I. Budaca and I. Silisteanu, J. Phys.: Conf. Series 413, 012027 (2013)
21. A.I. Budaca and I. Silisteanu, J. Phys.: Conf. Series 337, 012022 (2012)
22. I. Silisteanu and C.I. Anghel, Rom. J. Phys. 59, 724 (2014)
23. S.P. Maydanyuk, P.-M. Zang, and S.V. Belchikov, Nucl. Phys. A 940, 89 (2015)
24. V.Yu. Denisov, O.I. Davidovskaya, and I.Yu. Sedykh, Phys. Rev. C 92, 014602 (2015)
25. O.V. Kiren, S.B. Gudennavar, and S.G. Bubbly, Rom. J. Phys. 57, 1335 (2012)
26. K.P. Santhosh et al., J. Phys. G: Nucl. Part. Phys. 36, 115101 (2009)

27. K.P. Santhosh and B. Priyanka, Phys. Rev. C 90, 054614 (2014)
28. Yibin Qian and Zhongzhou Ren, Phys. Rev. C 90, 064308 (2014)
29. Yibin Qian and Zhongzhou Ren, Phys. Lett. B 738, 87 (2014)
30. R. Budaca, A. Sandulescu, and M. Mirea, Mod. Phys. Lett. A 30, 1550129 (2015)

10

Recent Advances in Beta Decay Measurements

Magali Estienne, Muriel Fallot, Lydie Giot, Loïc Le Meur and Amanda Porta

Subatech (CNRS/IN2P3), IMT Atlantique, Université de Nantes, 4 Rue Alfred Kastler, 44307 Nantes, France

ABSTRACT

Three observables of interest for present and future reactors depend on the β decay data of the fission products: the reactor decay heat, antineutrinos from reactors and delayed neutron emission. Concerning the decay heat, significant discrepancies still exist between summation calculations in – their two main ingredients: the decay data and the fission yields – performed using the most recent evaluated databases available. It has been recently shown that the associated uncertainties are dominated by the ones on the decay data. But the results subtantially differ taking into account

Citation: Magali Estienne, Muriel Fallot, Lydie Giot, Loïc Le Meur and Amanda Porta. "Recent advances in beta decay measurements" EPJ Nuclear Sci. Technol., 4 (2018) 24 DOI: https://doi.org/10.1051/epjn/2018034

or not the correlations between the fission products. So far the uncertainty propagation does not include as well systematic effects on nuclear data such as the Pandemonium effect which impacts a large number of nuclei contributing to the decay heat. The list of nuclei deserving new TAGS measurements has been updated recently in the frame of IAEA working groups. The issues listed above impact in thc same way the predicted energy spectra of the antineutrinos from reactors computed with the summation method, the interest of which has been recently reinforced by the Daya Bay latest publication. Nuclear data should definitely contribute to refine and better control these calculations. Lastly, a lot of nuclear data related to delayed neutrons are missing in nuclear databases. Despite the progresses already done these last years with new measurements now requiring to be included in evaluated databases, the experimental efforts which still need to be done are significant. These different issues will be addressed here before to comment on recent experimental results and on their impacts on the quoted observables. Some perspectives will also be presented. Solving the issues listed above will require to bring together experimental, simulation, evaluation and theoretical activities.

INTRODUCTION

In Pressurized Water Reactors, the thermal power is mainly induced by four isotopes constituting the fuel assembly. The two fissile and fertile nuclei ^{235}U and ^{238}U present in the fresh fuel undergo fission or capture processes inducing the production of the fissile nuclei ^{239}Pu and ^{241}Pu. The evolution of their compositions with the time t (reactor burnup) is typical: while the ^{235}U composition dominantly decreases with t compared to ^{238}U, ^{239}Pu is massively produced in comparison to ^{241}Pu. In the fission process which gives the thermal energy according to equation (1), we are essentially interested by the production of the two fission products FP1 and FP2.

$$n + {}^{235}\mathrm{U} \rightarrow {}^{236}\mathrm{U}^* \rightarrow FP1 + FP2 + neutrons. \qquad (1)$$

These latters most oftenly neutron-rich nuclei undergo β or β-n decays (when the daughter nucleus is produced in an excited state above the neutron emission threshold S_n). In such radioactive processes, a neutron converts into a proton and emits an electron and an electron antineutrino. Gammas are also emitted in the deexcitation of the produced daughter. Their independent measurement allows to assess β decay physics properties. In these proceedings, we will essentially focus on the physics that can be addressed via the γ channel. Before the 90s, high-resolution γ-ray spectroscopy through the use of germanium detectors was the conventional detection technique used for γ measurement. It offers excellent energy resolution but an efficiency which strongly decreases at high energy and usually limited acceptance. As direct consequences, weak γ-ray cascades or high energy γ-rays are likely difficult to detect and the decay scheme of the parent nucleus may be incomplete. There is a danger of overlooking the existence of β-feeding into the high energy nuclear levels of daugther nuclei especially with decay schemes with large Q-values. In other words, it might lead to an overestimate of the high energy part of the FP β spectra along with an underestimate of the γ energies. This phenomenon is commonly called "Pandemonium effect" highlighted by Hardy [1]. As a consequence, the content of nuclear databases (NDB) can be strongly biased in some cases, as illustrated in Figure 1 in reference [2]. In this figure, two antineutrino energy spectra reconstructed from the endpoints and branching ratios either tabulated in JEFF 3.1 or measured by the TAGS collaboration (Pandemonium free – see Sect. 3.1 for details on the detection technique) for ^{105}Mo are compared. The average energy of 1.524 MeV of the latter is to be compared to the 2.475 MeV of the former one. Their difference of the order of 50% is to be compared to error tabulated in JEFF of the order of the percent. The uncertainty has thus been clearly underestimated in this case. This illustrates the necessity to manipulate and use NDB with the utmost care.

Indeed β decay has important implications for present and future reactors. They are threefold: (i) The released γ-ray and β particle contribute to the decay heat (DH) which is critical for reactor safety and economy; (ii) The antineutrinos which escape can be detected. This impacts both fundamental physics (neutrino oscillation, mass hierarchy and reactor anomaly) and reactor applications (monitoring and non-proliferation); (iii) The study of β-n emitters provides an access to the delayed neutron fractions important for the operation and control of the chain reactions of reactors. In Section 2, the current status of the researchs carried out in those branches will be given and analyzed in the point of view of nuclear data. Specific attention will be focused on the treatment of uncertainties. Section 3 will be devoted to a discussion on recent and expected new results from the TAGS collaboration. At this occasion, the TAGS detection technique will be presented. Our conclusions and perspectives will be given in Section 4.

CURRENT STATUS OF *Β* DECAY APPLICATIONS

Decay Heat

The decay heat (DH) is defined as the energy released by the nuclear fuel after the shut-down of the chain reaction in a reactor. It concerns the three α, β and γ radioactivities whose emitters are either the FPs (which are the dominant contributors to the DH over the first 100 years after the reactor stop) or heavy nuclei such as the actinides (which contribution becomes not negligeable above 100 years after the reactor stop). As the DH represents a residual power of 6 to 12% of the nominal power of the reactor just after its shut-down, the evaluation of the reaction safety and various economic aspects of nuclear power generation require its good knowledge. The DH can be directly accessed and studied via integral measurements (for a recent review see [3]). However, it is also essential to be able to estimate it analytically and this is achievable thanks to the only predictive method for future reactors: the summation method [4]. It consists in the summation of all the fission product and actinide contributions inventoried for specific conditions of reactor operation and subsequent cooling period. In the case of fission pulses, the time ranges up to $\sim 10^6$ s and the fission of pure isotopes is considered. In these conditions, the contribution of actinides can be neglected and the decay heat can be expressed with equation (2),

$$f(t) = \sum_i (\overline{E}_{\beta,i} + \overline{E}_{\gamma,i}) \lambda_i N_i(t) \qquad (2)$$

where $\overline{E}_{\beta,i}$ and $\overline{E}_{\gamma,i}$ are respectively the mean energies of the β and γ decays of the ith FP, λ_i is its total decay constant and $N_i(t)$ the associated fission yield. The computation thus requires the access to these observables in NDB for as much fission products as possible with the best accuracy as possible. In the 70s, this formula was successfully computed with the use of nuclear DB taken worldwide, however important discrepancies were observed comparing the DH calculations and benchmark experiments mainly because of the Pandemonium effect. These differences were partially compensated by the inclusion of average β and γ energies derived from the Gross Theory of β decay which was able to compensate the missing β-strengths or the missing nuclei information [5,6]. But, since then this temporary solution was step-by-step replaced by the use of measured data with a new detection technique: the total absorption γ-ray spectroscopy TAGS that will be presented in Section 3.1. Since then, several collaborations and some worldwide efforts around TAGS are born with the support of international agencies as the IAEA, the NEA, etc. Figure 2 in reference [7] highlights how the use of non Pandemonium free measurements stored in NDB can strongly bias the computation of the DH and the success of the TAGS to correct for this effect. During the 1st TAGS campain of the Valencian's group of IFIC at the facility of JYFL of Jÿvaskÿla (Finland), Algora et al. measured the β-feeding of the 7 nuclei $^{102;104;105;106;107}$Tc, ^{105}Mo, and ^{101}Nb important contributors to the DH, 5 out of 7 indeed were suffering strongly from Pandemonium effect. The summation calculation of the DH has then been repeated substituting their measurements to the ones previously stored in ENDF/B-VII which came from high resolution measurements. These nuclei strongly contributed to improve the calculation reducing impressively the gap between Tobias integral measurement compilation [8] and the summation calculation and solving a large part of the long-standing discrepancy in the γ component of the decay-heat data of ^{239}Pu in the 4–3000 s range.

Figure 1 presents a comparison of some FISPACT-II simulations of the total and the electromagnetic decay heats performed with a variety of nuclear databases (ENDF/B [9], JEFF [10] and JENDL [11], the American, European and Japanese libraries respectively containing neutron-induced fission yield (nFY) and a variety of decay data (DD) and the General Description of Fission Observables model (GEF) [12]) to integral measurements on a thermal fission of ^{235}U [3]. The measurements performed from fission decay heat experiments include the ones using the Oak Ridge Research Reactor by Dickens et al. [13], a series of experiments performed under the direction of Schier from the university of Massachusetts Lowell [14–16] and a separate meta-analysis performed by Tobias which includes a combination of results from multiple laboratories [8]. Even if this set of comparative studies demonstrates the ability to calculate the decay heat within the experimental uncertainties of the various integral measurements, some efforts still need to be done in order to improve the quality of both nuclear databases and integral measurements to reach agreements between simulations and experiments. As an example the three following observations can be made regarding the γ component of the DH of ^{235}U: i) Integral measurements are not in agreement for several cooling times of the most well-known nuclide; ii) There are still discrepancies between data and simulations using different DB for γ heat; iii) The comparison between data and simulations is better for the total heat than for the γ one in favour of potential problems in the decay data β/γ feeds.

Concerning the determination of the total uncertainty associated to the calculation of the DH, a lot of work still needs to be done to improve our knowledge about β decay properties of FP and to estimate the associated uncertainties. In his PhD thesis, Benoît concludes that the uncertainty on its decay heat calculation mainly comes from the uncertainties on the fission yields [17] in opposition to the work performed by Katakura which points out that the main contribution to decay heat uncertainties comes from the evaluation of decay energies below 5×10^3 s [18]. Concerning the mean decay energy in the first approach [17], a mean error of 3.35% was considered for 369 nuclei tabulated in JEFF3.1, and 10% of uncertainties have been taken for 75 unknown nuclei. These are mainly nuclei with half life <10 s. Whereas in the second approach [18] (probably more realistic having in mind the bias coming from Pandemonium effect in nuclear databases discussed in Sect. 1) 100% error was considered on nuclei for which no uncertainty was given in databases. However, a really low uncertainty of 0.001% on 122 branching ratios over 250 has been considered in the first approach and no sensitivity study has been performed on the BRs in the second approach leaving an open question about the impact of uncertainties on BRs. Moreover different covariant matrices of the fission yields have been included in both works showing the necessity to take them into account in the complete computation of the DH uncertainty and emphasizing the need for some additional reflection on how to compute the covariance matrices.

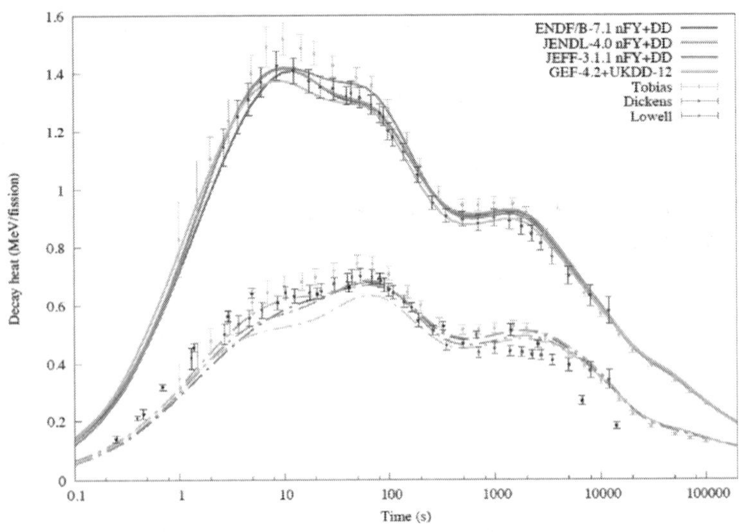

Figure 1: Total (solid) and γ (dash) decay heat from thermal pulse on ^{235}U as a function of time (s) [3].

Reactor Antineutrinos

Reactor antineutrinos are a second direct application of β decay both from fundamental and applied point of views.

In the field of the fundamental physics of neutrinos, the last 5 years have been essential for the first measurements of the θ_{13} mixing angle by the three experiments Double Chooz (DC) [19], Daya Bay (DBay) [20] and RENO [21]. Indeed, the precise measurement of the oscillation parameter required an independent computation of the antineutrino spectra. The conversion method, quasi independent from nuclear databases and in principle more precise than the summation method was chosen at the time by the different experiments for that purpose [22].

The spectral studies carried out for θ_{13} determination are at the origin of the so-called "reactor anomaly" [23]. They are responsible for the 6% deficit of the absolute value of the measured flux compared to the best prediction converted ILL data which has been first observed by short distance experiments and later confirmed by DC, DBay and Reno. One of the hypotheses which has been proposed to explain this deficit was the potential existence of one or several sterile neutrino(s), hypothesis which has motivated the deployments of a large amount of short baseline experiments [24]. However, an other hypothesis that could be considered concerns the converted spectra and to which extent they can be regarded as trustful. This hypothesis should not be neglected taking into account that: i) The three experiments observed a distortion in their full antineutrino energy spectra between 5 and 7 MeV not reproduced so far by the converted spectra; ii) The last results from DBay which points-out a potential problem in the converted antineutrino spectra from ^{235}U measured β spectrum at ILL [25,26]. The quite large number of recent publications on spectral studies in the frame of neutrino physics putting into question the integral β measurement of ^{235}U of Schreckenbach et al. and its conversion and the potential existence of sterile neutrinos strongly motivate the necessity to improve nuclear databases used by the summation method, the alternative to the conversion one [26]. Antineutrino spectra predictions will also be essential for the next generation reactor neutrino experiments like JUNO [27] or to compute the background for other multipurpose experiments.

As regard neutrinos and applied physics, it has been suggested to use the discrepancy between antineutrino flux and energies from U and Pu isotopes to infer reactor fuel isotopic composition and power [28]. Indeed, due to the nuclear structure properties of the 4 isotopes present in a reactor core, the distribution of their fission products is different implying different antineutrino emitted energy spectra after they undergo a β decay. On an average 6 antineutrinos are produced per fission and about 2 above the inverse β decay (IBD) reaction threshold, the reaction commonly used for their detection. Moreover, due to the time evolution of the fuel isotopic composition, the associated antineutrino flux also evolves with time both in norm and shape making of antineutrino detection an elegant tool to monitor the core fissile content. This second topic associated to neutrinos strongly motivates the necessity to perform new measurements of nuclei (of interest for the field) which β decay. The IAEA Nuclear Data Section has acknowledged this necessity by including a number of nuclei of interest for reactor antineutrino spectra in their priority lists of measurements [29].

The Summation Method

As for the decay heat, the antineutrino spectrum associated with one of the four fissioning isotopes in a moderated reactor can be computed as the sum of the contributions of all fission products using the full information available per nucleus in the NDB [22,30]. This method is useful for several reasons. Not only it is the only one that can be adapted to the computation of the antineutrino $(\overline{v})$ emission associated to various reactor designs, but also it allows for the computation of antineutrino spectra for which no integral β spectrum has been measured yet taking into account off-equilibrium effects and allowing the use of different energy binnings of interest for reactor neutrino experimental analyses.

The $\beta/\overline{v}$ spectrum per fission of a fissible isotope k, $S_k(E)$, can be broken-up into the sum of all fission product $\beta/\overline{v}$ spectra weighted by their activity A_{fp} according to equation (3). The $\beta/\overline{v}$ spectrum of one fission product being defined as the sum over the b branches of all β decay spectra (or associated $\overline{v}$ spectra), S^b_{fp}, of the parent nucleus to the daughter nucleus weighted by their respective branching ratios.

$$S_k(E) = \sum_{fp=1}^{N_{fp}} A_{fp} \times \sum_{b=1}^{N_b} BR^b_{fp} \times S^b_{fp}(Z_{fp},A_{fp},E^b_{0fp},E). \quad (3)$$

To calculate these quantities the MURE code is used to extract the percentage of fissions and the fission product activity [31]. The β spectrum S^b_{fp} of each b branch for the fp[th] fission product is derived from the Fermi theory taking into account Huber›s prescriptions for the treatment of corrections to be applied to the calculation [32]. Equation (3) illustrates the strong dependence of the summation method to nuclear data as BR^b_{fp}, Z_{fp}, A_{fp} and E^b_{0fp} are extracted from NDB for each spectral prediction as described in [30]. A given spectral computation thus requires the use of databases as exhaustive and accurate as possible, including existing measurements Pandemonium free. It also motivates the necessity to perform new measurements of nuclei with incomplete decay schemes for which the β-feeding is clearly biased as explained in Section 1.

Results after Pandemonium Bias Correction

Figure 2 shows our predictions obtained in 2012 with the summation method of the antineutrino energy spectra emitted by the ^{239}Pu (left) and ^{235}U (right) in a PWR. The details of such computation can be found elsewhere [30]. We emphasize here that to obtain these new spectra, we substituted 7 nuclei, ^{102,104,105,106,107}Tc, ^{101}Nb and ^{105}Mo, first simulated with the JEFF DB by the ones, Pandemonium free measured during the first TAGS campaign at the JYFL of Jÿvaskÿla (same nuclei as in Sect. 2.1). In the corresponding publication, the impact of these nuclei on the predicted spectra has been quantified and commented by looking at the ratio of the new predicted spectra and the ones obtained with the same data set but the latest TAGS data. The Pandemonium effect strongly affects the global $\overline{\nu}$ energy spectra of ^{235,238}U and ^{239,241}Pu. A noticeable deviation from unity (maximum 8% decrease) was obtained for ^{239,241}Pu. We observed a maximum of 3.5% deviation for ^{238}U. As regards ^{235}U, as these nuclei have a small contribution to the total spectrum, the effect was smaller than 1.5% at ~3 MeV.

In these proceedings we would like to highlight the two inserts included in both panels of Figure 2 which correspond to the ratios of the spectra predicted with the summation method over the spectra predicted by the conversion one for ^{239}Pu (left) and ^{235}U (right). This powerful comparison with the computations of Huber over the range 2–8 MeV shows reasonable agreement in terms of normalization and shape, despite the steeply falling shape of the spectra with energy. However, we observe a small increasing trend (reaching at maximum 10% discrepancy up to 7 MeV) of the ratio with energy for ^{239}Pu suggesting that the NDB used are not yet fully Pandemonium free. This strongly motivates the necessity to measure more Pandemonium nuclei with the TAGS technique. Moreover for ^{235}U, we obtained a ratio below 1 over the full energy range varying from less than 5% to 10% discrepancies. This observation is compatible with the latest PRL from Daya Bay which pointed out a potential problem in the converted spectrum of ^{235}U and supports the necessity to perfect our summation predictions with cleaner nuclear data.

At last, we would also like to stress the necessity to put uncertainties on our predictions which would definitively allow us to draw quantitative conclusions on comparison to neutrino experiments. For that purpose, one would need to have covariances between fission yields but also covariances for the decay data.

In [33], besides publishing our priority list of Pandemonium nuclei to be measured with TAGS, we have studied the impact of the recently measured ^{92}Rb on the decay heat and the antineutrino energy spectra of $^{235,238}U$ and $^{239,241}Pu$. Figure 3 illustrates the results of this new measurement on the spectra by comparing the ratio of the new predicted spectra and the ones obtained with the same data set but the latest TAGS data for ^{92}Rb (red dashed-dotted line). As expected, the main effect is in the 4 to 8 MeV antineutrino energy range, with a maximum between 7 and 8 MeV, and amounts to 4.5% for ^{235}U, 3.5% for ^{239}Pu, 2% for ^{241}Pu and 1.5% for ^{238}U. These discrepancies are due to the difference in the shapes of the antineutrino spectra built with the newly measured β feedings with respect to the antineutrino spectra converted from Rudstam's measurements. The ratio is displayed as well with green dotted lines, and is nearly superposed on the ratio built when using Rudstam data in the first place. The change becomes even more dramatic (black dashed lines) if one compares with summation method spectra in which an older version of the ENDF data was used, as in [35]. A gray horizontal bar is placed above the antineutrino energy scale to indicate the region of the distortion observed by the reactor antineutrino experiments with respect to converted spectra. This shows the relevance of the present ^{92}Rb decay data in the calculations. It also motivates the needs for more measurements to improve DB and to precisely quantify the uncertainties associated to the summation calculations.

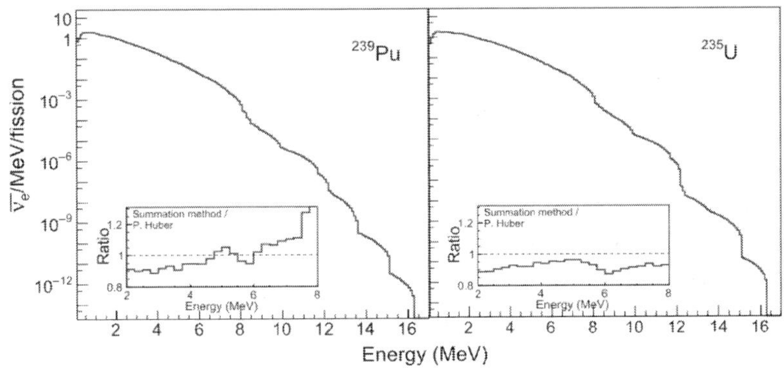

Figure 2: Reconstructed antineutrino energy spectra, including the latest TAGS data from [7] for ^{239}Pu (left) and ^{235}U (right). In the insets are the ratios of the spectra to the ones computed by Huber (converted spectra) [30].

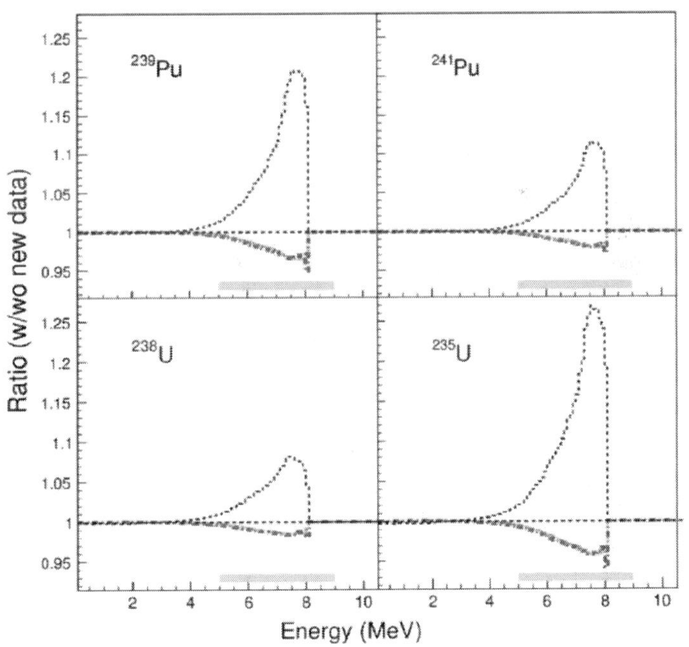

Figure 3: Ratio between the antineutrino spectra calculated using the results presented in [33] with respect to the data on ^{92}Rb decay used in [30] (thick red dashed-dotted line), in [34] (green dotted line) and in [35] (black dashed line) for ^{239}Pu (top left), ^{241}Pu (top right), ^{238}U (bottom left) and ^{235}U (bottom right).

β-Delayed Neutrons (BDN)

In some β decays, if the daugther nucleus is produced in an excited state of energy above the neutron emission threshold, the corresponding nuclei can deexcite with the emission of a neutron producing a new nucleus of characteristics (Z + 1, N − 2). This neutron is called delayed neutron with respect to the fast neutrons emitted during the fission process. A good control of these β-delayed neutron emission being essential for the reactor operation, it is crucial to have access to all the parameters characterizing a β-delayed neutron emission ideally via a dedicated database. Indeed the last compilation of β-delayed neutron data was published in 2002. Since then, many measurements were performed and published but no effort had been dedicated to list the whole data available worldwide. This has become an other concern of the IAEA which nuclear data section has organized a coordinated research project on the topic, with the aim of making a new database with up to date data of delayed neutron emission [36] and the aim of deducing from this new compilation a list of recommandations for new experimental efforts.

In addition, the IAEA has organized regular TAGS consultant meetings during the last decades [29], allowing to update the table of priority nuclei for the decay heat calculations and the reactor antineutrino predictions. The most recent table has been published in [33], and combines the important contributors to the decay heat, reactor antineutrinos, mentioning if they are β-delayed neutron emitters.

In parallel, the CEA Cadarache has started an experimental program in order to complement the databases in neutron delayed emission data [37].

RECENT AND UPCOMING RESULTS IN THE EUROPEAN TAGS COLLABORATION

The Total Absorption Spectroscopy: A Solution to the *Pandemonium* Effect

In order to avoid the detection issue pointed-out in Section 1 regarding the use of HPGe detectors to measure γ-ray emissions, the Total Absorption γ-ray Spectroscopy (TAGS) technique was proposed in the 90s and successfully used since then to detect β intensity to states at high excitation in the daughter nucleus [38,39]. A total absorption spectrometer (TAS) consists in a ~4π calorimeter for the detection of the γ cascades rather than individual γ-rays. The detection of the total energy allows the determination of the feeding probability of excited levels populated in the β decay. More details on the way the analysis is performed can be found elsewhere [40,41].

So far, two segmented TAS have been developed by the Valencia/Surrey research groups and exploited by the European TAGS collaboration already during two campains of measurement first at the IGISOL facility of the University of Jÿvaskÿla in 2009 and then after the upgrade of IGISOL in 2014. For the first campain, Rocinante, the Valencia-Surrey Total Absorption Spectrometer was used [42]. It consists in a cylindrical 12-fold segmented BaF_2 detector with a length and external diameter of 25 cm, and a longitudinal hole of 5 cm diameter. It was coupled to a 0.5 mm thick Si detector with a β-detection efficiency of about 25% to get rid of the backgrounds. The set-up offers a total efficiency of more than 80% up to 10 MeV. At that period, 7 nuclei of interest for decay heat purpose (including 4 β-delayed emitters) have been measured and published [7,33,42]. Their impacts on the decay heat and antineutrino energy spectra calculations have partly been presented and discussed in Sections 2.1 and 2.2.2 of these proceedings.

For the second campain, the new Decay Total Absorption γ-ray Spectrometer (DTAS) has been successfully installed and used [43]. It consists in 18 modules of NaI(Tl) crystals. The DTAS spectrometer was completed with a HPGe detector. During this campain, 12 nuclei of Subatech Nantes team proposal of direct interest for antineutrino spectra predictions associated to 11 nuclei of Valencia team proposal of direct interest for decay heat were measured. The whole set of nuclei are currently under study or in the process of being published in the course of 2018 [44,45]. They will allow to update or complete nuclear databases with a list of nuclei henceforth Pandemonium free.

Beta-delayed Neutron Emission Properties, A "By-product" of TAS Measurements

Concerning BDN, even if the TAGS measurements are not dedicated to direct measurement of neutrons, they can bring indirect information on some parameters of the properties of β-decays with delayed neutron emission. As an illustration, the 1st TAGS campain at the JYFL has allowed to evidence an enhanced γ emission above the neutron separation threshold in several nuclei. These emissions are not reproduced by the statistical model. These results have implications on nuclear astrophysics processes such as the r-process, as the ratio of γ to neutron widths used in the calculation of neutron capture cross sections can be extracted from the β decay data measurements [46].

The use of TAGS data in complement to neutron measurements (P_n values) brings a valuable cross-check to ensure that there is no inconsistensies in the β decay properties of fission products in databases.

Recent New Bias in Nuclear Data

Quite recently in a NEA/WPEC25 meeting, large discrepancies have been pointed-out between the mean γ energies associated to the β-decay of several nuclei deduced from TAGS Greenwood and Rudstam [4,47]. This point has been recently investigated by the TAGS Valencian's group who made a careful comparison of these mean energies between the data Pandemonium free from Greenwood and their own measurements, and the data measured by Tengblad/Rudstam [42]. This study is summarized in Figure 4 where the difference between the mean γ energies of (Greenwood/Valencia) and Rudstam is depicted as a function of the Q_β for 30 nuclei.

The comparison shows an uncertainty of the mean energy of 50% in many cases. Actually, the decay of ^{91}Rb which was assumed to be Pandemonium free by Rudstam et al. was used at the time of the measurements as a normalization point in their systematic studies. After applying a new normalization based on their own measurement of the ^{91}Rb on the results of Rudstam et al., Rice et al. show that even if the mean value of the differences of the mean γ energies is reduced from −360 keV to −180 keV, there is a remaining systematic difference between the two sets of data. Moreover, they still obtain a large spread of the observed differences. This example illustrates once more the necessity to multiply experiments to complete or correct nuclear databases as much as possible. One can also wonder how such kind of discrepancies can impact the evaluation of nuclei and indirectly the decay heat determination (or other applications).

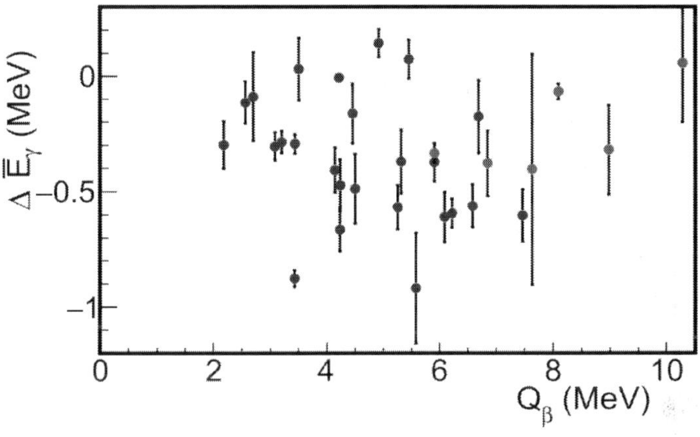

Figure 4: Differences between the mean energies reported in the work of Rudstam et al. and the deduced mean γ energies from the work of Greenwood et al. (blue points) and from the work of the TAGS Valencia/Nantes›s groups (red points) [42].

CONCLUSIONS AND PERSPECTIVES

In these proceedings, we have shown that measurements of many nuclei of interest for decay heat, antineutrinos, and a few β-delayed n emitters have been performed during the past years or are programmed in a near future using the Total Absorption γ-ray Spectroscopy technique. This direction still needs to be followed as it is mandatory to better characterize fission product fundamental β-decay properties and especially to improve NDB and all the subsequent applications. The results of the second TAGS campaign in Jyväskylä are already partly available and will be completed in 2018. We then expect to assess the overall impact of all the TAS measurements performed during the last decade on decay heat, reactor antineutrinos and β delayed neutron observables by end 2018. Some effort will also be dedicated to include the new measured data in generation IV reactor simulations. The next step regarding the summation method calculations for both DH and antineutrinos will be the adjunction of uncertainties on the predictions. A proper propagation of uncertainties including FY correlations and decay data correlations is mandatory for that purpose. This task is very challenging since no covariance matrices associated to the decay data exist and the methodology to obtain the covariance matrices of the fission yields is still debated. In addition models are still needed for several tens of nuclei and the shape of forbidden non-unique transitions is not well under control impacting the reactor antineutrino calculations as well as nuclear structure and astrophysics. Solving the issues listed above will require to bring together experimental, simulation, evaluation and theoretical activities.

AUTHOR CONTRIBUTION STATEMENT

All the authors have contributed to the scientific content of the paper and approved the final manuscript.

ACKNOWLEDGMENTS

Several results obtained by the European TAGS collaboration have been used to illustrate the authors statements in these proceedings. The authors would like to acknowledge their colleagues from the TAGS collaboration. The authors acknowledge the nuclear data section of the IAEA and especially Paraskevi (Vivian) Dimitriou for fostering research in the field through the TAGS consultant meetings and the CRP about β-delayed neutron emission. This work was supported by the CHANDA European project, the In2p3 institute of CNRS, and the NEEDS challenge through the NACRE project.

REFERENCES

1. J.C. Hardy et al., Phys. Lett. B **71**, 307 (1977)
2. M. Estienne et al., Nucl. Data Sheets **120**, 149 (2014)
3. M. Fleming, J.-C. Sublet, report 2015 Fispact II, CCFE-R(15)28
4. T. Yoshida et al., OECD/NEA Working Party for International Evaluation Co-operation, Vol. 1425 (2007) 25. Nuclear Science NEA/WPEC-25 (2007)
5. K. Takahashi, M. Yamada, T. Kondoh, At. Data Nucl. Data Tables, **12**, 101 (1973)
6. T. Yoshida, R. Nakasima, J. Nucl. Sci. Technol. **18**, 393 (1981)
7. A. Algora et al., Phys. Rev. Lett. **105**, 202501 (2010)
8. A. Tobias, Technical report, Central Electricity Generating Board, 1989
9. M. Chadwick et al., Nucl. Data Sheets **112**, 2887 (2011)
10. A. Koning et al., Technical Report NEA No. 6190, OECD NEA, 2006
11. K. Shibata et al., J. Nucl. Sci. Technol. **48**, 1 (2011)
12. K.-H. Schmidt, B. Jurado, C. Amouroux, General view on nuclear fission (2014), pp. 1–208, www.khs-erzhausen.de/GEFY.html
13. J.K. Dickens et al., Technical Report ORNL/NUREG-14, Oak Ridge National Laboratory, 1979
14. S. Li, Beta Decay Heat Following 235U, 238U and 239Pu Neutron Fission, Ph.D. thesis, University of Massachusetts Lowell, Department of Physics, 1997
15. H.V. Nguyen, Gamma-ray Spectra and Decay Heat Following 235U Thermal Neutron Fission, Ph.D. thesis, University of Massachusetts Lowell, Department of Physics, 1997
16. E.H. Seabury, Gamma-ray Decay Heat Measurements Following 238U(n,f) and 239Pu(n,f), Ph.D. thesis, University of Massachusetts Lowell, Department of Physics, 1997
17. J.-C. Benoît, Développement d'un code de propagation des incertitudes des données nucléaires sur la puissance résiduelle dans les réacteurs à neutrons rapides, PhD thesis, University of Paris Sud, CEA, 2012
18. J. Katakura, J. Nucl. Sci. Technol. **50**, 99 (2013)
19. Y. Abe et al. (Double Chooz Collaboration), Phys. Rev. Lett. **108**, 131801 (2012)
20. F.P. An et al. (Daya Bay Collaboration), Phys. Rev. Lett. **108**, 171803

(2012)

21. J.K. Ahn et al. (RENO Collaboration), Phys. Rev. Lett. **108**, 191802 (2012)
22. Th.A. Mueller et al., Phys. Rev. C **83**, 054615 (2011) [NASA ADS]
23. G. Mention, M. Fechner, Th. Lasserre, Th. Mucller, D. Lhuillier, M. Cribier, A. Letourneau, Phys. Rev. D **83**, 073006 (2011)
24. K.N. Abazajian et al., http://arxiv.org/abs/1204.5379
25. F.P. An et al. (Daya Bay Collaboration), Phys. Rev. Lett. **118**, 251801 (2017)
26. M. Fallot, Physics **10**, 66 (2017)
27. F. An et al. (Juno collaboration), J. Phys. G **43**, 030401 (2016)
28. L.A. Mikaelian, Proc. Int. Conf. Neutrino-77, v.2 (1977) p. 383
29. P. Dimitriou, A.L. Nichols, IAEA report INDC(NDS)-0676, IAEA, Vienna, Austria, 2015
30. M. Fallot et al., Phys. Rev. Lett. **109**, 202504 (2012)
31. MURE, MCNP utility for reactor evolution (2009), http://www.nea.fr/tools/abstract/detail/nea-1845, O. Meplan, Technical Reports No. LPSC 0912 and No. IPNO-09-01, 2009
32. P. Huber, Phys. Rev. C **84**, 024617 (2011)
33. A.-A. Zakari-Issoufou et al., Phys. Rev. Lett. **115**, 102503 (2015)
34. A.A. Sonzogni, T.D. Johnson, E.A. McCutchan, Phys. Rev. C **91**, 011301(R) (2015)
35. D.A. Dwye, T.J. Langford, Phys. Rev. Lett. **114**, 012502 (2015)
36. https://nucleus.iaea.org/sites/connect-members/bDN/Pages/default.aspx
37. D. Foligno, P. Leconte, EPJ. Nucl. Sci. Technol. **4**, 31 (2018)
38. A. Algora et al., Nucl. Phys. A **654**, 727c (1999)
39. Z. Hu et al., Phys. Rev. C **60**, 024315 (1999)
40. D. Cano-Ott et al., Nucl. Instr. Meth. A **430**, 488 (1999)
41. D. Cano-Ott et al., Nucl. Instr. Meth. A **430**, 333 (1999)
42. S. Rice et al., Phys. Rev. C **96**, 014320 (2017), arXiv1704.01915
43. V. Guadilla et al., Nucl. Inst. Meth. B **376**, 334 (2016)
44. V. Guadilla, TAGS measurements for neutrino physics and applications, Ph.D. thesis, University of Valencia, 2017

45. L. Le Meur, Etude des propriétés de désintégration bêta de noyaux exotiques à l'aide de la technique de Spectroscopie par Absorption Totale (TAS), Ph.D. thesis, University of Nantes, in preparation
46. J.L. Tain et al., Phys. Rev. Lett. **115**, 062502 (2015)
47. A.L. Nichols et al., INDC(NDS)-0499, 2006

Table 1: Legendre coefficients in W(θ) (2) for gamma rays from inelastic scattering of 14 MeV neutrons

Experiment	a_2	a_4
14.9 MeV (Zhou [8])	0.21 ± 0.02	-
14.2 MeV (Abbondanno [9])	0.20 ± 0.09	0.11 ± 0.14
14.1 MeV (This work)	0.19 ± 0.02	0.02 ± 0.03

CONCLUSION

- The inelastic scattering of neutrons with an energy of 14.1 MeV on natural silicon sample was investigated by using the method of tagged neutrons. The angular distribution of the γ-rays with energy of 1.779 MeV from 28Si(n, n') $^{28}Si^{*}$ was obtained.
- Within the limits of statistical errors, the calculated values of Legendre coefficient for the angular anisotropy is found to be in good agreement with the results published earlier [8, 9].
- Unlike the results for the angular distribution of gamma-rays from $^{12}C(n, n'\gamma)\,^{12}C$ with 14.1 MeV neutrons [10], it was confirmed that the contribution of the components of the 4th degree Legendre polynomials is insignificant for gamma-radiation from the first excited state $E^{*}(2^{+}) = 1.779$ MeV in ^{28}Si.

It is planned to do some model calculations of the angular distribution of gamma rays from the inelastic neutron scattering of 14.1 MeV on 28Si nuclei in order to reveal the possible mechanism of this reaction.

REFERENCES

1. S. Takeuchi, M. Matsushita, N. Aoi et al., Phys. Rev. Lett 109, 182501 (2012)
2. J. Fridmann, I. Wiedenhöver, A. Gade et al., Nature 435, 922 (2005)
3. A. Mutschler, A. Lemasson, O. Sorlin et al., Nature Physics 13, 152 (2017)
4. M.L. Markova, T.Yu. Tretyakova, N.A. Fedorov, Phys. Atom. Nucl. 80(2017) (in press)
5. V.M. Bystritsky, N.I. Zamyatin, E.V. Zubarev et al., Phys. of Part. and Nucl. Lett., 10, 442 (2013).
6. I.N. Ruskov, Yu.N. Kopatch, V.M. Bystritsky et al., EPJ Web of Conferences 146, 03024 (2017)
7. ADCM - An universal Digital Pulse Processing system for nuclear physics experiments: http://afi.jinr.ru/ADCM16-LTC.
8. H. Zhou, F. Deng, Q. Zhao et al., Phys.Rev. C 82, 047602 (2010)
9. H. Zhou, F. Deng, W. Cheng et al., Nucl. Inst. and Meth. A 648, 192 (2011)
10. U. Abbondanno, R. Giacomich, M. Lagonegro et al., J. Nucl. Energy. 27 227 (1973)
11. V.M. Bystritsky, D.N. Grozdanov, A.O. Zontikov et al., Phys. Part. Nucl. Lett. 13 504 (2016)

SECTION 5:
NUCLEAR REACTIONS FISSION

12

Experimental and Theoretical Cross Section Data of Deuteron Induced Nuclear Reactions on Platinum

F. Tárkányi[1], F. Ditrói[1], S. Takács[1], A. Hermanne[2] & A. V. Ignatyuk[3]

[1] Institute for Nuclear Research, Hungarian Academy of Sciences (ATOMKI), Debrecen, Hungary
[2] Cyclotron Laboratory, Vrije Universiteit Brussel (VUB), Brussels, Belgium
[3] Institute of Physics and Power Engineering (IPPE), Obninsk, Russia 249020

ABSTRACT

Additional experimental activation cross sections for deuteron induced reactions on platinum were measured in the 35–49 MeV energy range for the $^{nat}Pt(d,xn)^{191,192,193,194,195,196m2,196g,198m,198g,199}Au$, $^{nat}Pt(d,x)^{189,191,193m,195m,197g}Pt$ and $^{nat}Pt(d,x)^{188,189,190m2,190,192,194m2}Ir$ reactions by using the stacked foil irradiation technique and gamma ray spectrometry. The available experimental results

Citation: Tárkányi, F., Ditrói, F., Takács, S. et al. "Experimental and theoretical cross section data of deuteron induced nuclear reactions on platinum". J Radioanal Nucl Chem 321, 747–764 (2019). https://doi.org/10.1007/s10967-019-06624-4

are compared with the calculated values based on ALICE-D and EMPIRE-D model codes and with the theoretical predictions in the TALYS based TENDL-2017 library.

INTRODUCTION

Experimental activation cross section data of proton and deuteron induced reactions on platinum for production of medically relevant radioisotopes, for thin layer activation technology and for generally used activation data libraries [1,2,3,4,5,6] were reported by us earlier. The goal of this study was to clear the disagreement in our earlier measurement, and to compare all data with different theoretical models. Furthermore, we could determine new cross section for some reactions.

In our last investigation up to 50 MeV [6] we could deduce only scattered data for some activation products and for some reactions significant disagreements were found with earlier experimental data. During a recent investigation of activation data on other target elements we got the possibility to obtain additional results on platinum in the 35–49 MeV energy range. In the present report, we compare the new data with the literature values and in order to investigate the prediction capability of theoretical results we made calculations with the ALICE and EMPIRE codes, modified to better represent deuteron induced reactions.

EXPERIMENTAL DETAILS AND DATA ANALYSIS

Elemental experimental cross sections were determined by using the activation method, stacked foil irradiation technique and HPGe gamma ray spectrometry. Cross-section data were deduced relative to the excitation functions of the $^{27}Al(d,x)^{22,24}Na$ monitor reactions [7] re-measured in the whole covered energy range.

The irradiation was made at an external beam line of the Cyclone 90 cyclotron of the Université Catholique in Louvain la Neuve (LLN) for 60 min with a 50 MeV energy extracted beam having 100 nA beam intensity. The targets were irradiated in a short Faraday cup.

The irradiated stack contained a sequence of 7 blocks of Hf (10.54 μm), Al (49.54 μm), Al (49.54 μm), Pt (19.29 μm), Al (49.54 μm), Al (49.54 μm), CuMnNi alloy (24.73 μm), Al (49.54 μm), Al (49.54 μm) followed by 13 blocks of Hf (10.54 μm) Al (10.85 μm), Al (49.54 μm), CuMnNi alloy

(24.73 μm) and Al (10.85 μm). The 7 Pt targets covered the 49–35 MeV energy range.

Four series of gamma-ray spectra were measured to follow the decay, starting at 7.9–8.7 h, 21.3–25.1 h, 187.2–198.2 h and 2811.1–3022.9 h after the end of bombardment, respectively. Further experimental details are given in our previous work on deuteron induced reactions on manganese [8], which was irradiated in the same stack.

Gamma spectra were evaluated by automatic fitting algorithm included in the Genie 2000 package or in an iterative process using the Forgamma [9, 10] code. For data evaluation, NUDAT 2.6 [11] decay data were used (Table 1). As natural platinum consists of several stable isotopes with following abundance (%): ^{190}Pt–0.01, ^{192}Pt–0.79, ^{194}Pt–32.9, ^{195}Pt–33.8, ^{196}Pt–25.3, ^{198}Pt–7.2, so called elemental cross sections were deduced.

Table 1: Decay characteristic of the investigated reaction products and Q-values of the contributing processes

Nuclide spin/parity decay level energy	Half-life	$E_{\gamma(keV)}$	I (%)	Contributing process	*Q*-value (MeV)
^{191g}Au 3/2$^+$ ε: 100%	3.18 h	277.86 399.84 478.04 586.44 674.22	6.4 4.2 3.5 15.0 6.0	^{190}Pt(d,n) ^{192}Pt(d,3n) ^{194}Pt(d,5n) ^{195}Pt(d,6n) ^{196}Pt(d,7n) ^{198}Pt(d,9n)	1.55 −13.56 −28.17 −34.28 −42.20 −55.60
^{192}Au 1$^-$ ε: 100%	4.94 h	295.96 308.46 316.51 612.46	23 3.5 59 4.4	^{192}Pt(d,2n) ^{194}Pt(d,4n) ^{195}Pt(d,5n) ^{196}Pt(d,6n) ^{198}Pt(d,8n)	−6.52 −21.14 −27.24 −35.16 −48.56
^{193g}Au 3/2$^+$ ε: 100%	17.65 h	112.52 173.52 255.57 268.22 439.04	2.2 2.7 6.2 3.6 1.78	^{192}Pt(d,n) ^{194}Pt(d,3n) ^{195}Pt(d,4n) ^{196}Pt(d,5n) ^{198}Pt(d,7n)	−6.52 −12.43 −18.54 −26.46 −39.86
^{194}Au 1$^-$ ε: 100%	38.02 h	293.55 328.46 1468.88	10.58 60.4 6.61	^{194}Pt(d,2n) ^{195}Pt(d,3n) ^{196}Pt(d,4n) ^{198}Pt(d,6n)	−5.56 −11.66 −19.58 −32.98
^{195g}Au 3/2$^+$ ε: 100%	186.01 d	98.86 129.74	11.21 0.84	^{194}Pt(d,n) ^{195}Pt(d,2n) ^{196}Pt(d,3n) ^{198}Pt(d,5n)	2.87 −3.23 −11.15 −24.56

$^{196m2}Au$ 12^- IT: 100% 595.66 keV	9.6 h	147.81 168.37 188.27	43.5 7.8 30.0	$^{195}Pt(d,n)$ $^{196}Pt(d,2n)$ $^{198}Pt(d,4n)$	3.41 −4.51 −17.92
^{196g}Au 2^- β^-: 7.0% ε: 93.0%	6.1669 d	333.03 355.73	22.9 87	$^{195}Pt(d,n)$ $^{196}Pt(d,2n)$ $^{198}Pt(d,4n)$	3.41 −4.51 −17.92
^{198m}Au 12^- IT: 100% 811.7 keV	2.272 d	97.21 180.31 204.10 214.89 333.82	69 49 39 77.3 18	$^{198}Pt(d,2n)$	−3.33
^{198g}Au 2^- β^-: 100%	2.6947 d	411.80	95.62	$^{198}Pt(d,2n)$	−3.33
^{199}Au $3/2^+$ β^-: 100%	3.139 d	158.38 208.20	40.0 8.72	$^{198}Pt(d,n)$	4.254
^{188}Pt 0^+ **α: 2.6E** − 5% ε: 99.9999%	10.2 d	187.59 195.05 381.43 423.34	19.4 18.6 7.5 4.4	$^{190}Pt(d,p3n)$ $^{192}Pt(d,p5n)$ $^{194}Pt(d,p7n)$	−17.86 −32.97 −47.59
^{189}Pt $3/2^-$ ε: 100%	10.87 h	243.50 568.85 721.38	7.0 7.1 9.3	$^{190}Pt(d,p2n)$ $^{192}Pt(d,p4n)$ $^{194}Pt(d,p6n)$ $^{195}Pt(d,p7n)$	−11.14 −26.24 −40.86 −46.96
^{191}Pt $3/2^-$ ε: 100%	2.802 d	96.517 129.42 172.18 351.21 359.90 409.44 538.90	3.28 3.2 3.52 3.36 6.0 8.0 13.7	$^{190}Pt(d,p)$ $^{192}Pt(d,p2n)$ $^{194}Pt(d,p4n)$ $^{195}Pt(d,p5n)$ $^{196}Pt(d,p6n)$	4.22 −10.89 −25.50 −31.60 −39.53
^{193m}Pt $13/2^+$ IT: 100% 149.78 keV	4.33 d	135.50	0.112	$^{192}Pt(d,p)$ $^{194}Pt(d,p2n)$ $^{195}Pt(d,p3n)$ $^{196}Pt(d,p4n)$ $^{198}Pt(d,p6n)$	4.04 −10.58 −16.68 −24.60 −38.01
^{195m}Pt $13/2^+$ IT: 100% 259.29 keV	4.010 d	98.90 129.79	11.7 2.90	$^{194}Pt(d,p)$ $^{195}Pt(d,pn)$ $^{196}Pt(d,p2n)$ $^{198}Pt(d,p4n)$	3.88 −22.25 −10.15 −23.55

^{197m}Pt 13/2^{+} β^{-}: 3.3% IT: 96.7% 399.59 keV	95.41 min	346.5	11.1	^{196}Pt(d,p) ^{198}Pt(d,p2n)	3.62 −9.78
^{197g}Pt 1/2^{+} β^{-}: 100%	19.89 h	191.44	3.7	^{196}Pt(d,p) ^{198}Pt(d,p2n)	3.62 −9.78
^{188}Ir 1^{-} ε: 100%	41.5 h	155.05 477.99 633.02	30 14.7 18	^{190}Pt(d,2p2n) ^{192}Pt(d,2p4n) ^{194}Pt(d,2p6n)	−16.56 −31.67 −46.28
^{189}Ir 3/2^{+} ε: 100%	13.2 d	245.08	6.0	^{190}Pt(d,2pn) ^{192}Pt(d,2p3n) ^{194}Pt(d,2p5n) ^{195}Pt(d,2p6n)	−8.38 −23.49 −38.10 −44.21
^{190m}Ir (11)$^{-}$ IT: 8.6% ε: 91.4% 376.4 keV	3.087 h	186.7 361.2 502.5 616.5	64.2 86.72 89.38 90.14	^{190}Pt(d,2p) ^{192}Pt(d,2p2n) ^{194}Pt(d,2p4n) ^{195}Pt(d,2p5n) ^{196}Pt(d,2p6n)	−2.01 −17.12 −31.73 −37.84 −45.76
^{190}Ir 4^{-} ε: 100%	11.78 d	186.68 361.09 371.24 407.22 518.55 557.95 569.30 605.14	52 13.0 22.8 28.5 34.0 30.1 28.5 39.9	^{190}Pt(d,2p) ^{192}Pt(d,2p2n) ^{194}Pt(d,2p4n) ^{195}Pt(d,2p5n) ^{196}Pt(d,2p6n)	−2.01 −17.12 −31.73 −37.84 −45.76
^{192g}Ir 4^{+} ε: 4.76% β^{-}: 95.24%	73.829 d	295.95 308.46 316.51 468.07 604.41	28.71 29.70 82.86 47.84 8.216	^{192}Pt(d,2p) ^{194}Pt(d,2p2n) ^{195}Pt(d,2p3n) ^{196}Pt(d,2p4n) ^{198}Pt(d,2p6n)	−2.90 −17.51 −23.61 −31.54 −44.94
^{194m2}Ir (10,11) β^{-}: 100% 190.0 keV	171 d	328.5 338.8 390.8 482.6 562.4 600.5 687.8	93.0 55.0 35.0 97.0 35.0 62.0 59.0	^{194}Pt(d,2p) ^{195}Pt(d,2pn) ^{196}Pt(d,2p2n) ^{198}Pt(d,2p4n)	−3.67 −9.78 −17.70 −31.10

The Q values of the contributing reactions (Q value calculator [12]) are also shown in Table 1 to indicate the reaction thresholds.

Uncertainty on cross-sections was determined according to the recommendation given in Ref. [13] by taking the sum in quadrature of all individual linear contributions: beam current (7%), target thickness or homogeneity (5%), detector efficiency (5%), photo peak area determination and counting statistics (1–20%).

The median beam energy in each target foil was obtained by a degradation calculation based on the calibrated primary energy, the stopping powers of Andersen [14] and the fitted monitor reactions (Fig. 1) [15]. Uncertainty of energy was estimated by taking into account cumulative effects during the energy degradation (primary energy, target thickness, energy straggling) and the applied correction to monitor reactions.

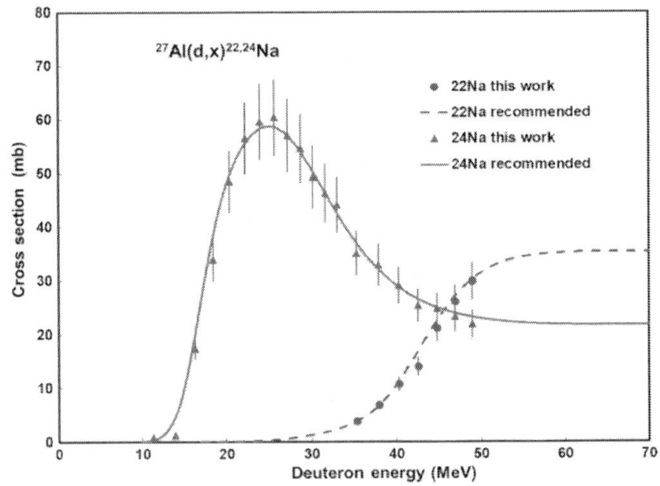

Figure 1: Simultaneously measured experimental data of $^{27}Al(d,x)^{22,24}Na$ monitor reactions in comparison with the recommended values [7]

Abundance of isotopes of natural Pt (%): ^{190}Pt–0.01, ^{192}Pt–0.79, ^{194}Pt–32.9, ^{195}Pt–33.8, ^{196}Pt–25.3, ^{198}Pt–7.2.

The *Q*-values shown in Table 2 refer to formation of the ground state. Decrease *Q*-values for isomeric states with level energy of the isomer.

Table 2: The measured cross sections for the $^{nat}Pt(d,xn)^{191,192,193,194,195,196m2,196g,198m,198g,199}Au$, $^{nat}Pt(d,x)^{189,191,193m,195m,197g}Pt$ and $^{nat}Pt(d,x)^{189,190m,190,192,194m2}Ir$ reactions

E (MeV)	ΔE (MeV)	^{199}Au		^{198m}Au		^{198g}Au		^{196m2}Au		^{196g}Au		^{195}Au		^{194}Au	
		σ (mb)	Δσ (mb)	σ (mb)	Δσ (mb)	σ (mb)	Δσ (mb)	σ (mb)	Δσ (mb)	σ (mb)	Δσ (mb)	σ (mb)	Δσ (mb)	σ (mb)	Δσ (mb)
48.5	0.3	6.98	0.98	1.12	0.19	2.61	0.34	5.38	0.73	32	4.13	74.80	9.64	184.62	23.80
46.5	0.3	4.88	0.64	0.80	0.13	2.35	0.31	5.37	0.75	31.13	4.02	77.01	9.93	161.87	20.87
44.3	0.4	6.28	0.95	1.46	0.25	2.72	0.36	5.80	0.78	36.22	4.68	95.75	12.34	181.08	23.34
42.1	0.4	6.26	0.90	1.11	0.18	3.49	0.45	7.57	1.01	44.57	5.75	118.04	15.21	212.52	27.39
39.8	0.5	6.38	0.90	1.18	0.18	3.68	0.48	9.54	1.26	49.41	6.37	122.26	15.76	240.05	30.94
37.3	0.5	8.03	1.12	1.37	0.20	4.01	0.52	12.49	1.63	59.5	7.67	123.93	15.97	307.49	39.63
34.8	0.5	4.12	0.60	1.38	0.20	3.71	0.48	14.97	1.95	65.75	8.48	101.08	13.05	313.60	40.41
E (MeV)	ΔE (MeV)	^{193}Au		^{192}Au		^{191}Au		^{197g}Pt		^{195m}Pt		^{193m}Pt		^{191}Pt(cum)	
		σ (mb)	Δσ (mb)	σ (mb)	Δσ (mb)	σ (mb)	Δσ (mb)	σ (mb)	Δσ (mb)	σ (mb)	Δσ (mb)	σ (mb)	Δσ (mb)	σ (mb)	Δσ (mb)
48.5	0.3	207.38	26.73	399.97	51.63	363.64	47.20	28.67	3.76	63.75	7.16	202.34	29.54	341.19	38.42
46.5	0.3	218.67	28.25	358.37	46.28	286.30	37.27	27.81	4.29	60.54	6.80	140.49	22.85	261.64	29.52
44.3	0.4	266.44	34.40	403.86	52.14	254.97	33.29	27.92	3.70	67.39	7.58			235.37	26.56
42.1	0.4	283.00	36.57	450.46	58.14	215.68	28.21	25.19	3.73	77.89	8.75	99.75	20.86	205.35	23.20
39.8	0.5	321.22	41.47	440.24	56.82	142.41	18.78	33.06	4.26	78.67	8.84	132.34	23.39	136.42	15.48
37.3	0.5	330.04	42.68	434.29	56.06	66.76	9.20	35.15	4.49	77.81	8.74	63.59	14.57	76.57	8.83
34.8	0.5	284.24	36.70	310.24	40.07	7.80	2.22	27.01	3.63	67.34	7.57			85.78	9.77

E (MeV)	ΔE (MeV)	^{189}Pt		^{194m2}Ir		^{192}Ir		^{190m2}Ir		^{190}Ir		^{189}Ir	
		σ (mb)	Δσ (mb)	σ (mb)	Δσ (mb)	σ (mb)	Δσ (mb)	σ (mb)	Δσ (mb)	σ (mb)	Δσ (mb)	σ (mb)	mb
48.5	0.3	10.81	1.91	0.30	0.04	2.82	0.32	0.64	0.17	4.29	0.49	11.99	1.46
46.5	0.3	9.21	1.61	0.21	0.03	2.51	0.28	0.90	0.19	3.70	0.43	9.10	1.14
44.3	0.4	9.79	1.78	0.15	0.03	2.70	0.31	0.83	0.20	3.36	0.40	7.22	1.09
42.1	0.4			0.24	0.03	2.97	0.33	0.67	0.18	3.58	0.42	6.08	0.86
39.8	0.5			0.15	0.02	2.91	0.33			2.92	0.35	4.43	0.72
37.3	0.5					3.15	0.36			3.04	0.35	1.93	0.47
34.8	0.5					2.55	0.29			1.84	0.22		

When complex particles are emitted instead of individual protons and neutrons the *Q*-values have to be decreased by the respective binding energies of the compound particles: pn $\rightarrow$ d + 2.2 MeV, p2n $\rightarrow$ t + 8.5 MeV, 2pn $\rightarrow$ ^{3}He + 7.7 MeV, 2p2n $\rightarrow$ α + 28.3 MeV.

THEORETICAL CALCULATIONS

The cross sections of the investigated reactions were calculated using the pre-compound model codes ALICE-IPPE [16] and EMPIRE-II [17] modified for deuterons by Ignatyuk (D-versions) [18]. These modified codes named ALICE-D and EMPIRE-D were developed by including a simulation of direct (d, p) and (d, t) transitions with the general relations for a nucleon transfer probability in the continuum through an energy dependent enhancement factor for the corresponding transitions. The theoretical curves were determined using one recommended input data-set [19] without any optimization or adjustment of parameters to the individual reactions or stable target isotopes. Independent data for isomers with ALICE-D code were obtained by using the isomeric ratios calculated with EMPIRE-D.

The experimental data are also compared with the cross section data reported in the TENDL-2017 [20] on-line nuclear reaction data library. The TENDL-2017 library is based on both default and adjusted TALYS 1.9 model calculations [21].

We had many problems with these calculations. At the beginning the optical potential for deuterons was taken in accordance with the default RIPL recommendations [22]. As for this potential all Empire results were much higher than he ALICE results, we decided to make new calculations with the Perry potential [23], which we used earlier for the ^{197}Au target [24]. This resulted in certainly more consistency. However, the main problems relate to the levels schemes, which are poor for the most of Pt and Ir isotopes. There are too many low-value spin- and a strong deficit for high-spin levels. So, the calculated isomeric ratios are too low for most cases. In principle some corrections for the spin distributions can be made to get better agreement.

RESULTS AND DISCUSSION

The measured experimental cross-section data are shown in Figs. 2, 3, 4, 5, 6, 7, 8, 9, 10, 11, 12, 13, 14, 15, 16, 17, 18, 19, 20, 21, 22 and 23. The numerical values are presented in Table 2. The investigated radio-products can be produced directly, through the internal transition of a long-lived isomeric state and through the decay of an isobaric parent. We made theoretical model calculations also for production of a few radioisotopes measured by us earlier, but no new experimental data are presented in this study. Agreement of the available experimental data and the theoretical data can clearly be seen in the figures of the present work. The comparisons result in different conclusions for each reaction, which is hard to explain. Some more detailed discussions are available in our previous work [6]. Therefore, in the summary we present only general conclusion on the predictive capability of the model calculations.

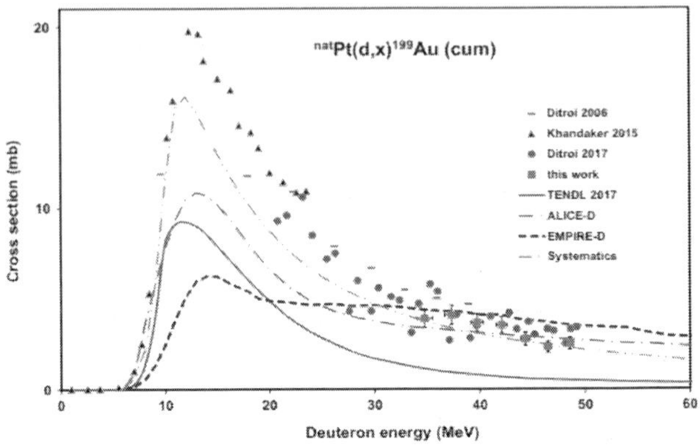

Figure 2: Excitation function of the $^{nat}Pt(d,x)^{199}Au$(cum) reaction in comparison with literature and theoretical values.

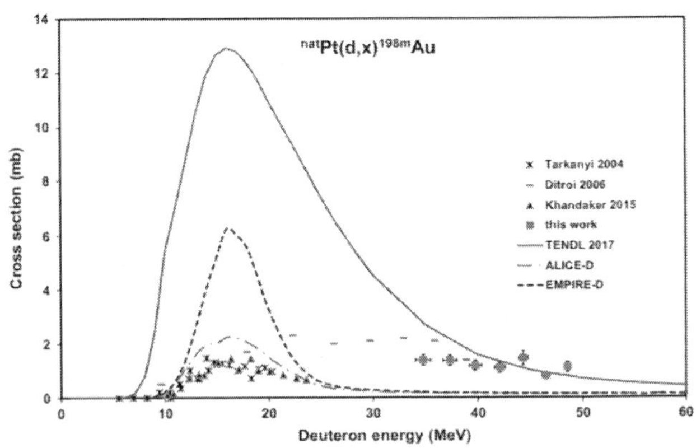

Figure 3: Excitation function of the $^{nat}Pt(d,x)^{198m}Au$ reaction in comparison with literature and theoretical values.

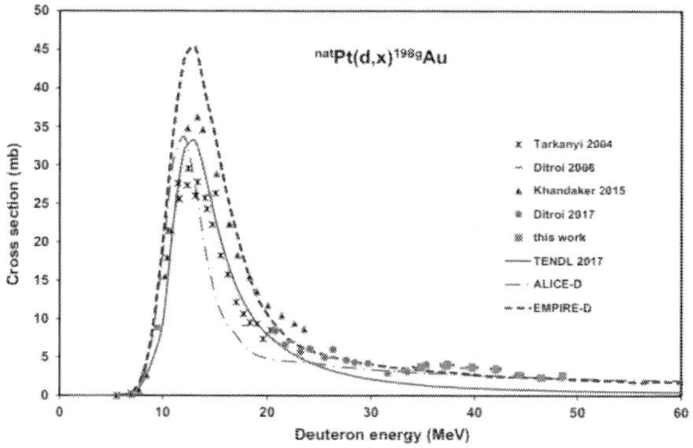

Figure 4: Excitation function of the $^{nat}Pt(d,xn)^{198g}Au$ reaction in comparison with literature and theoretical values.

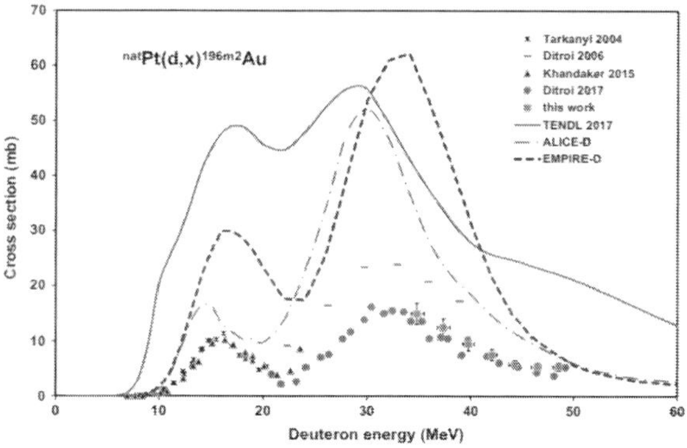

Figure 5: Excitation function of the $^{nat}Pt(d,xn)^{196m2}Au$ reaction in comparison with literature and theoretical values.

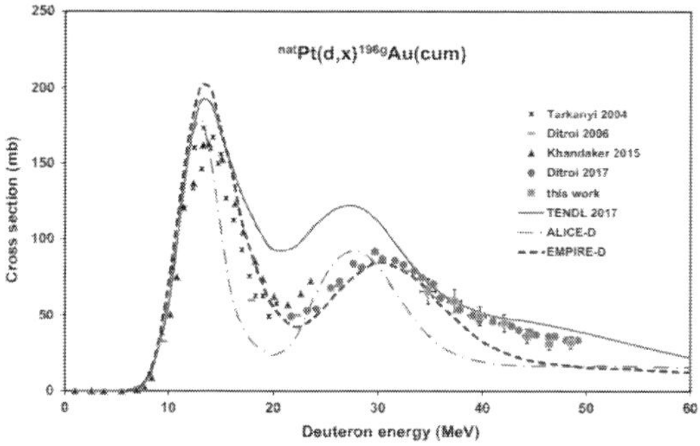

Figure 6: Excitation function of the $^{nat}Pt(d,xn)^{196g}Au$ (cum) reaction in comparison with literature and theoretical values.

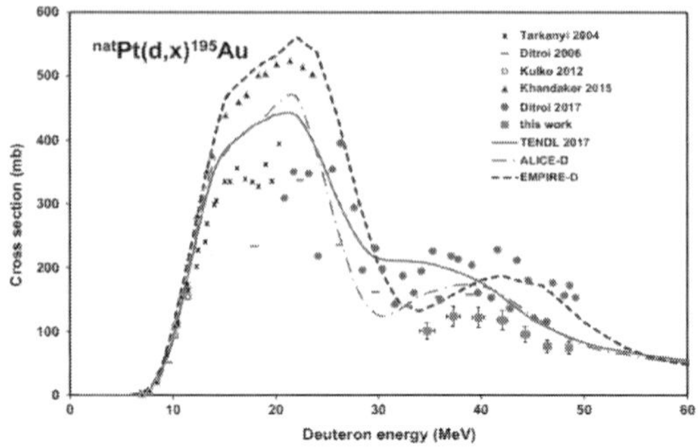

Figure 7: Excitation function of the $^{nat}Pt(d,xn)^{195}Au$ (cum) reaction in comparison with literature and theoretical values.

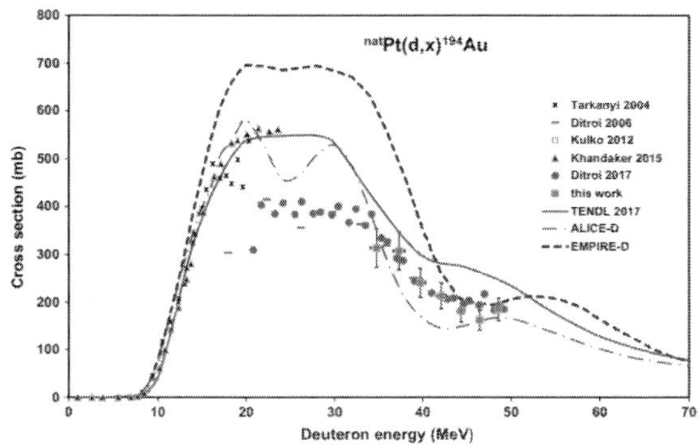

Figure 8: Excitation function of the $^{nat}Pt(d,xn)^{194}Au$ reaction in comparison with literature and theoretical values.

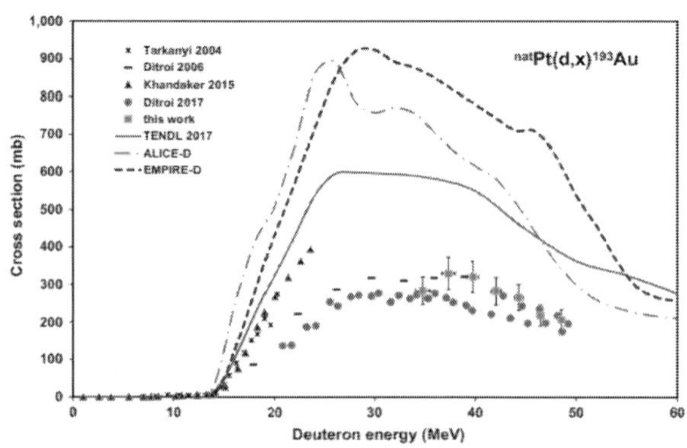

Figure 9: Excitation function of the $^{nat}Pt(d,xn)^{193}Au$(cum) reaction in comparison with literature and theoretical values.

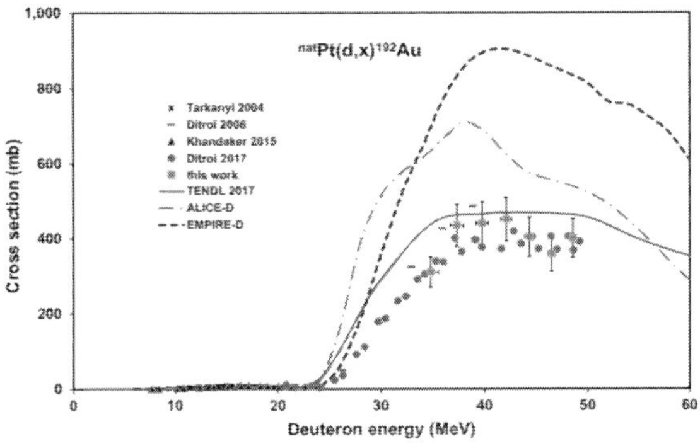

Figure 10: Excitation function of the $^{nat}Pt(d,xn)^{192}Au$ reaction.

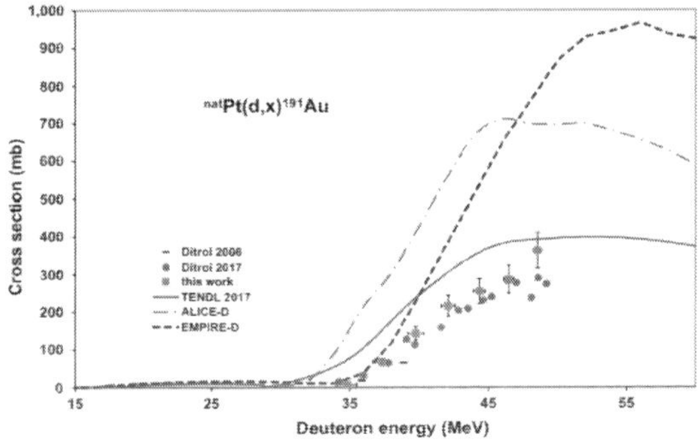

Figure 11: Excitation function of the $^{nat}Pt(d,xn)^{191}Au$ reaction in comparison with literature and theoretical values.

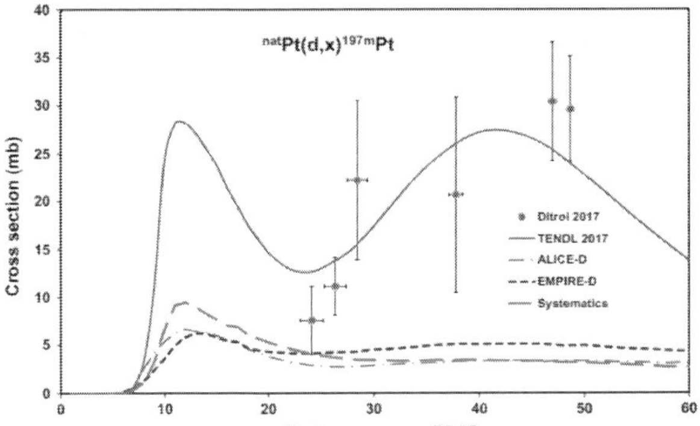

Figure 12: Excitation function of the $^{nat}Pt(d,xn)^{197m}Pt$ reaction in comparison with theoretical values.

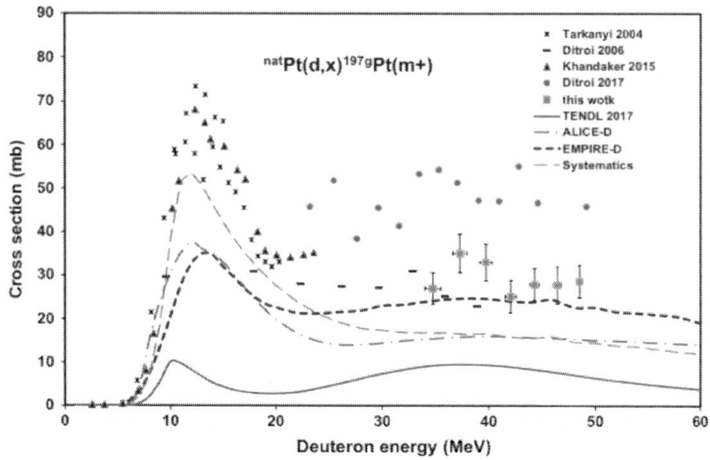

Figure 13: Excitation function of the $^{nat}Pt(d,x)^{197g}Pt(m+)$ (cum) reaction in comparison with literature and theoretical values.

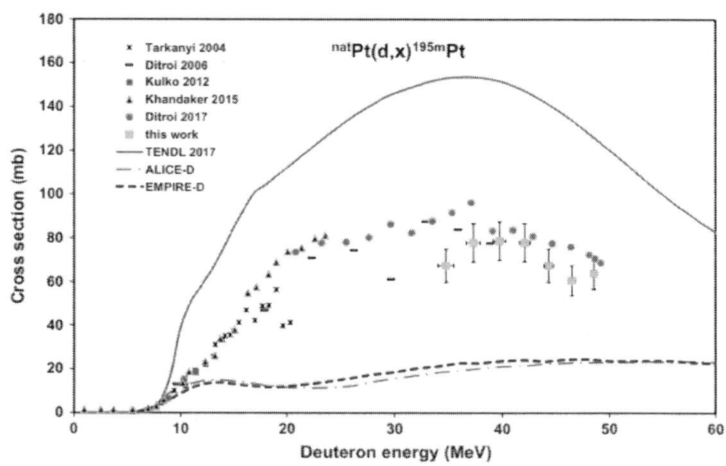

Figure 14: Excitation function of the $^{nat}Pt(d,xn)^{195m}Pt$ reaction in comparison with literature and theoretical values.

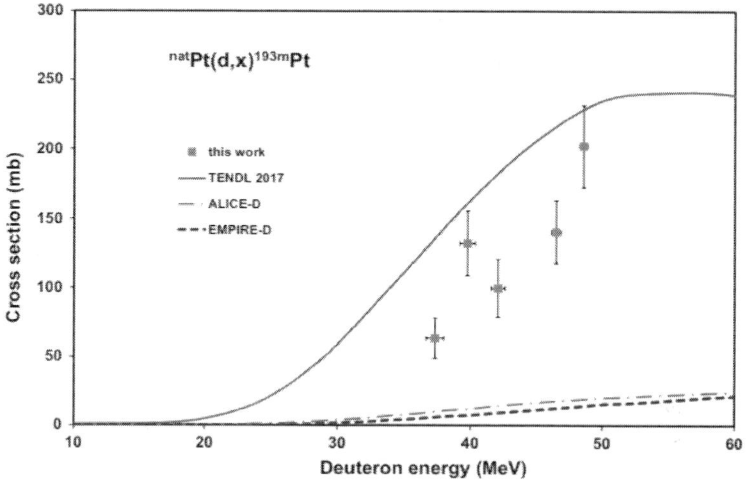

Figure 15: Excitation function of the $^{nat}Pt(d,xn)^{193m}Pt$ reaction in comparison with theoretical values.

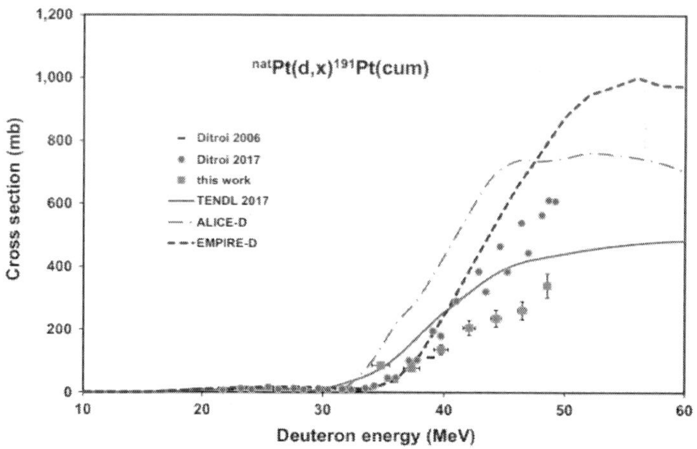

Figure 16: Excitation function of the $^{nat}Pt(d,xn)^{191}Pt$ reaction in comparison with literature and theoretical values.

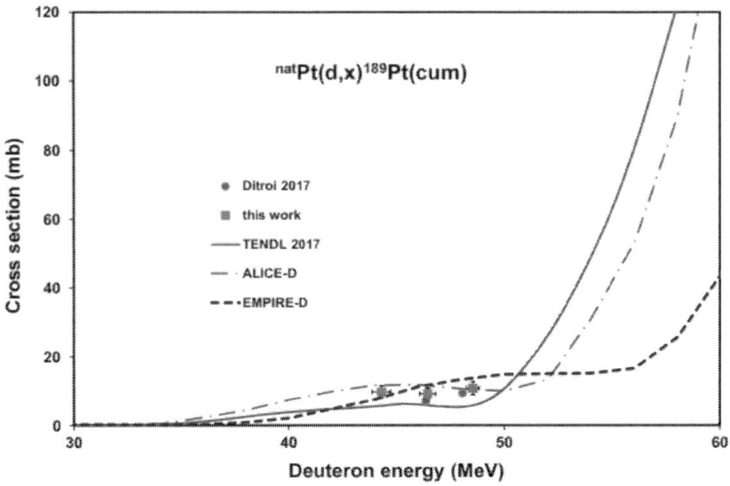

Figure 17: Excitation function of the $^{nat}Pt(d,xn)^{189}Pt$ reaction in comparison with literature and theoretical values.

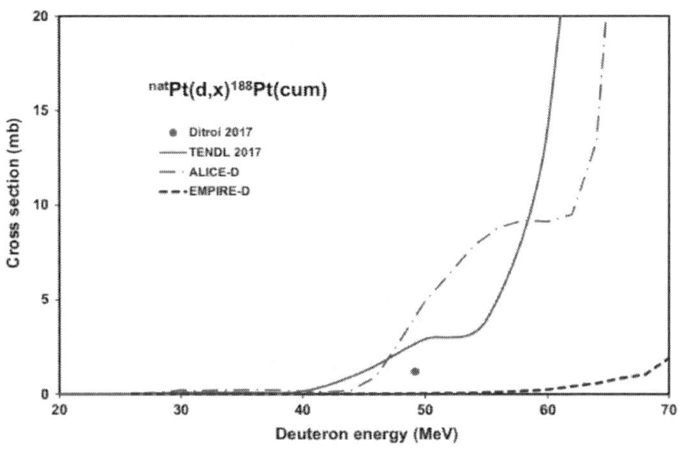

Figure 18: Excitation function of the $^{nat}Pt(d,xn)^{188}Pt$ reaction in comparison with theoretical values.

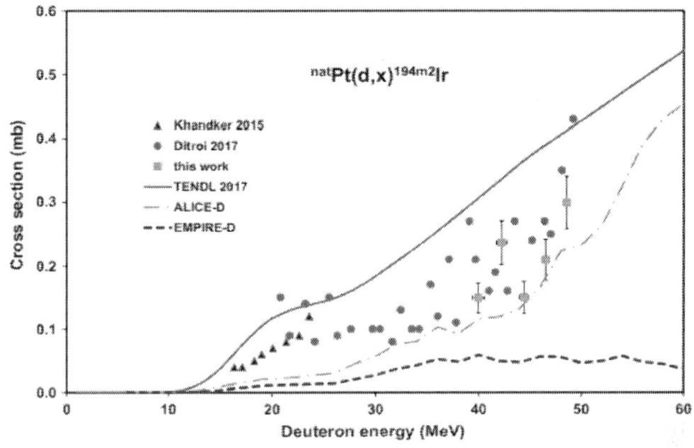

Figure 19: Excitation function of the $^{nat}Pt(d,xn)^{194m2}Ir$ reaction in comparison with literature and theoretical values.

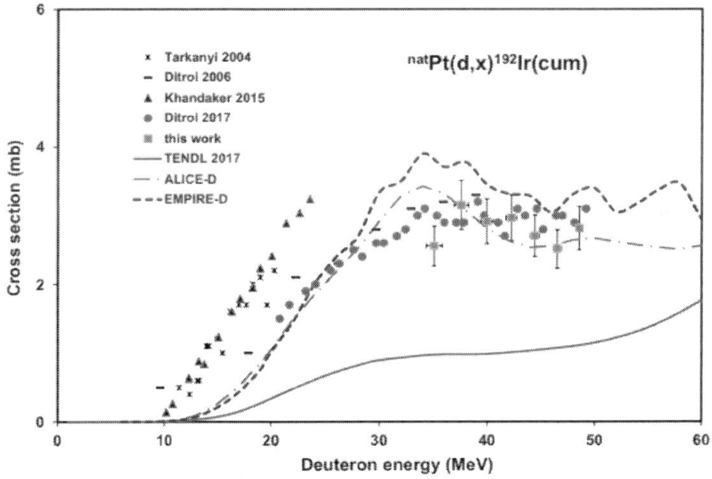

Figure 20: Excitation function of the $^{nat}Pt(d,xn)^{192}Ir$ reaction in comparison with literature and theoretical values.

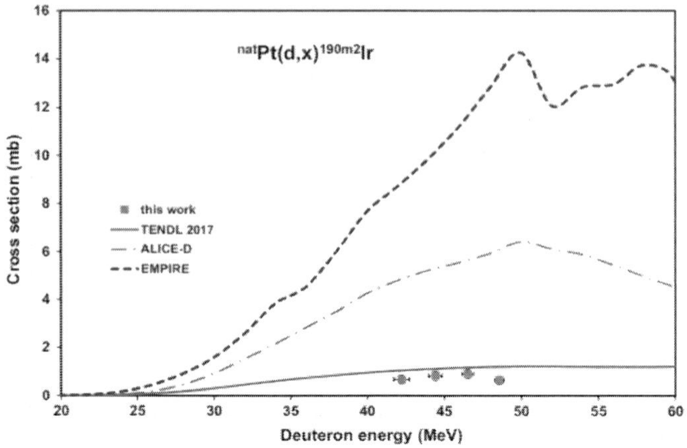

Figure 21: Excitation function of the $^{nat}Pt(d,xn)^{190m2}Ir$ reaction in comparison with theoretical values.

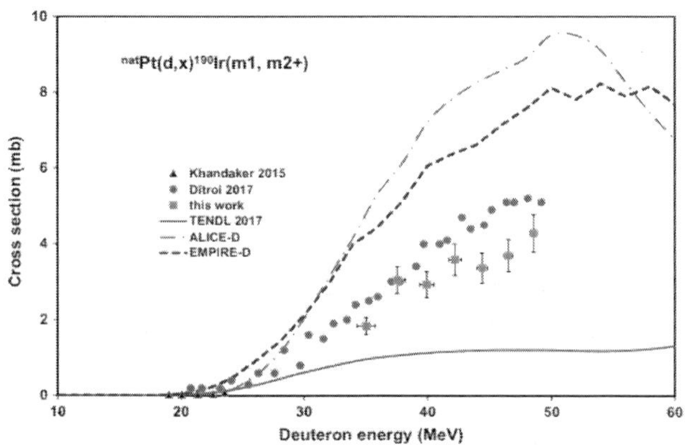

Figure 22: Excitation function of the $^{nat}Pt(d,xn)^{190}Ir$ reaction in comparison with literature and theoretical values.

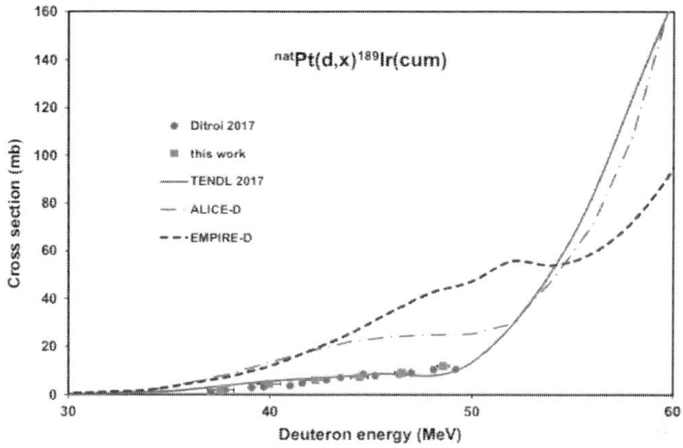

Figure 23: Excitation function of the $^{nat}Pt(d,xn)^{189}Ir$ reaction in comparison with literature and theoretical values.

Radioisotopes of Gold

^{nat} Pt(d,x) ^{199} Au(cum) reaction

The measured cross-sections of the ^{199}Au ($T_{1/2}$ = 3.139 d) are cumulative, produced directly through the $^{198}Pt(d,n)$ reaction and contribution of the $^{-}$-decay of the shorter-lived parent ^{199}Pt ($T_{1/2}$ = 30.8 min) produced through the $^{198}Pt(d,p)$ process (Fig. 2). A good agreement with the earlier experimental studies [6, 7] in the overlapping energy interval is seen. Because none of the theoretical model code gave acceptable predictions for the experimental values an estimated prediction from the systematics has also be presented in Fig. 2, which is proved to be the best approximation both in trend and in values.

^{nat} Pt(d,x) ^{198m} Au reaction

We could obtain cross section data for both longer-lived states of ^{198}Au. The cross sections for the ^{198m}Au ($T_{1/2}$ = 2.272 d, IT 100%) isomeric state are shown in Fig. 3. An acceptable continuation to higher energy of the data of Ditroi et al. [4] is seen. All the three model codes predict a large peak around 16–17 MeV, which was not confirmed by any of the experiments.

^{nat}*Pt(d,xn)* ^{198g}*Au reaction*

Independent, low, cross sections for direct production of the ground state of ^{198}Au ($T_{1/2}$=2.6947 d) are shown in Fig. 4 and are in acceptable agreement with the earlier values. The trends given by all model codes are good. The best value is estimated by the EMPIRE-D above 20 MeV, but it overestiimates strongly around the maximum.

^{nat}*Pt(d,xn)* ^{196m2}*Au reaction*

Direct production cross-sections for the $T_{1/2}$=9.6 h isomeric state ^{196m2}Au are shown in Fig. 5 and agree well with the Ditroi et al. [6] data. All theoretical model codes recognize the two maxima, but all overestimate the experimental values.

^{nat}*Pt(d,xn)* ^{196g}*Au(cum) reaction*

The cross sections for cumulative formation of the ground state ^{196g}Au ($T_{1/2}$=6.1669 d) measured after the complete decay of the two higher laying, short-lived, isomeric states (m2–$T_{1/2}$=9.6 h, m1–$T_{1/2}$=8.1 s) are shown in Fig. 6. Very good agreement with Ditroi et al. [6] was observed. All model codes describe the shape of the excitation function almost correctly, the best value approximation is given by the EMPIRE-D code.

^{nat}*Pt(d,xn)* ^{195}*Au(cum) reaction*

The cross-sections for formation of ^{195}Au ($T_{1/2}$=186.01 d) contain the decay of the $T_{1/2}$=30.5 s isomeric state (Fig. 7) and are lower than the rather scattered data of Ditroi et al. [6]. The theoretical codes follow the trend of the experimental curves, but give different value predictions.

^{nat}*Pt(d,xn)* ^{194}*Au reaction*

The excitation functions for production of the directly produced ^{194}Au ($T_{1/2}$=38.02 h) are shown in Fig. 8 and are in perfect agreement with Ditroi et al. [6]. The theoretical model codes give different values around the maxima, their approximation is acceptable only below 20 MeV.

^{nat}*Pt(d,xn)* ^{193g}*Au(cum) reaction*

The measured cross-sections of ^{193g}Au ($T_{1/2}$=17.65 h) include the internal decay of the $T_{1/2}$=3.9 s isomeric state (Fig. 9) additional to the direct production. The new measurements are about 20% higher than a part of the values in Ditroi et al. [6] but seem to agree better with the 2006 values [4]. All the theoretical model codes strongly overestimate the experimental values.

^{nat}Pt(d,xn) ^{192}Au reaction

The activation cross sections for production of ^{192}Au ($T_{1/2}$=4.94 h) are shown in Fig. 10 and are between our earlier values given in Refs. [4] and [6]. The best prediction is given by the TENDL-2017, both other model codes strongly overestimate the experimental values.

^{nat}Pt(d,xn) ^{191}Au reaction

The cross section data of ^{191}Au ($T_{1/2}$=3.18 h) include the contribution from the decay of the simultaneously produced short half-life isomeric state ($T_{1/2}$=0.92 s, 100% IT) (Fig. 11). Good overall agreement with Ditroi et al. [6]. Only TENDL-2017 follows correctly the trend of the experimental values with considerable overestimation.

Radioisotopes of Platinum

^{nat}Pt(d,x) ^{197m}Pt reaction

We could not deduce cross section data for production of the ^{197m}Pt ($T_{1/2}$=95.41 min), due to the long cooling time before measuring the first spectra. The earlier experimental data and the theoretical predictions are shown in Fig. 12. In spite of the large experimental uncertainties the best approximation is given by the TENDL-2017.

^{nat}Pt(d,x) ^{197g}Pt(m+) (cum) reaction

The measured production cross-section of ^{197g}Pt ($T_{1/2}$= 19.8915 h) is including the direct production, the contribution from internal transition ($T_{1/2}$=95.41 min, IT: 96.7) of the isomeric state and from the β^- decay of the short-lived ^{197m,g}Ir isomers ($T_{1/2}$=5.8 min and 8.9 min, respectively) (Fig. 13). Our new measurements are in better agreement with the Ditroi et al. [4] values than with the results in Ditroi et al. [6]. All the three theoretical model codes show the trend of the experimental values, but strongly underestimate them. Only the prediction of EMPIRE-D is closed to the experimental results above 25 MeV. Even the systematics cannot reconstruct the experimental results.

^{nat}*Pt(d,x)* ^{195m}*Pt reaction*

The excitation function of the high spin isomeric state ($T_{1/2}$ = 4.02 d) of ^{195}Pt is shown in Fig. 14 and confirm overall the previously measured values of [4] and [6]. All theoretical predictions follow the trend of the experimental values, but TENDL strongly overestimates, while ALICE-D and EMPIRE-D giving approximately the same underestimated values.

^{nat}*Pt(d,x)* ^{193m}*Pt reaction*

The cross sections for production of ^{193m}Pt ($T_{1/2}$ = 4.33 d, IT: 100%) are shown in Fig. 15. They include the contribution of 0.03% EC decay of ^{193m}Au ($T_{1/2}$ = 3,9 s). No earlier experimental data were found. Only the TENDL-2017 prediction gives an acceptable trend but with slight overestimation.

^{nat}*Pt(d,x)* ^{191}*Pt(cum) reaction*

The cumulative cross-sections for production of ^{191}Pt ($T_{1/2}$ = 2.802 d) includes direct production and formation through decay of ^{191}Au ($T_{1/2}$ = 3.18 h) and is shown in Fig. 16. Above 40 MeV the new values are significantly lower than the Ditroi et al. [6] data. The three theoretical model calculations give different results, the best approximation is given by the EMPIRE-D up to 40 MeV, but above this energy the TENDL-2017 predictions are better.

^{nat}*Pt(d,x)* ^{189}*Pt(cum) reaction*

The cross-sections for production of ^{189}Pt ($T_{1/2}$ = 11 h) are cumulative, measured after the complete decay of the two longer-lived states of the parent ^{189}Au ($T_{1/2}$ = 4.6 min and 28.3 min) radioisotope (Fig. 17). Good agreement of the few data points with the earlier measurement [6] was found. In this energy range all the three theoretical codes give acceptable predictions.

^{nat}*Pt(d,x)* ^{188}*Pt(cum) reaction*

Only a single cross section point for formation of ^{188}Pt ($T_{1/2}$ = 10.2 d) is available, reported in earlier work [6] (Fig. 18). The cross-section contains the contribution of the direct production and the simultaneous decay of the ^{188}Au ($T_{1/2}$ = 8.8 min) parent. The theoretical model codes give very different predictions.

Radioisotopes of Iridium

^{nat}Pt(d,x) ^{194m2}Ir reaction

The measured spectra allow to get cross sections only for the second, long-lived ($T_{1/2}$ = 171 d) metastable state (Fig. 19) out of the three longer-lived states of ^{194}Ir. Overall agreement between the new values and Ditroi et al. [6] was seen. TENDL-2017 and ALICE-D give similar trends, but with over- and underestimation. The EMPIRE-D prediction is much lower.

^{nat}Pt(d,x) ^{192}Ir(cum) reaction

Our experimental data (Fig. 20) include the direct production of the ground state ($T_{1/2}$ = 3.8 d) and the decay of the short-lived isomeric state ^{192m1}Ir ($T_{1/2}$ = 1.45 min). Overall agreement between the new values and Ditroi et al. [6] was seen. ALICE-D gives good approximation of the measured values in the whole energy region, while EMPIRE-D slightly overestimates. TENDL-2017 gives strong underestimated prediction.

^{nat}Pt(d,x) ^{190m2}Ir reaction

The production cross sections of the second, high spin isomeric state ^{190m2}Ir ($T_{1/2}$ = 3.25 h) are shown in Fig. 21 and agree well with the TENDL-2017 prediction while the values of the other codes are 4–10 times too high. No earlier experimental data are available in the literature.

^{nat}Pt(d,x) ^{190}Ir(m+) reaction

The measured cross sections of the ground state ($T_{1/2}$ = 11.8 d) include, in addition to the direct production, the decay of the first metastable state ^{190m1}Ir ($T_{1/2}$ = 1.2 h, IT 100%) and the second high-spin isomeric state ^{190m2}Ir ($T_{1/2}$ = 3.25 h, IT 5.6%) (Fig. 22). Our new values are about 10% lower than Ditroi et al. [6]. TENDL-2017 underestimates, while EMPIRE-D and ALICE-D overestimate the experimental values.

^{nat}Pt(d,x) ^{189}Ir(cum) reaction

The cross-section determined for ^{189}Ir ($T_{1/2}$ = 13.2 d) is cumulative and contains the direct production and the contribution from the decay of ^{189}Pt ($T_{1/2}$ = 10.87 h) (Fig. 23). Perfect agreement with Ditroi et al. [6] was found. In this energy region the TENDL-2017 prediction is acceptable, while the ALICE-D and EMPIRE-D results overestimate the experimental values.

***nat* Pt(d,x) *188* Ir(cum) reaction**

The cross-section determined for ^{188}Ir ($T_{1/2}$=41.5 h) is cumulative and contains the direct production and the contribution from the decay of ^{188}Pt ($T_{1/2}$ = 10.2 d) (Fig. 24). In this energy region the TENDL-2017 prediction is acceptable with slight overestimation, while the ALICE-D and EMPIRE-D results overestimate the experimental values.

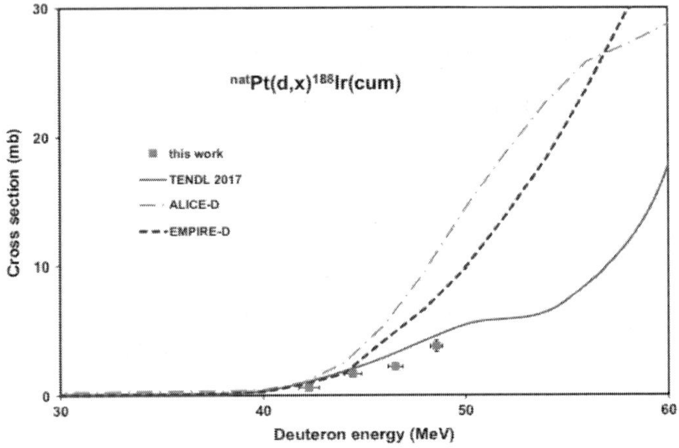

Figure 24: Excitation function of the ^{nat}Pt(d,xn)^{188}Ir reaction in comparison with literature and theoretical values.

SUMMARY AND CONCLUSION

We report on new experimental cross sections leading to formation of ^{nat}Pt (d,xn)^{191,192,193,194,195,196m2,196g,198m,198g,199}Au, ^{nat}Pt(d,x)^{189,191,193m,195m,197g}Pt and ^{nat}Pt (d,x)^{188,189,190m2,190,192,194m2}Ir in the 35–49 MeV energy range. For production of ^{193m}Pt and ^{188,190m}Ir no earlier experimental data were found. A partial or good agreement was found with the results of our previous studies in the overlapping energy regions, but some cases there are large disagreements. The comparison with the theoretical predictions of TENDL-2017 shows moderate agreement or large disagreement especially for isomeric cross sections. The situation is nearly same in the case of ALICE-D and EMPIRE-D, which underline the importance of experimental data that can help to still improve the codes and models, especially for deuteron induced reactions.

13

Production of Medically Useful Bromine Isotopes via Alpha-particle induced Nuclear Reactions

Katharina Breunig[1], Bernhard Scholten[1], Ingo Spahn[1], Alex Hermanne[2], Stefan Spellerberg[1], Heinz H. Coenen[1], and Bernd Neumaier[1]

[1] Forschungszentrum Julich, Institute of Neuroscience and Medicine, INM-5: Nuclear Chemistry, 52428 J ¨ ulich, Germany ¨
[2] Vrije Universiteit Brussel (VUB), Cyclotron Laboratory, 1050 Brussels, Belgium

ABSTRACT

The cross sections of α-particle induced reactions on arsenic leading to the formation of $^{76,77,78}Br$ were measured from their respective thresholds up to 37 MeV. Thin sediments of elemental arsenic powder were irradiated together with Al degrader and Cu monitor foils using the established stacked-foil technique. For determination of the effective α-particle energies and of the effective beam current through the stacks the cross-section ratios of the monitor nuclides $^{67}Ga/^{66}Ga$ were used. This should help resolve

Citation: Katharina Breunig, Bernhard Scholten, Ingo Spahn, Alex Hermanne, Stefan Spellerberg, Heinz H. Coenen and Bernd Neumaier. "Production of medically useful bromine isotopes via alpha-particle induced nuclear reactions" EPJ Web Conf., 146 (2017) 08006 DOI: https://doi.org/10.1051/epjconf/201714608006

discrepancies in existing literature data. Comparison of the data with the available excitation functions shows some slight energy shifts as well as some differences in curve shapes. The calculated thick target yields indicate, that ^{77}Br can be produced in the energy range $E_\alpha = 25 \rightarrow 17$ MeV free of isotopic impurities in quantities sufficient for medical application.

INTRODUCTION

Halogens have been shown to be useful for labelling molecules for diagnosis and therapy in nuclear medicine. One of the main radionuclides is ^{18}F with 110 min half-life for Positron Emission Tomography (PET) studies, besides several radioiodine isotopes used in diagnosis and therapy.

Fluorine has the strongest carbon-halogen bond, apart from that use of ^{18}F is limited to PET and the radionuclide has a too short half-life to study slow metabolic processes, i.e. of proteins and peptides. Iodine on the other side has isotopes with different decay characteristics suitable for a broader spectrum of medical applications but the chemical bonding is sometimes too weak and the labelled molecules are then not stable enough in vivo.

Bromine radionuclides can be an alternative because their bonding in molecules is stronger than that of iodine; furthermore have interesting decay modes for medical applications. [cf. 1-3] The "non-standard" positron emitters ^{75}Br ($T_{1/2}$ = 96.7 min) and ^{76}Br ($T_{1/2}$ = 16.2 h) could be used for medical application.

The ^{77}Br, decaying by electron capture with a half-life of 57 h, can be used in Auger-therapy with the opportunity to follow the radioactivity distribution in the body using Single Photon Emission Computed Tomography (SPECT). The shorter lived ^{80m}Br ($T_{1/2}$ = 4.4 h), decaying by isomeric transition, can also be used for Auger-therapy while the ß–- emitter ^{82}Br ($T_{1/2}$ = 35.2 h) has potential for therapy.

The labelling of a molecule with radiobromine can be done, in principle, analogous to the labelling with radioiodine, for which many radiotracers are known and established in nuclear medicine for diagnosis and therapy as well (for early reviews cf. [4,5]). Carrier free bromine radionuclides can be produced at accelerators and many reaction channels have been examined. Proton and deuteron induced reactions on highly enriched selenium or krypton isotopes offer the production of pure isotopes of nearly any interesting bromine radionuclide with high yield but have the disadvantage of expensive target material and the need of reusable targets.

Indirect production of the lighter bromine isotopes ^{76}Br or ^{77}Br can be done by proton and deuteron irradiations of natural bromine via the product ^{76}Kr and ^{77}Kr which can be separated from the bromine target material, decaying subsequently to the desired radiobromines.

^{3}He particle and alpha particle induced reactions on the monoisotopic arsenic were examined, too, but they are also limited to a few bromine isotopes and furthermore accelerators dedicated to production of medical radioisotopes generally do not offer ^{3}He beams (for early reviews cf. [6,7]). A detailed review of alpha induced reaction cross sections and production of medical radionuclides was recently published be us in Radiochimica Acta [8]. Monoisotopic arsenic as target in radiobromine production has the advantage of low cost, easy target preparation and simple chemical separation without the need of recovering the target material.

Therefore our aim was to re-examine the α-particle induced reactions which lead to the bromine isotopes ^{76}Br, ^{77}Br and ^{78}Br, whose cross-section data are discrepant in literature. All reactions were studied from their respective thresholds up to 37 MeV.

EXPERIMENTAL

Targets and Irradiations

Thin arsenic targets, as mandatory for cross-section measurements, were prepared by sedimentation from an ethanolic slurry on Al backings.

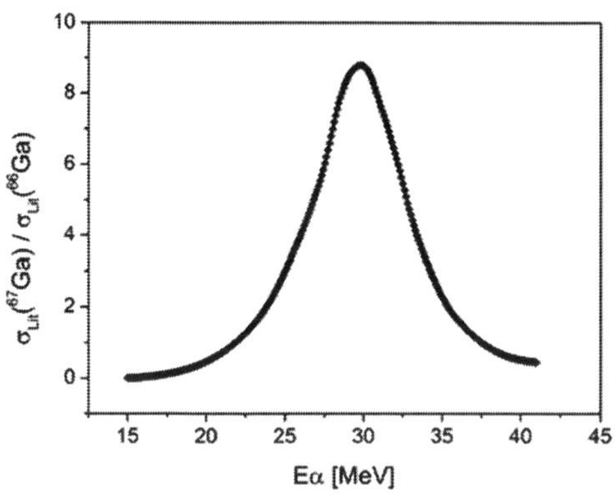

Figure 1: Cross-section ratio of $^{67}Ga/^{66}Ga$ of the nuclear monitor reaction Cu(α,x).

The dried sediment was covered by thin Al foil, and the sandwiches thus obtained, were stacked together with Al degraders and Cu monitors, placed in front of each arsenic sample.

The nuclear reactions on Cu leading to ^{66}Ga and ^{67}Ga served as monitors for the determination of alpha beam flux and energy. The energy degradation within the foil stack was pre-calculated using the code STACK.

The irradiation was performed at the cyclotron CGR560 of the Vrije Universiteit Brussel with 38 MeV α-particles in stacked-foil arrangement, using a beam current of about 70 nA for 30 min. Only for crosssection determination of the short-lived ^{78}Br (6.5 min), the irradiation time was limited to 6 min in separate experiments. In total 30 irradiations were done.

Determination of Activity and Beam Flux

After activation, the foils within the stacks were separated and the activity of all the Al-As-Al target sandwiches and of the monitor foils was measured γ -spectrometrically using well calibrated HPGe-detectors. The decay of the radionuclides was followed by recurring measurements.

While during irradiation the beam flux was measured by a Faraday Cup, the exact beam flux was determined by calculation from the monitor reactions on Cu with ^{66}Ga and ^{67}Ga as the activation products of interest.

Alpha-particles have a much higher stopping power than protons and deuterons and also a stronger straggling. Therefore small errors or uncertainties in the stack preparation as well as increasing energy degradation with decreasing α-particle energy within the stack cause higher uncertainties in beam flux calculation.

Therefore we determined reliably the α-particle energy in each foil using the cross-section ratio ^{67}Ga to ^{66}Ga, shown in Fig. 1, calculated from the evaluated cross sections taken from the IAEA-Website [9].

A comparison of the re-measured $Cu(\alpha,x)^{66}Ga$ cross sections, once uncorrected, obtained from the flux detected by the Faraday Cup, and on the other hand corrected, based on the cross-section ratios of the two Ga isotopes, to that of the recommended excitation function from IAEA-TECDOC-1211 [9] showed that only the energy corrected cross sections are in very good agreement with the recommendation. This underlines the necessity of individual energy validation, and its absence in some earlier experiments can explain the deviations in the literature data of the following nuclear reactions especially when large energy degradations were applied.

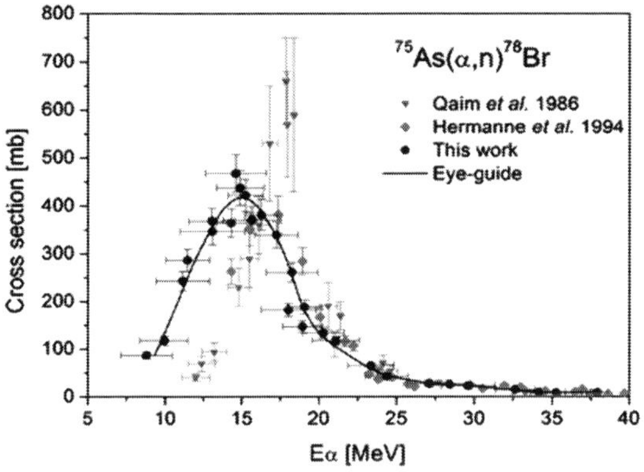

Figure 2: Cross sections of the $^{75}As(\alpha,n)^{78}Br$ reaction together with literature data [10,11].

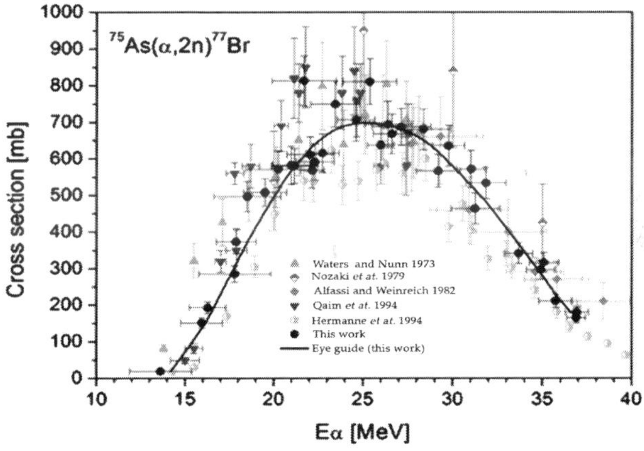

Figure 3: Cross section of the $^{75}As(\alpha,2n)^{77}Br$ reaction together with literature data [10–14].

RESULTS

The cross sections given by us are calculated using the above mentioned corrections of the individual beam energy in each As sample and the individual beam flux within a stack based on the monitor foils.

Cross Section of ^{75}As(α,n)^{78}Br Reaction

Figure 2 shows our results for the (α,n)-reaction on arsenic leading to ^{78}Br in comparison with the data of Qaim [10] and Hermanne [11], which are shifted 2 to 3 MeV to higher energies. The data of Qaim are somewhat higher with a narrower peak at the maximum of the reaction cross sections.

Cross Section of ^{75} As(α,2n)^{77}Br Reaction

The cross sections of the (α,2n)-reaction on arsenic leading to ^{77}Br are shown in Fig. 3, together with the literature data [10–14]. Within uncertainties the data are in fair agreement to our data. Nevertheless, small energy shift can be seen and also the values of cross sections differ up to about 30%.

Cross Section of ^{75} As(α,3n)^{76}Br Reaction

Our cross sections of the ^{75}As(α,3n)^{76}Br reaction together with the data of Nozaki, Paans, Alfassi and Hermanne [11,13–15] are given in Fig. 4. The data show energy shifts. The largest shift is in the data published by

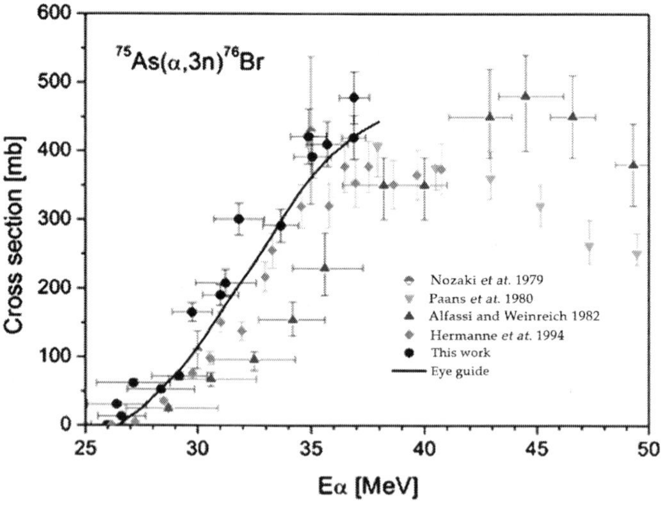

Figure 4: Cross section of the ^{75}As(α,3n)^{76}Br reaction together with literature data [11,13–15].

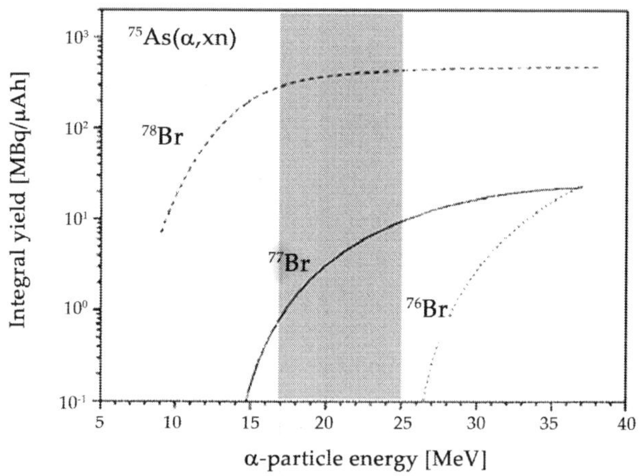

Figure 5: Thick target yields of the $^{75}As(\alpha,xn)$ reaction products.

Alfassi et al. moved about 6 MeV to higher energies, but their data support our cross-section magnitude. The other data are somewhat lower and slightly shifted also to higher alpha energies.

Thick Target Yields

Calculating the theoretical thick target yields, given in Fig. 5, on the basis of the respective eye guides, it is obvious that only ^{77}Br can be produced in a quality necessary for medical applications. While ^{78}Br with a half-life of only 6.48 min decays out very fast after the irradiation, the only longer lived bromine isotopes are ^{77}Br (57.0 h) and ^{76}Br (16.2 h). Selecting the energy range 25 to 17 MeV, ^{77}Br can be produced free of isotopic impurities with a yield of 9.1 MBq/μ Ah, sufficient for medical application.

The co-produced non-isotopic impurity ^{74}As can be chemically separated together with the bulk of arsenic from the radiobromine.

SUMMARY

We measured the excitation functions of alpha-particle induced reactions on arsenic leading to $^{76,77,78}Br$. Thereby we used individual corrections of beam energy and flux in each target foil on the basis of the ^{67}Ga to ^{66}Ga crosssection ratios in Cu monitor foils. This method strengthens the data base and clarifies discrepancies in literature data.

The alpha induced reaction on arsenic is a suitable route to produce ^{77}Br in sufficient amounts and in pure form for medical application.

This reaction will be a promising application at our Cyclone 30 XP accelerator using 30 MeV alpha particle beam to produce ^{77}Br either for medical application or for developing radiolabelling strategies.

The results are related to our R&D work on the production of radiohalogens.

REFERENCES

1. B. Maziere, C. Loc'h, Appl. Radiat. Isot. 37, 703–713 (1986)
2. D.J. Rowland, T.J. McCarthy, M.J. Welch, in Handbook of Radiopharmaceuticals, John Wiley & Sons, Ltd (2003), 441–465
3. V. Tolmachev, Current Radiopharmaceuticals 4, 76–89 (2011)
4. H.H. Coenen, S.M. Moerlein, G. Stocklin, ¨ Radiochim. Acta 34, 47–68 (1983)
5. H.H. Coenen, in: Progress in Radiopharmacy. Development in Nuclear Medicine 10 (Cox PH, Mather SJ, Sambson CB, Lazarus CR, eds.), Martinus Nijhoff Publishers, Dordrecht (1986), 196–220
6. S.M. Qaim, G. Stocklin, Radiochim. Acta ¨ 34, 25–40 (1983)
7. S.M. Qaim, Appl. Radiat. Isot. 37, 803–810 (1986)
8. S.M. Qaim, I. Spahn, B. Scholten, B. Neumaier, Radiochim. Acta 104, 601 (2016)
9. F. Tark ´ anyi, S. Tak ´ acs, K. Gul, A. Hermanne, ´ M.G. Mustafa, F.M. Nortier, P. Oblozinsky, S.M. Qaim, B. Scholten, Yu. N. Shubin, Y. Zhuang, Beam monitor reactions: in Charged particle cross section database for medical radioisotope production. IAEA-TECDOC-1211, International Atomic Energy Agency, Vienna 2001, p. 49. www-nds.iaea.org/medical/monitor reactions.html
10. S.M. Qaim, G. Blessing, H. Ollig, Radiochim. Acta 39, 57–60 (1986)
11. A. Hermanne, M. Sonck, J. Van Hoyweghen, D. Terriere, J. Mertens, Proceedings of the International Conference on Nuclear Science and Technology, Gatlinburg, Tennessee, 1039–1041 (1994)
12. S.L. Waters, A.D. Nunn, M.L. Thakur, J. Inorg. Nucl. Chem. 35, 3413–3416 (1973)
13. T. Nozaki, M. Iwamoto, Y. Itoh, Appl. Radiat. Isot. 30, 79–83 (1979)
14. Z.B. Alfassi, R. Weinreich, Radiochim. Acta 30, 67–71 (1982)
15. A.M.J. Paans, J. Welleweerd, W. Vaalburg, S. Reiffers, M. G. Woldring, Appl. Radiat. Isot. 31, 267–273 (1980)

SECTION 6:
RADIATION DETECTORS

14

Silicon Carbide Microstrip Radiation Detectors

Donatella Puglisi [1,2] **and Giuseppe Bertuccio** [1,3]

[1]Department of Electronics, Information and Bioengineering, Politecnico di Milano, Campus Como, 22100 Como, Italy
[2]Department of Physics, Chemistry and Biology, Sensor and Actuator Systems, Linköping University, 58183 Linköping, Sweden
[3]Italian National Institute of Nuclear Physics (INFN), Section Milano, 20133 Milan, Italy

ABSTRACT

Compared with the most commonly used silicon and germanium, which need to work at cryogenic or low temperatures to decrease their noise levels, wide-bandgap compound semiconductors such as silicon carbide allow the operation of radiation detectors at room temperature, with high performance, and without the use of any bulky and expensive cooling equipment. In this work, we investigated the electrical and spectroscopic performance of an innovative position-sensitive semiconductor radiation detector in epitaxial 4H-SiC. The full depletion of the epitaxial layer (124 μm, 5.2×10^{13} cm^{-3})

Citation: Puglisi, D.; Bertuccio, G. "Silicon Carbide Microstrip Radiation Detectors". Micromachines 2019, 10, 835. https://doi.org/10.3390/mi10120835

MATERIALS AND METHODS

Two different designs of silicon carbide microstrip detectors have been realized on top of two-inch high-purity epitaxial 4H-SiC wafer produced by LPE Epitaxial Technology Center [**23**]. Each detector consists of 32 strips with a length of 2 mm, a width of either 25 μm (SM1) or 50 μm (SM3), and a pitch of either 55 μm (SM1) or 100 μm (SM3)—see Figure 1. Each of these strips can be read out independently by a front-end electronics channel, and therefore behaves as a separate detector. A cross-sectional view of the 4H-SiC microstrip structure is shown in Figure 2. The SiC epitaxial layer, which is the active region of the detector, has a maximum thickness of 124 μm, as experimentally measured (Figure 3).

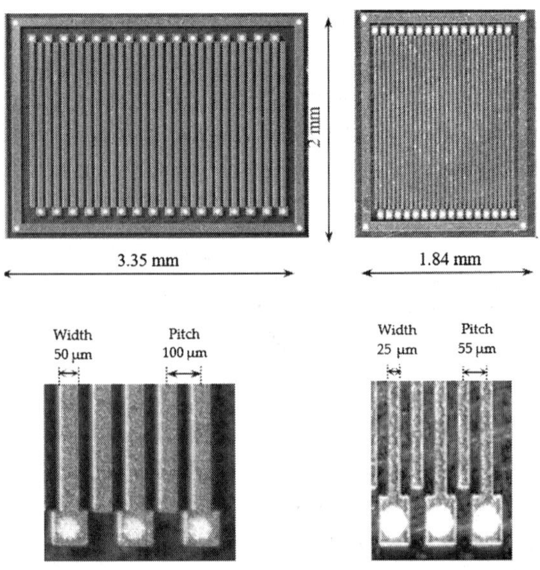

Figure 1: Photographs of the two microstrip detectors used in this work, together with a detail of their peripheral regions with bonding pads.

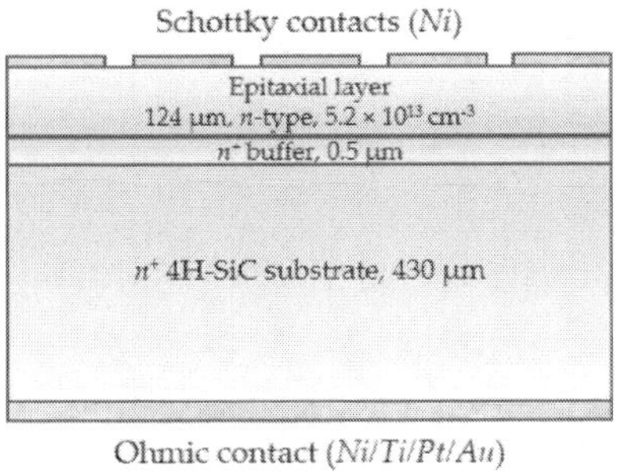

Figure 2: Cross-sectional view of the 4H-SiC microstrip structure.

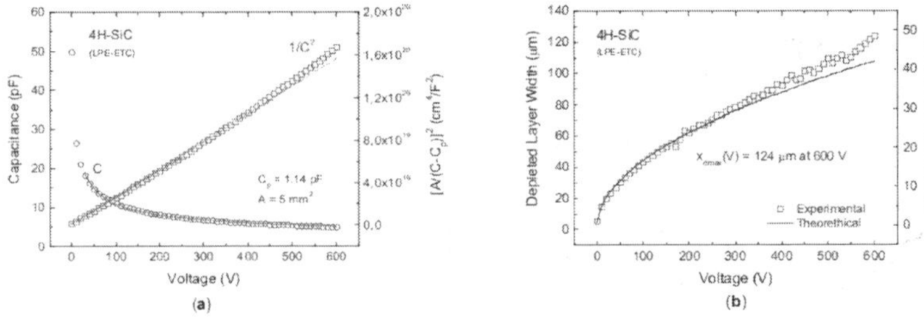

Figure 3: (a) Capacitance–voltage (C–V) and $1/C^2$–V per unit area characteristics. From the fit curve, the mean value of the donor concentration is expected to be $(5.56 \pm 0.05) \times 10^{13}$ cm^{-3}; (b) depleted layer and mean electric field as a function of applied voltage, as derived from C–V measurements. The theoretical result is obtained using 5.56×10^{13} cm^{-3} as the mean value of the doping concentration.

The application of a reverse bias at the common back electrode (ohmic contact) creates a depletion region, x_d, depending on the applied bias, V_R, and the residual doping (donor) concentration, N_D, within the material, according to

$$x_d = \sqrt{\frac{2\varepsilon_0\varepsilon_r}{qN_D}\left(\psi_{bi} - V_R - \frac{kT}{q}\right)} \quad (1)$$

where ψ_{bi} is the built-in potential and the term $\frac{kT}{q}$ arises from the contribution of the majority-carrier distribution tail [24].

The residual doping concentration depends on the homogeneity of the epitaxial layer. The donor concentration profile, $N_D(x)$, was determined as a function of the depleted layer width from capacitance-voltage measurements, as described in Section 3.

RESULTS

Electrical Characterization

Capacitance–Voltage Characterization

Capacitance–voltage (C–V) measurements were carried out up to 600 V at 25 °C, in order to determine the donor-concentration profile of the epitaxial layer (Figure 3a). The detector was placed in a test fixture Agilent 16065A connected to an Agilent 4284A Precision LCR Meter (Santa Clara, CA, USA). A Keithley 2410 voltage source (Cleveland, OH, USA), operating in the four-wire connection mode, was used to bias the device and measure the applied voltage. The measurement was performed with a 100 mV AC signal at 100 kHz. The donor-concentration profile as a function of the depleted layer width was determined from the slope of a $1/C^2$–V curve, according to [24], see Figure 3a. Please note that the C–V measurements were performed using a 4H-SiC Schottky diode with area $A = 5$ mm^2, produced from the same wafer. A full depletion of 124 μm was reached, polarizing the detector up to 600 V (Figure 3b). A mean value of $\langle N_D \rangle = (5.20 \pm 0.06) \times 10^{13}$ cm^{-3} was determined (Figure 4).

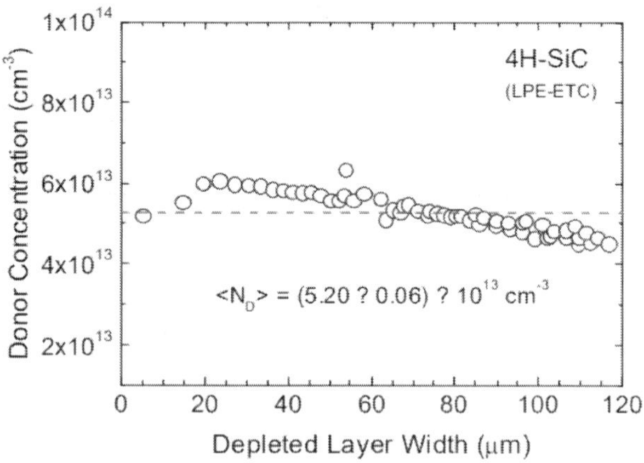

Figure 4. Donor concentration profile as a function of the depleted layer width. The full depletion of 124 μm was reached at 600 V. A mean value of $\langle N_D \rangle = 5.2 \times 10^{13}$ cm^{-3} was determined.

Statistical Leakage–Current Distribution

Current–voltage (I–V) measurements were carried out on each of the 32 strips of two different SiC detectors, SM1 and SM3, at room temperature (Figure 5). The two detectors were biased at 100 V and 200 V from the back ohmic contact using a Keithley 2410 source meter, whereas the current of each strip was measured connecting a Keithley 6430 electrometer to the front, rectifying the Schottky contact. The guard electrode surrounding the microstrips was kept to ground to collect the parasitic current generated at the device's chip edges. Figure 5 shows the current and current density values for each of the 32 strips of the two different microstrip detectors as measured at 25 °C and 200 V. Such bias voltage generates an inner electric field of about 30 kV/cm. Ultra-low current mean values of 2.2 fA and 7.6 fA were measured for the two microstrips under test, corresponding to current densities of 4.4 pA/cm^2 and 15.2 pA/cm^2.

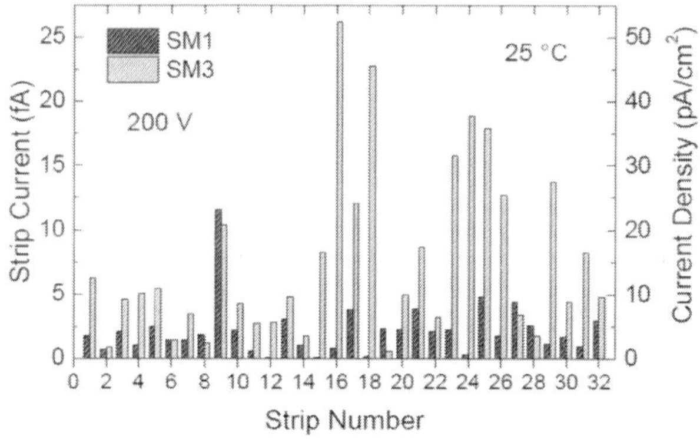

Figure 5: Current and current density measured at 25 °C and 200 V on the 32 strips of the two microstrip detectors, SM1 and SM3. Current values of few fA (current densities of low pA/cm^2) were measured on all strips.

Temperature Dependence

The temperature dependence of the strip current as a function of reverse-bias voltage is shown in Figure 6. From now on, only results obtained with the detector SM1 are presented. In order to perform this measurement, the device was attached to a Teflon circuit board using a silver conductive glue. Electrical contacts were established using 25 μm gold wire-bonding connections. Measurements were acquired, biasing up to 200 V the back contact with the Keithley 2410 source meter and reading the currents from the Keithley 6430 electrometer. Tests were carried out inside an environmental chamber, setting the temperature at 27 °C, 47 °C, 67 °C, 87 °C, and 107 °C, and monitoring it by means of a thermocouple placed near to the device. During each measurement, the temperature changes were monitored within ±0.1 °C. The current density was calculated considering a strip active area of 5×10^{-4} cm^{-2}.

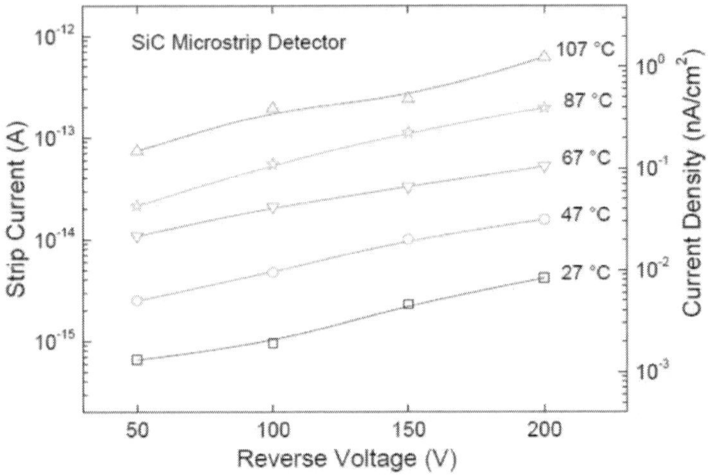

Figure 6: Current and current density dependence from temperature in the range 27 to 107 °C.

Such a thermally activated process is described by the Arrhenius plot that, according to the emission theory, is expressed by

$$I = I_0 \cdot \exp\left[\left(-\frac{E_A}{kT}\right)\cdot\left(1-\frac{T}{T_0}\right)\right] \tag{2}$$

where I_0 is the saturation current, T_0 is the room temperature, and E_A is the activation energy [5,25].

Figure 7 shows the Arrhenius plots of the leakage current as a function of $1000/T$ at four different reverse voltages, i.e., 50 V, 100 V, 150 V, and 200 V. The activation energy is given by the slope of linear fit of data. Values from 0.57 eV to 0.65 eV were calculated in the voltage range 50 V to 200 V. According to the literature, these values refer to major deep levels ($Z_{1/2}$ center) within the bandgap [26,27,28].

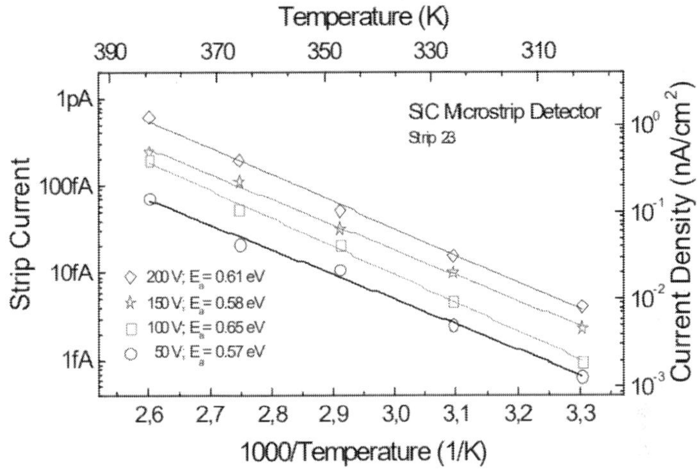

Figure 7: Arrhenius plots of the leakage current (and current density) versus the reverse of temperature at four different voltages. The activation energy values are from 0.57 eV to 0.65 eV in the voltage range 50 to 200 V.

Interstrip Resistance Measurements

Current–voltage measurements shown in previous sections refer to the characterization of single strips, considering that the measured current arrived only from the back contact of the device, and possible latent currents from adjacent strips were negligible. In order to determine the bias limit condition so that two adjacent strips can be considered isolated, we measured the interstrip resistance.

Interstrip resistance measurements were carried out by measuring the current between two consecutive strips, keeping the back contact at 100 V and the guard at 0 V. One of the two strips (microstrip 2 in Figure 8) is biased from −5 V to +5 V while the current of the other strip (microstrip 1 in Figure 8), kept at 0 V, is measured by an electrometer. We repeated the test on three couples of strips. The negative slope shown in Figure 8 is due to the application of the bias voltage to microstrip 2 while measuring the current at microstrip 1. The mean value of resistance between two adjacent strips of SM1 resulted in 5.3 TΩ.

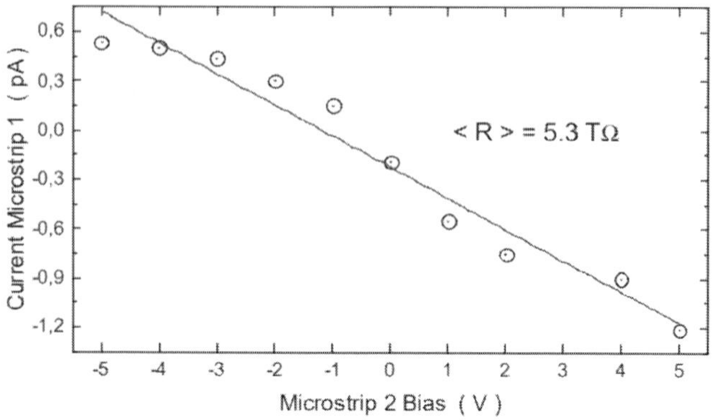

Figure 8: Mean value of resistance between two adjacent strips.

X-Ray Spectroscopy

Room Temperature

The detector was irradiated with a 397 kBq ^{241}Am source placed at a few cm from the detector surface. Figure 9a shows a ^{241}Am spectrum acquired using a SiC microstrip detector at 21 °C. The spectrum was acquired at 200 V reverse-bias condition using 12.8 μs peaking time in the triangular signal processing. The pulser line width is 214 eV full width at half maximum (FWHM) corresponding to an equivalent noise charge of 11.6 electrons root mean square (rms). Figure 9b shows a detail of the same measurement up to 28 keV. Several X-ray lines from Mn, Cu, Np, and Ag can be clearly distinguished with a very good resolution, i.e., enough to separate the *K* and *L* lines of neighboring elements. Conventionally, the energy resolution, that is the FWHM, is specified for the Mn *Kα* peak at 5.9 keV, which is 213 eV for our SiC microstrip detector at room temperature (Figure 9). It is notable that Si(Li) and silicon drift detectors can achieve 130–150 eV FWHM, and Ge detectors can even achieve 115 eV FWHM for the Mn *Kα* peak at 5.9 keV, but with liquid-nitrogen cooling [29].

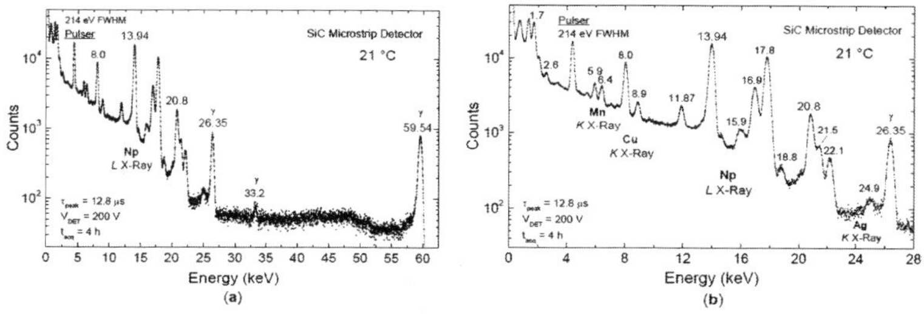

Figure 9: (a) X-ray spectrum from a ^{241}Am source acquired at 21 °C using the SiC microstrip detector SM1 and an ultra-low noise front-end (PRE5 no. 3); (b) detailed X-ray spectroscopy in the energy range 0 to 28 keV.

The analysis of linearity calculated on seven well-resolved peak lines at 8.0, 11.87, 13.94, 17.8, 20.8, 26.35 and 59.54 keV is shown in Figure 10. The percentage error from linearity is below ±0.05%.

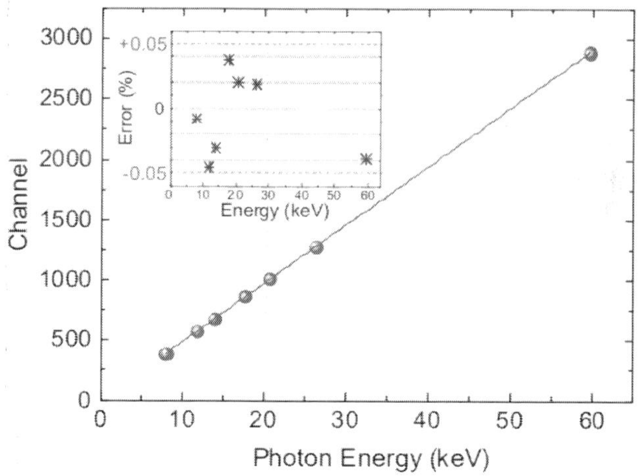

Figure 10: Analysis of linearity calculated on seven well-resolved peak lines, as shown in Figure 9. The percentage error from linearity is below ±0.05%.

Dependence of X-Ray Response on Detector Bias

The dependence of X-ray response on detector voltage was explored by biasing the detector from 10 V to 200 V at 25 °C and using a peaking time of 12.8 μs. A comparison between two spectra acquired at 10 V (bottom, blue) and 200 V (top, red) is shown in Figure 11. The pulser centroid is stable at both 10 V and 200 V. The peak at 13.94 keV shows a small shift of six channels (from 669 at 10 V to 675 at 200 V) which corresponds to 124 eV (Figure 12). The two peaks, as compared in Figure 12a, show a Gaussian symmetry without any tails. This experiment shows that SiC detectors can be operated in a wide range of bias voltages without suffering from a strong performance loss. Figure 12b shows the detection rate of the 13.94 keV photon peak at six different applied reverse-bias voltages, V_b. As expected, the photon rate increases with the square root of V_b due to the widening of the active region (depletion layer).

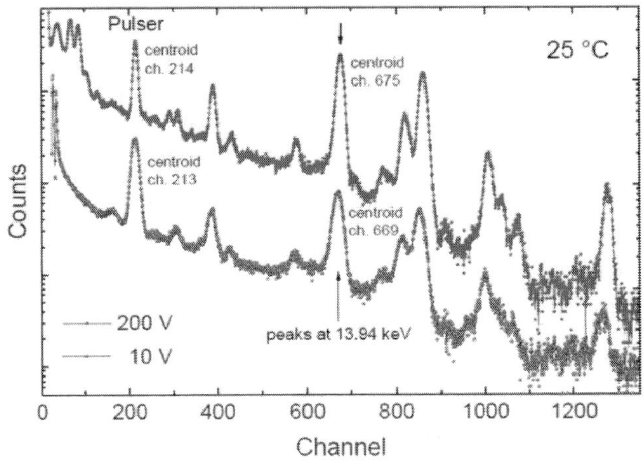

Figure 11: X-ray spectra from a ^{241}Am source acquired at 25 °C and at 10 V (bottom) and 200 V (top).

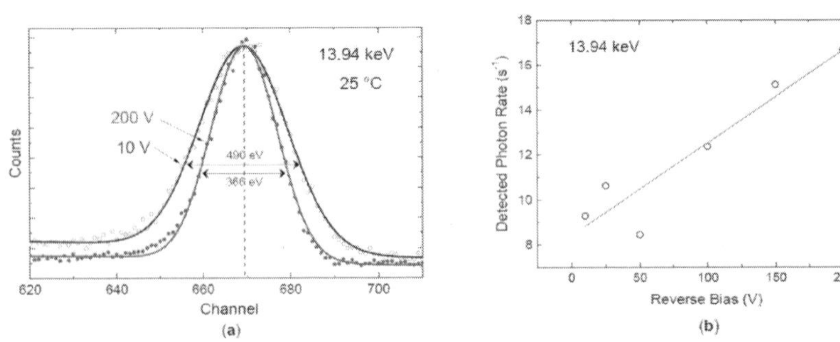

Figure 12: (a) Comparison between the two peaks at 13.94 eV obtained at 10 V and 200 V of applied voltage. The Gaussian symmetry without tails can be noticed; (b) Detected 13.94 keV photon rate as a function of the applied reverse bias.

High Statistics and Temperature Dependence

Figure 13 shows the results obtained by acquiring an X-γ-ray spectrum for 10 h at 80 V reverse-bias condition, and maintaining the thermostatic chamber at a constant temperature of 30 °C. Figure 13a compares spectra acquired after almost 2 h and after 10 h. The very small broadening of the pulser and emission lines should be noted, which demonstrates very good stability of the detector response to X-ray exposure. The analysis of linearity on *K* Cu and *L* Np X-ray monoenergetic lines shows a very small linearity error within ±0.04% after 10 h of acquisition (Figure 13b).

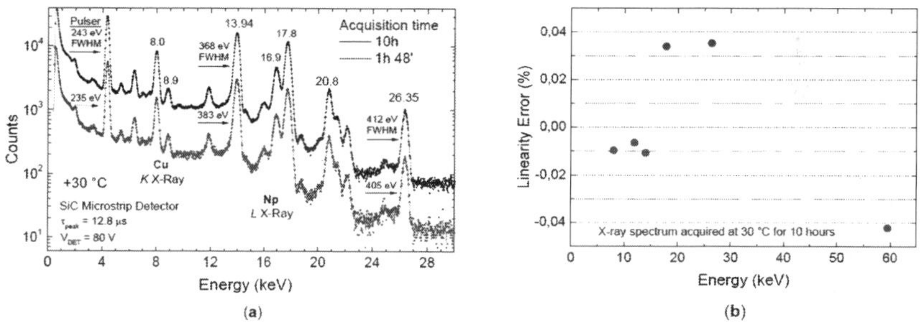

Figure 13: (a) X-ray spectra from ^{241}Am source after almost 2 and 10 h of acquisition at 30 °C in a thermostatic chamber. The very small broadening of pulser and emission lines demonstrates very good stability of the detector response to X-ray exposure; (b) linearity error after 10 h of acquisition based on *K* Cu and *L* Np X-ray monoenergetic lines.

Figure 14 shows a comparison between X-ray spectra acquired at three different temperatures, i.e., −20 °C, +30 °C, and +80 °C. As expected, the width of pulser and emission lines increased by increasing the operating temperature: the FWHM of the pulser changed from 205 eV at −20 °C, to 215 eV at +30 °C, and 249 eV at +80 °C. It is worth noticing the small broadening of the lines at +80 °C, which demonstrates the suitability of our microstrip detector to be used at high temperatures with very good stability of the detector response.

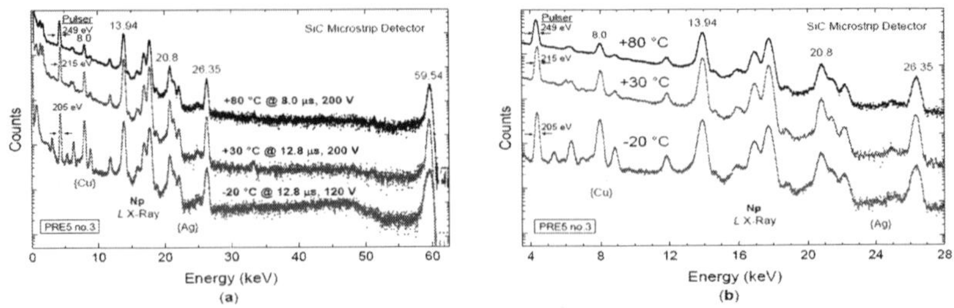

Figure 14: (a) Comparison between three X-ray spectra acquired at −20 °C, +30 °C, and +80 °C in the range 0–60 keV; (b) Detail of the X-ray spectra in the range 4–28 keV. Note a small broadening of the pulser line by increasing the operating temperature.

DISCUSSION

Two SiC-based microstrip detectors were fully characterized in a wide temperature range by means of electrical and spectroscopic measurements. A high stability of the detector response as a function of operating temperature as well as of applied voltage was widely demonstrated.

The very low leakage currents (current densities) from about 2 fA (4 pA/cm^2) at 25 °C to 620 fA (1.2 nA/cm^2) at 107 °C are among the best values measured on SiC detectors, and more than one order of magnitude lower than most silicon detectors [30,31]. Since the shot noise of the leakage current is a significant noise contribution in a radiation spectroscopy system, we can say that our SiC detectors allow the achievement of high signal-to-noise ratios. A good isolation between adjacent strips was demonstrated by the high value of the measured interstrip resistance of 5.3 TΩ, confirming that possible latent currents from adjacent strips can be considered negligible.

A very good doping uniformity of the whole epitaxial layer was also demonstrated. For the first time, a full depletion of 124 μm was reached, polarizing the detector at 600 V, and a mean value of $\langle N_D \rangle = 5.2 \times 10^{13}$ cm^{-3} was determined. In comparison with previous studies, this result is the best observed [32].

The X-ray spectra acquired from a ^{241}Am source at different voltages, temperatures, and exposure times showed high stability and a high spectroscopic resolution under all tested experimental conditions.

Different voltages were used to verify the effect of the applied bias voltage on the device performance. The spectroscopic response of our SiC detector does not significantly depend on the bias voltage, as shown in Figure 11, where two extreme bias voltages (10 V and 200 V) were used. No tails in the spectral lines were observed, which means that no significant charge trapping occurred in these devices. Remarkably, no strong performance loss was observed at 10 V, and substantially no difference was observed under operation between 80 V and 200 V. The possibility of using a lower voltage without losing significant information is an advantage for those applications wherein lower power consumption is desirable. The better resolution obtained at 200 V is due to a lower capacitance. We avoided operating the SiC detector above 200 V to prevent the risk of possible damages due to accidental breakdown or electrostatic discharge. Also, the exposure time to the ^{241}Am source does not affect the spectroscopic performance of our SiC detector, as demonstrated by negligible differences in the peak resolution after almost 2 and 10 h of acquisition (Figure 13). This means that it is not necessary to wait for a long time before getting all the main information from the device.

Finally, it is worth noticing the high resolution and very good stability in the performance of our SiC microstrip detector between −20 °C and +80 °C (Figure 14), which pave the way for use in a wide range of applications that are prohibitive for other conventional semiconductor detectors.

CONCLUSIONS

Even considering an initial higher cost for the SiC material in comparison to Si or Ge, SiC-based semiconductor radiation detectors are advantageous for use in operating conditions under which conventional semiconductor detectors in Si or Ge cannot adequately perform. One overall advantage is the elimination of cryogenic or Peltier cooling systems that allows the fabrication of far more compact, more stable, lighter, and lower power radiation detector systems. This also implies, as a direct consequence, economic advantages that go beyond the mere costs for the material itself.

Our findings confirm the high quality and the good uniformity of the epitaxial layer used for manufacturing our SiC-based prototypes, as well as the suitability of such devices to be used as high resolution X-ray detectors over a wide range of operating temperatures.

AUTHOR CONTRIBUTIONS

Investigation, D.P. and G.B.; Supervision, G.B.; Writing – original draft, D.P. and G.B.; Writing – review & editing, D.P.

ACKNOWLEDGMENTS

All experiments were performed at the Semiconductor and Integrated Circuit Laboratory of the Department of Electronics, Information and Bioengineering, Politecnico di Milano, Como Campus, Italy. The authors would like to thank C. Lanzieri for device manufacturing process at Selex Integrated Systems (Rome), P. Ferrari for his contribution to the electrical characterization, S. Masci for the device bonding, S. Caccia and D. Macera for their collaboration for the readout electronics. The authors would like to dedicate this work to the everlasting memory of Andrea Puglisi.

REFERENCES

1. Van Heerden, P.J. The Crystal Counter: A new apparatus in nuclear physics for the investigation of β- and γ-rays. Part, I. *Physica* 1950, *16*, 505–516.
2. Van Heerden, P.J.; Milatz, J.M.W. The Crystal Counter: A new apparatus in nuclear physics for the investigation of β- and γ-rays. Part II. *Physica* 1950, *16*, 517–527.
3. Lerov, C.; Rancoita, P.-G. *Principles of Radiation Interaction in Matter and Detection*, 2nd ed.; Word Scientific: Singapore, 2009; p. 83. ISBN 978-981-4470-89-6.
4. Scholze, F.; Rabus, H.; Ulm, G. Mean energy required to produce an electron-hole pair in silicon for photons of energies between 50 and 1500 eV. *J. Appl. Phys.* 1998, *84*, 2926–2939.
5. Bertuccio, G.; Casiraghi, R. Study of silicon carbide for X-ray detection and spectroscopy. *IEEE Trans. Nucl. Sci.* 2003, *50*, 175–185.
6. Knoll, G.N. *Radiation Detection and Measurement*, 4th ed.; Wiley: New York, NY, USA, 2010; ISBN 978-0-470-13148-0.
7. Lutz, G. *Semiconductor Radiation Detectors: Device Physics*, 2nd printing of the 1st ed.; Springer: Berlin/Heidelberg, Germany, 2007; ISBN 978-3-540-71679-2.
8. McGregor, D.S.; Hermon, H. Room-temperature compound semiconductor radiation detectors. *Nucl. Instrum. Methods Phys. Res. A Accel. Spectrom. Detect. Assoc. Equip.* 1997, *395*, 101–124.
9. Bertuccio, G. Prospect for Energy Resolving X-ray Imaging with Compound Semiconductor Pixel Detectors. *Nucl. Instrum. Methods Phys. Res. A Accel. Spectrom. Detect. Assoc. Equip.* 2005, *546*, 232–241.
10. Schlesinger, T.E.; James, R.B. (Eds.) *Semiconductor for Room Temperature Nuclear Detectors Applications*, Semiconductor and Semimetal Series; 1st ed.; Academic Press: San Diego, CA, USA, 1995; Volume 43, ISBN 9780080571997.
11. Owens, A.; Bavdaz, M.; Peacock, A.; Poelaert, A. High resolution x-ray spectroscopy using GaAs arrays. *J. Appl. Phys.* 2001, *90*, 5376–5381.
12. Abbene, L.; Del Sordo, S.; Caroli, E.; Gerardi, G.; Raso, G.; Caccia, S.; Bertuccio, G. Hard X-ray Response of Pixellated CdZnTe detectors. *J. Appl. Phys.* 2009, *105*, 124508.

13. Bertuccio, G.; Pullia, A. Room Temperature X-Ray Spectroscopy with a Silicon Diode Detector and an Ultra Low Noise Preamplifier. *IEEE Trans. Nucl. Sci.* 1994, *41*, 1704–1709.
14. Pullia, A.; Bertuccio, G. Resolution Limits of Silicon Detectors and Electronics for Soft X Ray Spectroscopy at non Cryogenic Temperatures. *Nucl. Instrum. Methods Phys. Res. A Accel. Spectrom. Detect. Assoc. Equip.* 1996, *380*, 1–5.
15. Bertuccio, G.; Caccia, S.; Casiraghi, R.; Lanzieri, C. Possibility of sub-electron noise with room temperature silicon carbide pixel detectors. *IEEE Trans. Nucl. Sci.* 2006, *53*, 2421–2427.
16. Nava, F.; Bertuccio, G.; Cavallini, A.; Vittone, E. Silicon carbide and its use as a radiation detector material. *Meas. Sci. Technol.* 2008, *19*, 102001.
17. Kimoto, T.; Cooper, J.A. *Fundamentals of Silicon Carbide Technology*, 1st ed.; IEEE Press: Piscataway, NJ, USA; John Wiley & Sons: Hoboken, NJ, USA; Singapore Pte. Ltd.: Singapore, 2014; ISBN 978-1-118-31352-7.
18. Mandal, K.C.; Muzykov, P.G.; Terry, J.R. Highly sensitive x-ray detectors in the low-energy range on n-type 4H-SiC epitaxial layers. *Appl. Phys. Lett.* 2012, *101*, 051111.
19. Sadrozinski, H.F.-W. Silicon microstrip detectors in high luminosity application. *IEEE Trans. Nucl. Sci.* 1998, *45*, 295–302.
20. Rossi, G.; Morse, J.; Protic, D. Energy and position resolution of germanium microstrip detectors at X-ray energies from 15 to 100 keV. *IEEE Trans. Nucl. Sci.* 1999, *46*, 765–773.
21. Dusi, W.; Auricchio, N.; Brigliadori, L.; Cola, A.; Donati, A.; Landini, G.; Mengoni, D.; Perillo, E.; Siffert, P.; Ventura, G. A study of the spectroscopic performance of a CdTe microstrip detector. *IEEE Trans. Nucl. Sci.* 2003, *50*, 1026–1030.
22. Kaluza, A.; Ohms, T.; Rente, C.; Engels, R.; Reinartz, R.; Luth, H. Microstrip X-ray detector with a very high dynamic range based on LPE-GaAs. *IEEE Trans. Nucl. Sci.* 1998, *45*, 724–727.
23. LPE Epitaxial Technology, S.p.A. Available online: http://www.lpe-epi.com/ (accessed on 21 June 2019).
24. Sze, S.M.; Ng, K.K. *Physics of Semiconductor Devices*, 3rd ed.; John Wiley & Sons Inc.: Hoboken, NJ, USA, 2007; pp. 85–137. ISBN 978-0-471-14323-9.

25. Schroder, D.K. *Semiconductor Material and Device Characterization*, 3rd ed.; John Wiley & Sons Inc.: Hoboken, NJ, USA, 2006; p. 691. ISBN 978-0-471-73906-7.

26. Knoll, L.; Kranz, L.; Alfieri, G. Electrically active deep levels formed by thermal oxidation of n-type 4H-SiC. *J. App. Phys.* 2019, *125*, 205302.

27. Mannan, M.A.; Nguyen, K.V.; Pak, R.O.; Oner, C.; Mandal, K.C. Deep Levels in n-Type 4H-Silicon Carbide Epitaxial Layers Investigated by Deep-Level Transient Spectroscopy and Isochronal Annealing Studies. *IEEE Trans. Nucl. Sci.* 2016, *63*, 1083–1090.

28. Nava, F.; Castaldini, A.; Cavallini, A.; Errani, P.; Cindro, V. Radiation Detection Properties of 4H-SiC Schottky Diodes Irradiated up to 10^{16} n/cm^2 by 1 MeV Neutrons. *IEEE Trans. Nucl. Sci.* 2006, *53*, 2977–2982.

29. Statham, P.J. Recent Developments in Instrumentation for X-ray Microanalysis. In *Modern Developments and Applications in Microbeam Analysis*; Love, G., Nicholson, W.A.P., Armigliato, A., Eds.; Springer: Wien, Austria; New York, NY, USA, 1998; pp. 1–9. (Suppl. 15), ISBN 978-3-211-83106-9.

30. Ruddy, F.H.; Seidel, J.G.; Chen, H.; Dulloo, A.R.; Ryu, S.-H. High-Resolution Alpha-Particle Spectrometry Using 4H Silicon Carbide Semiconductor Detectors. *IEEE Trans. Nucl. Sci.* 2006, *53*, 1713–1718.

31. Bertuccio, G.; Caccia, S.; Nava, F.; Foti, G.; Puglisi, D.; Lanzieri, C.; Lavanga, S.; Abbondanza, G.; Crippa, D.; Preti, F. Ultra low noise epitaxial 4H-SiC X-ray detectors. *Mat. Sci. Forum* 2009, *615*, 845–848.

32. Bertuccio, G.; Puglisi, D.; Macera, D.; Di Liberto, R.; Lamborizio, M.; Mantovani, L. Silicon Carbide Detectors for in vivo Dosimetry. *IEEE Trans. Nucl. Sci.* 2014, *61*, 961–965.

flushed with Ar/CO_2 (90/10%) at atmospheric pressure with a flow of 5 l/h. One of the electrodes was used as cathode and connected directly to a negative high-voltage supply. The other electrode was used as anode and connected to ground through a protection resistor and an ammeter used to measure the collected current as shown in figure 2c. The ionisation chamber was placed in front of a copper (Cu) target X-ray tube which was operated with an acceleration voltage of 20 kV and varying tube currents. The collected anode current was measured as a function of the X-ray tube current for different electric fields between the two electrodes ranging from 250 V/cm to 2000 V/cm as shown in figure 3.

While saturation attributed to recombination was observed for low electric fields, the response of the ionisation chamber was approximately linear over the full range of investigated irradiation intensities with an electric field of 2000 V/cm between the electrodes.

Readout Strip Anode

The same printing approach of combining PLA and graphite-loaded PLA with a dual-head FFF printer was also used to create a 2D strip anode with readout strips in orthogonal directions. The readout anode is shown in figure 4a. The bottom strips were 3 mm wide while the top strips were 2 mm wide. Bottom as well as top strips were located at a pitch of 6 mm and were 0.5 mm thick. The insulating layer between the strips was 0.5 mm thick. The electrical contact was brought to the back of the anode by printing a conductive path from the front to the back of the structure.

The readout anode was used to read out signals from a triple-GEM stack in a Time Projection Chamber (TPC) [13, 14] setup as shown in figure 4b. It was placed below the last GEM in the stack with an induction gap of 1 cm.

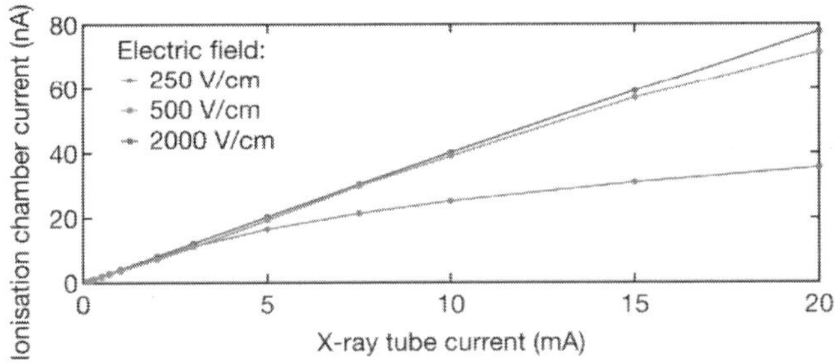

Figure 3: Anode current in 3D printed ionisation chamber as a function of X-ray tube current for different electric fields.

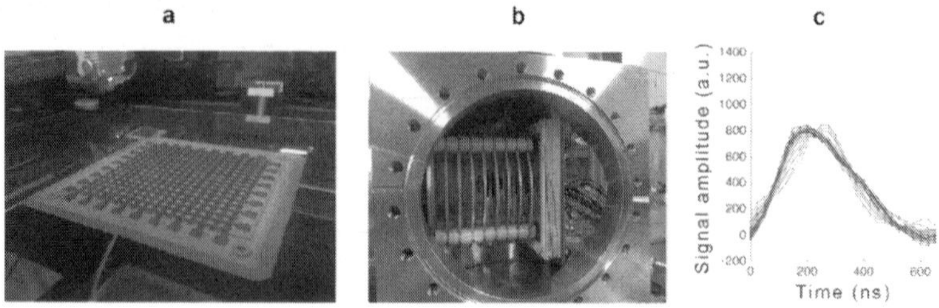

Figure 4: Readout strip anode with crossed strips in orthogonal directions. (a) 3D printed readout anode with an active area of 10×10 cm^2 . (b) TPC based on triple-GEM detector read out with 3D printed strip anode. (c) Signals of alpha particle event recorded from readout strips.

The strips were connected to an APV25 ASIC [15] and read out by the RD51 Scalable Readout System (SRS) [16]. The APV25 was triggered by a signal from a PhotoMultiplier Tube (PMT) recording secondary scintillation light pulses. The detector was operated in an Ar/CF_4 (80/20%) gas mixture at atmospheric pressure at a flow rate of 5 l/h. A gaseous radon (Rn) source was used to bring alpha particles into the gaseous volume and individual events were recorded. Figure 4c shows an example of an alpha particle event with waveforms from different strips displayed. The shift in the arrival time of the pulses can be used to deduce the orientation and angle of the recorded alpha particle track.

Although it was possible to record electronic signals from a 3D printed composite readout strip anode, printing performant readout anodes remains challenging. The low conductivity of the graphite-loaded PLA might decrease the recorded signal amplitudes and modify the timing characteristics of electronic signals. Although higher conductivity filaments are available, the achieved conductivities are orders of magnitude below those of metals. Charging up effects might also affect the stability of recorded signals and remain to be investigated for 3D printed readout structures. While these challenges might be addressed by future studies and optimisations, the resolution which can be achieved with FFF 3D printing limits the applicability of this manufacturing approach. With a nozzle size of approximately 0.4 mm, millimetre-scale structures can be achieved but strip widths below 0.5 mm and pitches below 1.5 mm are hard to achieve. This may be sufficient for some coarse applications but cannot be used to achieve fine readout structures for good spatial resolution. It is also not enough to 3D print amplifying structures capable of achieving electron avalanche multiplication.

INKJET 3D PRINTING

Thick Gaseous Electron Multiplier

High-resolution 3D printing with multiple materials was used to create a Thick Gaseous Electron Multiplier (THGEM) structure [17]. A THGEM geometry with a total active area of 5×5 cm^2 divided into two sectors with different hole diameters and pitches was designed. The CAD model featured cylindrical holes with diameters of 600 μm or 800 μm in the two sectors with a hole pitch of 1 mm or 1.5 mm, respectively. The nominal dielectric thickness was 400 μm and the nominal electrode thickness was 2 μm for both the top and bottom electrodes.

The employed 3D inkjet printing process is typically used for printing individual conductive traces on extended dielectric substrates for prototyping PCBs. The large conductor surfaces and relatively large thickness of the THGEM structure presented challenging conditions for the print process resulting in some adaptions of the printed geometry with respect to the CAD model. While these changes were not ideal for the operation of the THGEM structure, they might be improved by further process optimization.

Due to the printing process, the holes in the printed structure were conical instead of cylindrical and larger in diameter than desired. In the left sector of the THGEM shown in figure 5a, the hole diameter was increasing from 772 μm at the bottom to 928 μm at the top of the holes. In the right sector, the average hole diameters were increasing from 604 μm at the bottom to 790 μm at the top. The dielectric layer printed with polymer ink was 397 μm thick. The variation of the thickness of the printed structure across its surface was on the scale of micrometres. The thickness of electrodes printed with Ag nanoparticle ink was adjusted for low resistivity. While the bottom electrode was printed 2.8 μm thick, the top one was printed 21 μm thick as shown in figure 5c. The significantly larger thickness of the top electrode was necessary to achieve low resistance (<1 Ω between the two sides of the electrode) due to accumulated surface roughness during the printing process. The 3D inkjet printer deposits many layers consecutively to build up 3D structures from bottom to top and the resulting surface roughness of the dielectric layer required a thicker top electrode to ensure good electrical conductivity. The thickness variations of the structure can be seen in the microscopic image of the border between the two sectors in figure 5b.

Across each electrode, a resistance below 1 Ω was achieved. The top and bottom electrodes were separated by a resistance larger than 100 GΩ. The capacitance of each sector was approximately 110 pF. The device could withstand up to 1600 V potential difference in air. This is significantly lower compared to conventional THGEMs with a thickness of 400 μm, which can typically withstand well above 2000 V potential difference in air.

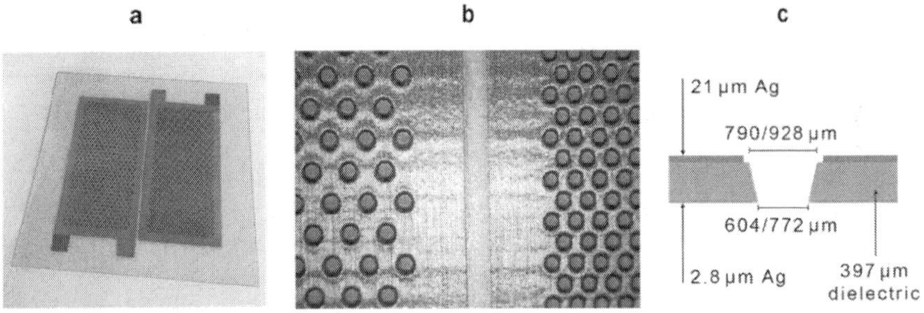

Figure 5: Printed THGEM detector. (a) Device with two sectors with different hole diameters and pitches. (b) Microscopic top view of the two sectors with different hole geometries along the border between them. (c) Measured dimensions of the THGEM structure. The two different numbers given for the top and bottom hole diameters correspond to the two sectors. Schematic not drawn to scale.

In contrast to the top electrode, the printed bottom electrode was very flat due to the fact that it was printed directly on the uniform print bed of the 3D inkjet printer. This difference in the visual appearance of the two electrodes can be seen in figure 6, where microscopic images of the top (a) and bottom (b) electrodes are shown.

Thin dielectric residuals were observed in the holes as shown in figure 6a, which are attributed to dielectric ink flowing down into the holes and accumulating on the print bed during the printing process before hardening. These residuals were effectively decreasing the free hole size in the printed THGEM. Manual removal of the residuals was possible, but they may be avoided in future prints by careful print parameter tuning.

As dielectric and conductive structures were printed consecutively with dedicated print heads, precise alignment between the two heads is important. Minor misalignment between the holes in the dielectric structure and in the printed electrodes was observed as shown in figure 6b. The magnitude of this misalignment varies across the active area of the device with almost perfect alignment achieved on one side and significant misalignment on the other side of the structure. This misalignment results in a crucial weakness in the high-voltage stability of the device. Due to the misalignment, the electrodes reach up to the hole in the dielectric without any rim remaining which is a weak point when high voltage is applied. Indeed, the regions with significant misalignment were found to be most susceptible to instabilities and discharges occurred predominantly where dielectric-conductor misalignment was most pronounced. Figure 6c shows a microscopic view of a region of the bottom electrode after discharges on several holes. Three holes with a strong discoloration around them are visible and the discharges appear to be focused on the side of the holes were the conductor reaches up the hole in the dielectric without any remaining rim. The maximum achievable voltage in an Ar/CO_2 (70/30%) gas mixture was limited to approximately 1100 V before discharges occurred frequently. Better alignment between dielectric and conductive structures may be achieved by further print process optimisations and might help to achieve significantly higher potential differences across the THGEM without the onset of discharges.

The THGEM was mounted in a gas volume to study its response as shown in figure 7a. To visualise signal amplification across the active area of the device, optical readout of the detector under X-ray irradiation was used.

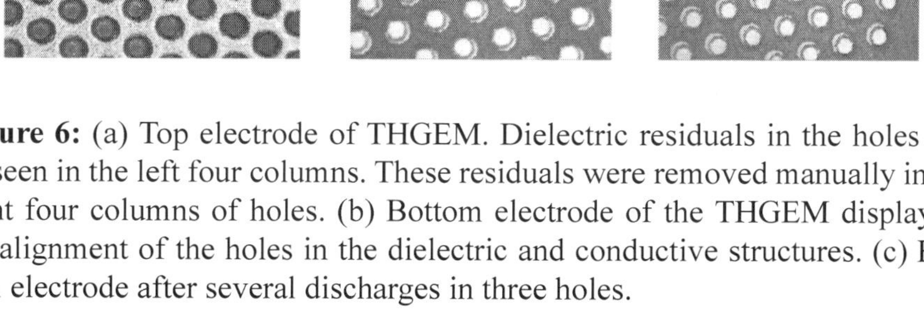

Figure 6: (a) Top electrode of THGEM. Dielectric residuals in the holes can be seen in the left four columns. These residuals were removed manually in the right four columns of holes. (b) Bottom electrode of the THGEM displaying misalignment of the holes in the dielectric and conductive structures. (c) Bottom electrode after several discharges in three holes.

The THGEM was placed between a thin foil cathode and a transparent ITO-based anode [18, 19]. The distances between the cathode and the THGEM as well as between the THGEM and the anode were 4 mm each. The gas volume was flushed with an Ar/CF4 (80/20%) gas mixture at atmospheric pressure at a flow rate of 5 l/h. A gas mixture containing CF4 was chosen for its high secondary scintillation light yield in the visible wavelength region [20] to allow for optical readout with CCD or CMOS cameras [21]. A low-noise CCD camera (QImaging Retiga R6 [22]) was placed outside of the gas volume at a distance of approximately 20 cm to record secondary scintillation light emitted during electron avalanche multiplication in the THGEM. The detector was irradiated with a Cu X-ray tube operated with an acceleration voltage of 20 kV and a tube current of 1 mA. The top electrode of the THGEM was biased with a negative high voltage power supply while the bottom electrode was grounded. A drift field of 50 V/cm and an induction field of 1000 V/cm were used. The potential difference across the THGEM was gradually increased up to 950 V where discharges started to occur. At each voltage, an image as shown in figure 7b was recorded with an exposure time of 10 s. A hardware binning setting of 8×8 pixels was used to collect signals of an 8×8 matrix of physical pixels and read them out as one virtual larger pixel, thus minimising the relative contribution of readout noise and increasing the signal-to-noise ratio while decreasing image resolution. Background images without X-ray irradiation were recorded and subtracted from the recorded images. The brightness of the resulting images was used as a quantification of the gain of the detector. Figure 7c shows the average pixel value intensity as a function of THGEM voltage where the onset of avalanche multiplication is visible around 300 V. Dividing the pixel value at the maximum THGEM voltage of 950 V by the one at low voltages, a gain factor of approximately 5 was determined. Instabilities of the detector at higher THGEM voltages prohibited higher gain factors.

To record spectra of the electronic signals from the THGEM detector, it was mounted as the first amplification stage in a triple-stage detector above two conventional GEM foils as shown in figure 8a. Signals were recorded from a pad anode with amplifying and signal shaping electronics and a Multi-Channel Analyser (MCA) to acquire histograms of signal amplitudes. In this setup, the detector was operated in an Ar/CO_2 (70/30%) gas mixture at atmospheric pressure at a flow rate of 5 l/h.

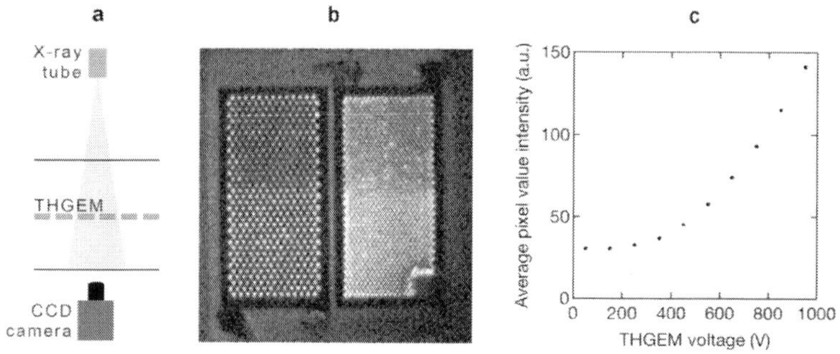

Figure 7: Optical readout of printed THGEM. (a) X-ray irradiation setup with THGEM and optical readout. Schematic not drawn to scale. (b) Optically read out image with a THGEM voltage of 950 V. The bottom right corner was disconnected after destructive discharges. The dielectric residuals in the holes shown in figure 6a were removed from the top halves of both sectors. The higher pixel values observed in the bottom halves of the sectors are attributed to the confinement of the holes by the remaining dielectric residuals in these regions. (c) Pixel value intensity as a function of THGEM voltage.

The THGEM was placed 4 mm below the cathode and 3 mm above the first of the two GEM foils. A transfer gap of 2 mm was used between the two GEM foils. The anode was placed 3 mm below the last GEM foil. In the drift region, an electric field of 50 V/cm was applied. In the transfer region between the THGEM and the GEM foil, a transfer field of 1500 V/cm was used. An induction field of 2000 V/cm between the last GEM foil and the anode was used. The two GEM foils were operated with a voltage difference of 430 V across each GEM. The detector was irradiated with a collimated ^{55}Fe source. The charging up behaviour of the THGEM structure was not addressed in this study and will need to be investigated to understand the dynamics of the detector response under irradiation. The ^{55}Fe source used to record energy spectra had a considerably lower rate than the X-ray tube used to record optically read out images as shown in figure 7. Thus, the time scales of charging up of the THGEM detector may be significantly different.

The acquired energy spectra from the detector for different THGEM voltages are shown in figure 8b. The peak at an amplitude around 250 is the signal from X-ray photons converting between the THGEM and the GEM foils. Events converted in this region are amplified only by the two GEM foils and their amplitude is independent of the THGEM voltage. Thus, this peak remains at the same amplitude for all THGEM voltages. The dominant peak at lower amplitudes is composed of X-rays interacting in the drift region between the cathode and the THGEM and its position thus depends on the THGEM voltage. For higher THGEM voltages, the peak is shifted to higher amplitudes. Since the gain of the two GEM foils was kept constant, this shift reflects signal amplification in the THGEM. The significant width of the spectra and their degradation with higher THGEM voltages is attributed to poor collection and extraction of electrons from the holes of the THGEM due to geometrical issues as shown in figure 6. In the case of optical readout, scintillation light produced in the holes can be recorded independently of electron extraction from the holes and this effect was therefore not observed in the measurement shown in figure 7c.

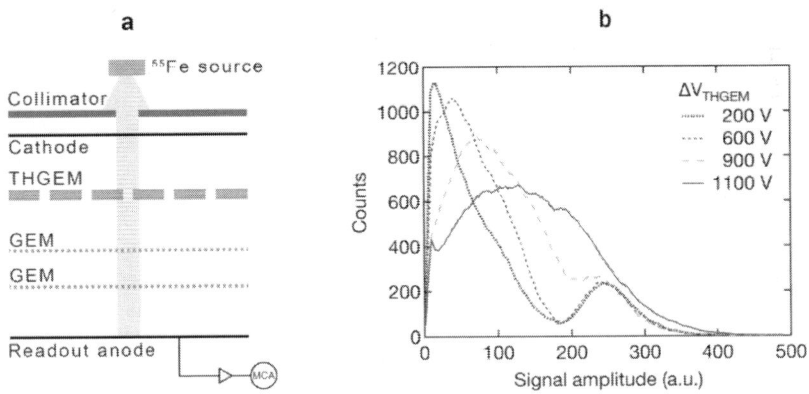

Figure 8: Electronic readout of printed THGEM. (a) X-ray irradiation setup with THGEM placed in a gas volume above two standard GEM foils. Signals from the anode were recorded with a Multi-Channel Analyser (MCA). Schematic not drawn to scale. (b) Histogram of anode signal amplitudes for different THGEM voltages.

The discrepancy between the optical and electronic measurements is attributed to the poor electron transfer through the THGEM. This inhibits the recording of energy spectra resolving the double peak structure of 55Fe in gaseous detectors and the quantification of the gain of this structure from electronic signals and will be addressed in further studies. Nevertheless, signal amplification can be observed in the recorded energy spectra and improved print quality may increase the collection and extraction efficiencies of the device.

CONCLUSION

Additive manufacturing can be used to create functional gaseous radiation detectors. Multi-material FFF 3D printing was employed to manufacture an ionisation chamber which displayed a linear response to X-ray radiation. In addition, a 2D strip anode was printed and used to read out electronic signals from a triple-GEM detector. FFF 3D printing may be used for coarse geometries including pad readout anodes with varying pad sizes as well as intricate 3D structures. However, it does not allow for sufficient resolution to create amplifying structures.

Multi-material inkjet 3D printing enables high-resolution functional structures and was used to print a THGEM structure. The THGEM was operated in Ar/CO_2 and Ar/CF_4 gas mixtures and read out by optical and electronic means. Signal amplification in the 3D printed structure could be observed. Process related issues such as misalignment of dielectric and conductive materials and dielectric residues in the holes limited the maximum voltage and thus the achievable gain in the device and are subject to further process optimization.

Additive manufacturing may enable advanced detector geometries not accessible by photolithographic techniques and allow for MPGD structures which are inherently stable against discharges.

Precise multi-material 3D printing of functional devices could significantly benefit radiation detector research by offering a fast and cost-effective approach to prototyping and can enable results-driven detector optimisation.

ACKNOWLEDGMENTS

We gratefully acknowledge the technical support and important contribution in manufacturing the THGEM device by inkjet 3D printing of Simon Fried, Jaim Nulman, Gal Begoun, Udi Zamwel and Robert Even from Nanodimension.

REFERENCES

1. F. Sauli, GEM: A new concept for electron amplification in gas detectors, Nucl. Instrum. Meth. A 386 (1997) 531.
2. Y. Giomataris, P. Rebourgeard, J.P. Robert and G. Charpak, MicrOMEGAs: A high granularity position sensitive gaseous detector for high particle flux environments, Nucl. Instrum. Meth. A 376 (1996) 29.
3. H.N. Chia and B.M. Wu, Recent advances in 3D printing of biomaterials, J. Biol. Eng. 9 (2015) 1.
4. K.V. Wong and A. Hernandez, A Review of Additive Manufacturing, ISRN Mech. Eng. 2012 (2012) 1.
5. Y. Mishnayot, M. Layani, I. Cooperstein, S. Magdassi and G. Ron, Three-dimensional printing of scintillating materials, Rev. Sci. Instrum. 85 (2014) 085102 arXiv:1406.4817..
6. J. Son et al., Improved 3D Printing Plastic Scintillator Fabrication, J. Korean Phys. Soc. 73 (2018) 887.
7. S. Fargher, C. Steer and L. Thompson, The Use of 3D Printing in the Development of Gaseous Radiation Detectors, EPJ Web Conf. 170 (2018) 01016.
8. M. Hohlmann, Printing out Particle Detectors with 3D-Printers, a Potentially Transformational Advance for HEP Instrumentation, in Proceedings, 2013 Community Summer Study on the Future of U.S. Particle Physics: Snowmass on the Mississippi (CSS2013): Minneapolis, MN, U.S.A., July 29 – August 6, 2013, arXiv:1309.0842.
9. Black Magic 3D, Black Magic 3D, https://www.blackmagic3d.com, (8 May 2018).
10. Nano Dimension, www.nano-di.com, (20 September 2019).
11. Leapfrog 3D Printers, Leapfrog 3D Printer s, https://www.lpfrg.com, (7 May 2018).
12. Protoplant, Proto-pasta, www.proto-pasta.com, (30 September 2019).
13. D.R. Nygren, The Time Projection Chamber — A New 4pi Detector for Charged Particles, eConf C740805 (1974) 58.
14. F.M. Brunbauer et al., Live event reconstruction in an optically read out GEM-based TPC, Nucl. Instrum. Meth. A 886 (2018) 24.

15. M.J. French et al., Design and results from the APV25, a deep submicron CMOS front-end chip for the CMS tracker, Nucl. Instrum. Meth. A 466 (2001) 359.
16. S. Martoiu, H. Müller, A. Tarazona and J. Toledo, Development of the scalable readout system for micro-pattern gas detectors and other applications, 2013 JINST 8 C03015.
17. R. Chechik, A. Breskin, C. Shalem and D. Mörmann, Thick GEM-like hole multipliers: Properties and possible applications, Nucl. Instrum. Meth. A 535 (2004) 303 physics/0404119..
18. T. Fujiwara, Y. Mitsuya, T. Yanagida, T. Saito, H. Toyokawa and H. Takahashi, High-photon-yield scintillation detector with Ar/CF 4 and glass gas electron multiplier, Jap. J. Appl. Phys. 55 (2016) 106401.
19. F.M. Brunbauer et al., Combined Optical and Electronic Readout For Event Reconstruction in a GEM-based TPC, IEEE Trans. Nucl. Sci. 65 (2018) 913.
20. M.M.F.R. Fraga, F.A.F. Fraga, S.T.G. Fetal, L.M.S. Margato, R. Ferreira-Marques and A.J. P.L. Policarpo, The GEM scintillation in He CF-4, Ar CF-4, Ar TEA and Xe TEA mixtures, Nucl. Instrum. Meth. A 504 (2003) 88.
21. F.M. Brunbauer et al., Radiation imaging with optically read out GEM-based detectors, 2018 JINST 13 T02006.
22. QImaging Corporation, Retiga R6, (2016).

SECTION 7:
RADIOCHEMICAL TECHNIQUES

16

New Ultra-sensitive Radioanalytical Technologies for New Science

Pavel P. Povinec

Department of Nuclear Physics and Biophysics, Faculty of Mathematics, Physics and Informatics, Comenius University, Mlynska dolina F1, 84248 Bratislava, Slovakia

ABSTRACT

Recent developments in radiometric and mass spectrometry technologies have been associated in the radiometric sector mainly with underground operations of large volume Ge detectors, while the mass-spectrometry sector, represented mainly by accelerator mass spectrometry and inductively coupled plasma mass spectrometry has become the most sensitive technique for ultra-low-level analyses of long-lived radionuclides. These new developments have had great impact on investigations of rare nuclear processes and applications of radionuclides in environmental, life and space sciences. New scientific investigations have been carried out therefore

Citation: Povinec, P.P. "New ultra-sensitive radioanalytical technologies for new science". J Radioanal Nucl Chem 316, 893–931 (2018). https://doi.org/10.1007/s10967-018-5787-3

which have not been possible before either because of lack of sensitivity or required large sample size.

INTRODUCTION

Radioanalytical technologies have always been a limiting factor for experiments in nuclear sciences comprising investigations of rare processes in nuclear physics and chemistry, in space research, in environmental radioactivity studies, in isotope oceanography and hydrology, in biomedical research and in many other branches of science. This has been mainly because of the fact that the available sensitivity was not high enough to get meaningful results, or the required sample size was too big to carry out such investigations [1,2,3,4]. Environmental radionuclide tracer studies began about 60 years ago with applications of cosmogenic (^{14}C and ^{3}H) and radiogenic (^{210}Pb, ^{226}Ra) radionuclides. This has been a rapidly developing field, which included studies on behaviour of radionuclides in the environment, use of natural and anthropogenic radionuclides as tracers of environmental processes, marine radioactivity studies, radiation protection, radioecology, etc., to mention at least a few of them. These studies have always been limited by the techniques available for sampling and analysis of radionuclides in the environment. As the levels of radionuclides observed recently in the environment have been very low, high sensitive radioanalytical systems have been required for carrying out new environmental investigations.

Traditionally radionuclide analyses have been carried out using alpha-, beta-, and gamma-spectrometers, jointly called radiometric counting systems. In the field of radiometric analytical technologies we moved from simple radiochemical radionuclide separation methods and gas counters to efficient robotic radiochemical techniques and sophisticated detectors working on line with powerful computers. The radiation detectors have often been situated underground, or they have been using anticosmic/antiCompton shielding to protect them against cosmic radiation. Such arrangements have considerably decreased the detectors background, and thus increased their sensitivity for analysis of radionuclides at very low levels in various types of samples.

Simultaneously, the philosophy of sampling and laboratory measurements has changed, where appropriate, to in situ analysis of radionuclides in the air, on land, in water and in the sediment, thus enabling developments of concentration maps and/or time series on radionuclide distributions in the investigated environments [1, 3, 5,6,7,8,9,10,11,12,13]. This has been a complimentary detection method to a traditional sampling and laboratory analysis, which can have several advantages, e.g. it can avoid a complicated and laborious sampling, pre-concentration and separation works, and long-term measurements are not necessary, especially when mapping of gamma-emitters in large surface areas would be required. The systems commonly used for underwater gamma-spectrometry are mostly based on NaI(Tl) detectors, with one exception when a methane cooled Ge detector was used [5, 6, 8]. The advantage of NaI(Tl)-based systems is firstly related to the high detection efficiency of NaI(Tl) crystals at much lower cost than equivalent Ge crystals. The drawbacks of NaI(Tl) systems are the high-power consumption for the operation of the photomultiplier tube, and a relatively poor energy resolution. The Ge-based systems have the advantage of good energy resolution, and hence excellent radionuclide identification capability. Other available types of detectors—crystals and semiconductors have not become competitive with NaI(Tl) and Ge detectors mainly because of their lower efficiency for gamma-rays in the energy interval of 30–3000 keV. The in situ technology has enabled to carry out not only radionuclide mapping experiments (for example using a towed detector system placed on a boat [9, 10]), but also to develop on line monitoring systems, either for monitoring of radionuclide releases from nuclear installations, investigations of changes in oceanic current systems [13], as well as for monitoring radon decay products in submarine groundwater discharge studies [14,15,16,17].

There have been several motivations for new developments of laboratory-based ultra-sensitive radioanalytical techniques in nuclear and environmental studies: (i) Levels of anthropogenic radionuclides after over 60 years of their injection to the environment have decreased considerably [1,2,3,4, 18,19,20]; (ii) Sample size required for radiometric analyses should be comparable to mass spectrometry analyses (e.g. in seawater profile sampling Rosette systems with 10–20 L bottles should be used instead of large 200 L (for ^{137}Cs) or 400 L (for Pu isotopes) sampling containers [1, 3, 4, 21,22,23,24,25,26]; (iii) Highly accurate, precise and traceable data are required for environmental and climate change studies, which would require detection limits below 1 nBq/g [1,2,3,4, 27,28,29,30,31]; (iv) New scientific ideas—such as investigations of rare nuclear processes and decays, investigations of cosmic dust, solar variations, supernova explosions, deep-sea bottom studies, DNA studies, environmental biotechnology, environmental nanotechnology, climate change studies, etc., have not been possible to realize till now as they have been requiring new ultra-sensitive radioanalytical technologies [32,33,34,35,36,37,38], with sensitivities good enough to analyse even very small samples.

The recent developments in low-level radioanalytical techniques have helped to improve the detection limits using: (i) High efficiency Ge detectors (up to 200% of relative efficiency compared to 76 mm in diameter and 76 mm long NaI(Tl) crystal) operating in underground laboratories which represent the most important achievement in radiometric analysis of radionuclides [1,2,3,4, 18,19,20,21,22,23,24,25,26, 35, 36]; (ii) Anticosmic shielding of detectors operating in surface and shallow-depth underground laboratories [39,40,41,42] which helped to improve detection limits comparable to deeper underground laboratories; (iii) Multidetector coincidence gamma-spectrometry systems for analysis of cascade and/or positron emitters [43,44,45,46,47,48,49,50,51]; (iv) Multidimensional gamma-spectrometry [52, 53]. Thanks to excellent energy resolution of Ge detectors (< 2 keV for 1.33 MeV gamma-rays), and possibility to analyse various gamma-emitters in material and environmental samples selectively and very often non-destructively, they dominated in the field of ultra-low-level gamma-spectrometry [54,55,56,57,58,59,60]. If such spectrometers can operate at least a few tens of meters underground, their performance can be superior in comparison with laboratories operating at the surface [39, 40, 47, 56, 60]. However, the underground gamma-spectrometry techniques are restricted to gamma-emitters only such as ^{7}Be, ^{40}K, ^{54}Mn, ^{60}Co, ^{137}Cs, ^{210}Pb, and many others. For analysis of beta-emitters found in the environment (e.g. ^{3}H, ^{14}C, ^{81}Kr, ^{85}Kr, ^{90}Sr, ^{133}Xe, ^{135}Xe, ^{241}Pu, and others), gas counting [61,62,63,64,65, 66,67,68,69,70,71,72,73,74,75,76,77] and liquid scintillation spectrometers have been used [37, 78,79,80]. In some applications, however, radiometric techniques of analysis of long-lived beta-emitters did not allow to carry out state of the art nuclear and environmental research either because of lack of sensitivity or requirement of large samples for analysis.

The most important break-through in the radioanalytical technologies has been, however, a general change in philosophy of radionuclide analysis. We moved from the concept of counting of radioactive decays (and thus waiting for them for a long time in the case of long-lived radionuclides) to the direct counting of atoms (as they would be stable) using high-sensitive mass spectrometers working either with low-energy ions (e.g. Inductively Coupled Plasma Mass Spectrometry—ICPMS [81,82,83,84,85], Resonance Ionization Mass Spectrometry—RIMS [86, 87], Thermal Ionization Mass Spectrometry—TIMS [88, 89], Secondary Ionization Mass Spectrometry—SIMS) [2, 90], or with ions accelerated up to hundreds of MeV in accelerator mass spectrometry (AMS) systems [91,92,93,94,95,96,97,98]. These developments have considerably improved the detection limits for analysis of radionuclides in all scientific applications requiring ultra-high radionuclide sensitivities and small sample size.

Two mass-spectrometry systems have recently been dominating in the field of ultra-sensitive radionuclide research, namely AMS and ICPMS, thanks to their ultra-high sensitivity, accuracy and precision, as well as to their ability to reach high output rates in analyses of long-lived radionuclides in all types of samples. The AMS dominated in analyses of several long-lived radionuclides such as ^{10}Be, ^{14}C, ^{26}Al, ^{41}Ca, ^{129}I, U and Pu isotopes, and others [91,92,93,94,95,96,97,98]. For analysis of some other long-lived radionuclides (e.g. ^{99}Tc, ^{129}I, U and Pu isotopes), the ICPMS has recently been representing, however, also very important breakthrough.

Another important trend which we should mention for applications of radionuclides in environmental sciences is connected with a transfer from a simple bulk sample analysis to specific compound analysis of stable and radioactive isotopes, very often resulting in use of coupled analytical systems, e.g. coupling of gas chromatography with AMS systems. By a combination of gas chromatographs, which select and separate necessary amounts of specific compounds, with an AMS which perform mass isotope analysis, it has been possible to open new windows for new isotope research [98,99,100,101]. This trend is having, however, more general applications, e.g. in geochemistry and cosmochemistry, where move from bulk radionuclide analyses of samples to specific radionuclide analysis of minerals would be feasible. More attention should also be given to analyses of radionuclides on particles, and generally on speciation studies of radionuclides in the environment [3].

Sample sizes have been constraining many new environmental investigations, which could be illustrated e.g. for surface seawater sampling and analysis: (i) 0.5 L sample size limit has already been achieved for ^{3}H analysis using in-growth mass spectrometry of ^{3}He [2, 102, 103] and in AMS analysis of ^{14}C and ^{129}I [2, 92, 93, 98, 102,103,104,105]; (ii) About 5 L seawater samples could be analysed for ^{137}Cs in deep-underground gamma-spectrometry laboratories such as Gran Sasso (Italy) and Modane (France); (iii) About 10 L samples could be analysed for ^{137}Cs in shallow and medium-depth underground gamma-spectrometry laboratories, operating at about 100 m water equivalent, w.e. (for comparison of laboratories operating in different rock environment we normalize them to the water depth) [60,

106,107,108]; (iv) About 10 L samples could be analysed for Pu isotopes by AMS [60, 96, 109, 110], and similarly also by ICPMS and TIMS [81, 88, 97, 110]. These developments in the radioanalytical technologies have had important impacts on sampling strategies as well. Due to over two orders of magnitude decrease in the sampling size it has been possible to sample e.g. even deep-sea water columns using conventional Rosette systems during one/two casts only. This enabled to carry out high resolution radionuclide water profile studies, which were not realizable before using conventional large volume water samplers (200–500 L), just because of sampling and financial constraints [1,2,3,4, 60, 106,107,108,109]. This move from simple radioanalytical techniques to the present sophisticated state of the art technologies has also been accompanied by a considerable change in the philosophy of environmental studies as well. We moved from institutional environmental investigations to global international projects carried out during recent years, e.g. WOCE (World Ocean Circulation Experiment), CLIVAR (Climate Variability and Predictability), PAGES (Past Global Changes), WOMARS (World Ocean Marine Radioactivity Studies), SHOTS (South Hemisphere Ocean Tracer Studies), GEOTRACES (Geochemical Traces in the oceans) to mention at least a few of them.

We shall discuss in this review developments in ultra-sensitive radionuclide analysing techniques, focusing on radiometric and mass spectrometry methods. We shall shortly follow developments in radiocarbon measuring techniques from the Libby counter through proportional gas counters and liquid scintillation spectrometers to more recent developments of AMS. While during the first 60 years of the radiocarbon measurements the beta-counting, specifically the gas counting was the dominant technique, in the present the dominant technology in radiocarbon science is AMS. We shall also focus on gamma-spectrometry, on the development of large volume Ge detectors very often operating in shallow and deep underground laboratories. Monte Carlo simulations of detector background characteristics have been important pre-requisite when designing low-level counting systems, operating hundreds of meters underground, where radioactive purity of construction materials and radon concentration in the air has become dominant factors controlling the detector background. Mass spectrometry technologies, mainly developments and applications of AMS and ICPMS for environmental studies and for radiopurity measurements of construction parts of large-scale underground nuclear physics experiments will also be discussed in detail. The radiometric and mass spectrometry technologies will be compared with each other and with neutron activation (NAA) methods, which, especially in radiopurity measurements, represent a renaissance of this method for ultra-low-level uranium and thorium analysis in construction materials. These new developments in mass and radiometric spectrometry for ultra-low-level radionuclide analyses have had great impact on investigations of rare nuclear processes and applications in environmental, life and space sciences. We shall present at least a few examples (SuperNEMO experiment, isotope groundwater hydrology and radionuclide tracing in the marine environment). As the topic is very wide, it has not been possible to cover all aspects of sampling and radionuclide analyses in detail; the emphasis has been on recent developments in the field. More information can be found in already published review papers [1,2,3,4, 60, 112, 114, 115]. Similarly, the list of references, although very comprehensive, could not cover all the work done in this field, but it is mostly listing our recent publications, including book and book chapters [1,2,3,4, 112,113,114,115]. A few monographs and book chapters published by other authors have also been included to help readers to find more specific information [91, 94,95,96, 116,117,118,119,120,121,122,123,124].

I also organised several conferences on low-level counting and spectrometry with published proceedings, where more detailed information can be found [125,126,127,128,129,130,131]. Later, the low-level conferences have been regularly organized by the ICRM (proceedings published in Applied Radiation and Isotopes). For the environmental aspects, I co-organised conferences on environmental radioactivity ENVIRA (proceedings published in Journal of Environmental Radioactivity and in proceedings books). Many conferences were organized on radioanalytical/radiochemistry aspects (e.g. MARC-Kona, RADCHEM-Marianske Lazne, RANC-Budapest, with proceedings published in Journal of Radioanalytical and Nuclear Chemistry). AMS conferences had proceedings regularly published in Nuclear Instruments and Methods Sect. B).

We hope by publishing this paper to offer readers an overview of recent developments in ultra-sensitive radioanalytical technologies and their applications in nuclear and environmental sciences, to encourage and advice-them how to build the state of the art radioanalytical laboratories for nuclear and environmental research. As the paper contains many abbreviations and technical terms, Table 1 lists at least some of them which have been used frequently.

Table 1: Abbreviations and technical terms used in the paper

Abbreviation/term	Explanation
TU	Tritium Unit: 1 TU = 10^{-18} $^{3}H/^{1}H$, Eq. 0.118 Bq L^{-1} **of water**
pMC	Percent Modern Carbon is used for recent environmental samples, calculated against a reference sample of ^{14}C activity from a known standard; 100 pMC is ^{14}C activity for radiocarbon age of 0 y BP (before present, i.e. before 1950)
$\Delta^{14}C$	$\Delta^{14}C = \delta^{14}C - 2(\delta^{13}C + 25)(1 + \delta^{14}C/1000)$ (‰) $\delta^{14}C = [(^{14}C_{sample} - {}^{14}C_{standard})/{}^{14}C_{standard}] \times 10^{3}$ **(‰)**
AMS	Accelerator mass spectrometry
ICPMS	Inductively-coupled plasma mass spectrometry
TIMS	Thermal ionization mass spectrometry
PIXE	Particle-induced X-ray emission
IAEA	International Atomic Energy Agency

DEVELOPMENT OF RADIOMETRIC TECHNOLOGIES

Gas Counting Systems

My scientific carrier started with development of methods for analysis of radiocarbon in atmospheric carbon dioxide [65], and at the beginning it was heavily influenced by W. Libby, D. Lal, K. Münnich, H. Oeschger and many other radiocarbon scientists. There have been several important breakthroughs in the radiocarbon technologies, starting from the Libby counter in which the cathode was covered with carbon sample [132], followed by proportional gas counters [133,134,135], liquid scintillation spectrometers [136, 137], and finally developing AMS [138,139,140], which shifted the analytical concept from counting of ^{14}C decay products (and waiting for beta-electrons) into direct counting of ^{14}C atoms present in a sample. As the half-life of ^{14}C is relatively long (5730 year), the number of ^{14}C atoms present in a sample compared to number of ^{14}C beta-decays observed during one day of counting is ~ 3 × 10^{6}, significantly in favour of AMS. Recently we moved into a stage when a bulk sample analysis has been replaced by compound specific analysis, e.g. in gas chromatography-AMS coupled analytical systems, which opened new dimensions in the radiocarbon science [98].

A traditional method of activity measurement using a low current ionization chamber would not give meaningful results for radiocarbon dating because of high background. It was fortunate that gas counters for direct counting of pulses originating in the radiation detector as a result of radioactive decay of nuclei and emission of beta-particles were already in use. Libby [132] introduced a detector shielding to decrease its background (20 cm of iron), but this was not enough because a further, at least by a factor of ten decrease was required to get a reasonable background. Libby recognized that the hard component of cosmic rays (muons), penetrating even the heavy iron shield should be at least partially eliminated. Such a trigger was used for the first time by Blackett and Occhialini [141] to identify tracks of positrons in a cloud chamber. Libby found a revolutionary solution in using a similar trigger, however, in an anticoincidence regime, which eliminated pulses coming from the central detector, if they were simultaneously registered by Geiger tubes surrounding the central detector (i.e. the anticosmic veto). This arrangement decreased the counter background by a factor of twenty, and was good enough to proceed with regular radiocarbon measurements. It has been clear from the beginning of radiocarbon measurements that for achieving larger radiocarbon data outputs, and to ensure better precision, the Libby counter should be replaced by another type of a detector. Because of low energy of ^{14}C beta-electrons (their maximum energy is only 156 keV), the best solution would be to incorporate a sample in a form of a gas directly used in the detector. Therefore, it was natural to use carbon dioxide, the first product of sample combustion, which contains ^{14}C atoms from the sample. After several trials, this has been achieved by de Vries and Barendsen [133], although it has been recognized that carbon dioxide must be well cleaned to get reasonable counting characteristics. Simultaneously Suess [134] developed a method for preparation of acetylene, and Burke and Meinschein [135] for preparation of methane as suitable ^{14}C counting gases. Later we found that the cleaning of CO_2 from impurities such as water vapours, nitrogen and sulphur oxides, and halogen compounds is required due to its high sensitivity to electronegative impurities, which decreased the transit time of electrons from a place of their origin to the anode by about a factor of ten when compared with CH_4 [71]. As we predicted that the best counting characteristics should have CH_4, we constructed an apparatus for preparation of CH_4 [142], following the Lal's design [143].

Further advantages of using CH_4 were that we could use the same method for ^{3}H as well as for ^{14}C measurements, which was naturally later enlarged to simultaneous ^{3}H and ^{14}C counting in the same CH_4 gas filling [70, 74]. This method has been very promising, and even nowadays it has been used for simultaneous ^{3}H and ^{14}C measurements in natural methane [144].

It has also been well known from building first ^{14}C counters that only radioactivity free construction materials should be used for construction of low-background gas counters (e.g. quartz, electrolytic copper). After a careful analysis of background components of proportional counters, it has been found that the most important contribution to the background of the central detector is the thickness and the material of the inner cathode. The inner cathode should be very thin so the contribution of δ-electrons originating in the cathode by interactions of gamma-rays with the cathode material should be minimized [145]. This became feasible with introduction of a multiwire guard counter, surrounding the inner detector, so only a thin cathode foil could separate both detectors [68, 145] (Fig. 1). A detector with a lowest background should have an optimized thickness of the inner cathode, just for stopping beta-electrons originating from radioactive decay of e.g. ^{14}C (in the case of ^{14}C beta-electrons the optimum thickness is ~ 10 mg/cm^2). The cathode should be made from radioactively pure foil (e.g. gold coated Mylar foil or thin copper). In the case of ^{3}H counting (because of small energy of beta-electrons, max. 18.6 keV), the internal cathode foil may be even replaced by wires [68]. If the sensitivity of the detector should be increased, the detector could operate under higher than atmospheric pressure.

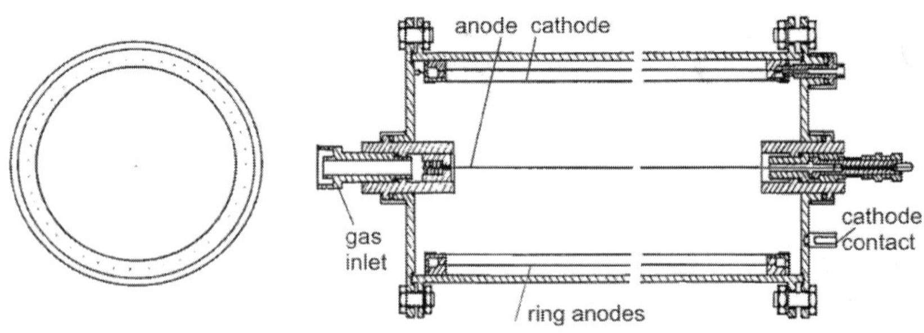

Figure 1: Low-background proportional counter with thin copper cathode for ^{3}H and ^{14}C counting [68].

A simple single wire inner counter has been later replaced by a system of cells forming thus a multielement detector (the Povinec detector) [146,147,148,149,150,151] (Fig. 2). Such a detection system can be used for tracking beta-electrons from decays of nuclei inside the detector. Except the 3H and 14C activity measurements, the multielement structure was also used for searching of double beta-decay of ^{136}Xe in the Gran Sasso underground laboratory (the first experiment carried out in the newly constructed Gran Sasso laboratory [151]). Even a more sophisticated high spatial resolution time projection chambers (possibly inserted into a magnetic field) could be constructed for tracking single beta-electrons [152]. However, these techniques would require large sample volumes. and with the invention of AMS for ^{14}C measurements, the radiometric detection systems, both gas, as well as liquid scintillation counters could not compete with AMS measurements. Radiocarbon laboratories with radiometric detectors have, however, significantly contributed to the radiocarbon science, especially in better understanding of the behaviour of radiocarbon in the environment. Most of the high precision ^{14}C results (a relative precision of $\Delta^{14}C$ in modern samples below 0.5%) for the development of the ^{14}C calibration curve were obtained using gas proportional and liquid scintillation detectors.

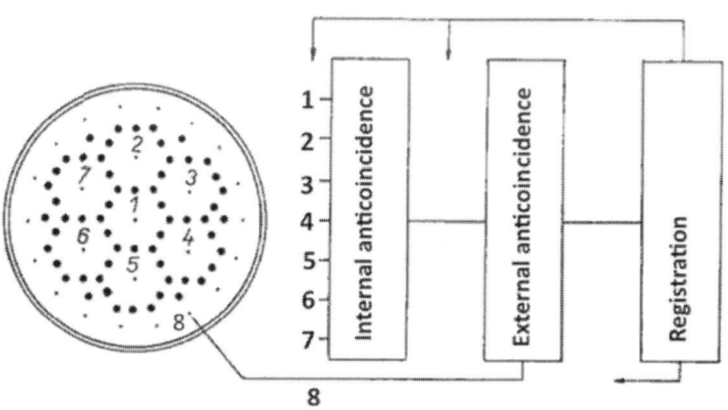

Figure 2: Cross-section of the 7-element proportional counter designed for ^{3}H counting (more elements can be added following the project requirements [147, 151]).

As the maximum energy of beta-electrons emitted by decaying ^{3}H and ^{14}C nuclei is only 18.6 and 156 keV, respectively, there is a room for decreasing a counter background by discriminating pulses which are above this energy threshold [153]. Using two channel electronics we can register e.g. ^{3}H pulses in the first channel and the background pulses above the ^{3}H

spectrum in the second channel, where about 50% of background pulses originate. Therefore, if we are measuring background in the second channel we may predict its value in the first channel even in the case when we were registering in this channels pulses from ^{3}H decays (the Povinec method). We may monitor in this way background variations during long-term measurements (more details can be found in [153]). This is possible to realize either by a multichannel analyser for pulse amplitude evaluation, or by a time–amplitude analysis. Such a registration system has also an advantage in simultaneous registration of pulses from a sample and background, which could help in discriminating false pulses originating either in various electromagnetic disturbances, or by sudden changes in cosmic-ray intensity. As in ^{3}H and ^{14}C analyses the counting times are typically over a few days, such background monitoring is useful [154]. Recently developed digital multichannel analysers have even better characteristics for long-term monitoring of the detector background, which is frequently used at present in commercial liquid scintillation spectrometers.

Radiochemical Separation Methods

Recent developments in radiometric and mass spectrometry technologies have been contributing significantly to new scientific investigations carried out in nuclear and environmental sciences. Sampling techniques have developed from simple devices operating with large volume samplers (from 100 to 500 L, Fig. 3) into the present Rosette multisampling system, and even to robotic systems based on ROVs (Remotely Operating Vehicles). Sophisticated sampling technologies have been accompanied with satellite views of areas for the optimisation of sampling. For example, in the marine environment, where the research work has been heavily depending on the new technologies, we have seen a replacement of time consuming and expensive large volume water sampling (500 L) from several km water depths by Rosette multisampling systems enabling high resolution water sampling of the water column within one or two casts only [1, 3, 4].

Figure 3: Large-volume (200 L) water samplers (top) compared with the Rosette multisampling system (24 bottles with 20 L volumes) (bottom) operating during deep-water sampling in the Pacific Ocean (IAEA'1997 cruise [22]).

There have also been great developments in radiochemical separation methods where we moved from a simple direct/nondestructive analysis of gamma-emitters to complicated radiochemical procedures separating radionuclides of interest from all possible matrices (water, sediment, biota, construction materials, etc.), and finally to robotic systems working on line with computers to minimize human working power in this, traditionally highly time-consuming works. As we have drastically decreased the levels of radionuclides to be analysed, especially in mass spectrometry methods, new problems have arrived, mostly associated with radioactive contamination of chemicals and tracers, and generally with radioactive contamination of laboratories (air, glassware, etc.). Therefore, clean rooms of class 100–1000 have to be used, if ultra-sensitive radionuclide analyses have to be carried out.

For analysis of all radionuclides of interest, developed radiochemical separation methods represent at present state of the art technologies, specifically focusing on type of the radionuclide, matrices of their occurrence and measuring methods to be applied for analyses. As this is a very widely developed field we shall not go into details, but readers can consult with specialised papers, reviews or monographs to get more information [1, 3, 60]. As a typical example, we present radiochemical procedures to be used for separation of ^{90}Sr, ^{137}Cs and Pu isotopes from seawater, and their subsequent analyses by beta-, gamma- and alpha-counting, and by AMS/ICPMS. After filtering (0.45 μm mesh) of collected seawater samples, on-board pre-concentration procedures (sequential extraction) were usually carried out to separate ^{90}Sr, ^{137}Cs and Pu isotopes from the collected seawater samples (MnO_2 co-precipitation, Fig. 4). ^{90}Sr is usually separated by co-precipitation with oxalic acid, and determined using the ^{90}Y in-growth method followed by beta-counting in gas or liquid scintillation counters [1, 3, 60]. ^{137}Cs is concentrated in seawater samples by adsorption onto AMP (ammonium molybdophosphate) using a method described in detail elsewhere [1, 60], and ^{137}Cs activities are determined either directly on AMP solution, or on separated Cs by low-level gamma-spectrometry with high efficiency Ge detectors. Thanks to these new developments it was possible to analyse small volume samples of seawater and thus to reach a high ^{137}Cs data density, which allowed to draw a detail picture on the spatial and depth distribution of ^{137}Cs in the Indian, Atlantic and Pacific Oceans [33, 107,108,109].

Figure 4: Large volume plastic containers (400 L) on the shipboard for scavenging of ^{90}Sr, ^{137}Cs, Pu isotopes and ^{241}Am from seawater samples (IAEA'1997 cruise [22]).

Transuranics are purified using anion-exchange resins and extraction chromatography [110, 111]. The samples are then either electrodeposited on stainless steel disks for alpha-spectrometry or used for ICPMS or AMS [60, 64, 81, 82, 110, 111, 155]. Figure 5 shows as an example a typical flow chart for separation of Pu from seawater and sediment samples and subsequent ICPMS, TIMS or AMS measurements [1, 3].

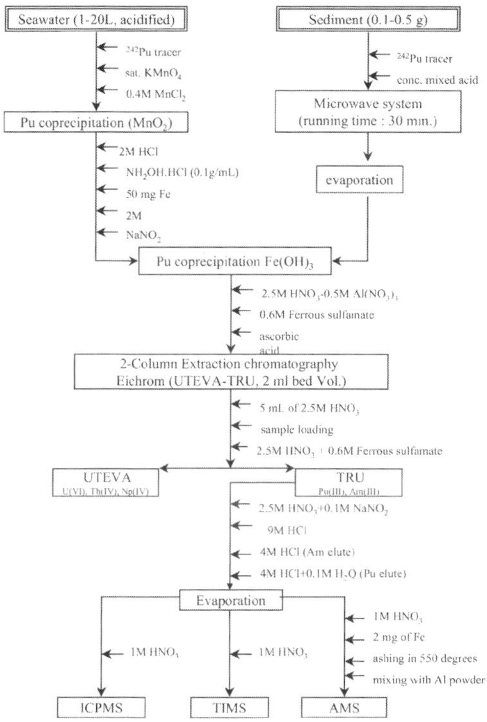

Figure 5: Separation of Pu from seawater and sediments samples and its subsequent analysis by ICPMS, TIMS or AMS.

High efficiency and excellent energy resolution of Ge detectors permit the analyses of gamma-emitters in composite samples selectively and very often non-destructively (e.g., in sea sediments). If such spectrometers can operate at least a few tens of meters underground, or they are protected against penetrating cosmic -ray muons by anticosmic shielding, their performance can be superior in comparison with laboratories located at the surface [1,2,3, 54, 56, 60]. However, this technique is restricted to gamma-emitters only (e.g., for ^{7}Be, ^{40}K, ^{54}Mn, ^{60}Co, ^{137}Cs, ^{210}Pb, etc.). Other radionuclides frequently found in the marine environment are the pure beta-emitters, like ^{3}H, ^{14}C, ^{32}Si, ^{32}P, ^{90}Sr, ^{241}Pu, etc., where mainly liquid scintillation spectrometry has made great improvements in recent years [112, 113, 156,157,158]. However, for some of these radionuclides mass spectrometry methods represent a real breakthrough in low-level counting, e.g., "^{3}He in-growth" mass spectrometry [159, 160] for ^{3}H, or accelerator mass spectrometry for ^{14}C [91, 94, 95].

Another important group of radionuclides is represented by alpha-emitters, both natural (like Ra, U and Th isotopes) as well as anthropogenic (like Pu and Am isotopes). These radionuclides have traditionally been analysed by semiconductor alpha-spectrometry. However, there were several limitations, e.g., in sensitivity, resolution and mass of samples used for analysis. Some of them have been partially overcome, e.g., the problems with resolution for analysis of ^{239}Pu and ^{240}Pu using high resolution alpha-spectrometers with suitable deconvolution software, however, recently especially ICPMS and AMS dominate in analysis of long-lived alpha-emitters in the environment, as these methods eliminate all the above-mentioned problems [2,3,4, 60, 96].

Gamma-spectrometers Operating in Surface Laboratories

Ge detectors with anticosmic shielding

Availability of large volume Ge detectors has been the most important developments in the radiometric sector. The reasons are in the excellent energy resolution and high efficiency of recently produced Ge detectors (up to about 200%). Muon-induced background becomes dominant for such large volume Ge detectors as the most prominent peaks observed (e.g. annihilation peak, neutron activation peaks) are due to cosmic -ray interactions with Ge detectors [2, 161]. Although the most effective way of increasing the sensitivity of a spectrometer is to increase counting efficiency and the amount of sample to be analysed, frequently, the only possible way is to decrease background of Ge detectors. The background components in a typical low-level Ge detector, not situated deep underground, are cosmic radiation (cosmic muons, neutrons, photons, and material activation products), radioactivity of construction materials, radon and its progenies. For a present-day, carefully designed low-level Ge spectrometer, the dominating background component is cosmic radiation, mainly cosmic-ray muons (Table 2). In a single Ge spectrometer there is no protection against cosmic muons, therefore, a spectrometer with anticosmic shielding will greatly reduce the background. The anticosmic shield can be made of gas or plastic scintillation detectors, which surround the lead/iron/copper shields housing the Ge detector [2, 39, 40, 47, 54, 161]. Another possibility is to use an antiCompton spectrometer, which is a powerful tool for reducing the detector's background as it combines both anticosmic and antiCompton background suppression [117].

Table 2: Detector background components of a low-level Ge detector of 1 kg mass located in a lead shield of 10 cm thick

Background component	**Integral counting rate (s^{-1})**
Environmental radionuclides	30–300
Muons	0.3–2
Cosmic neutrons	0.03–0.2
Radionuclides in the shield	0.01–0.2
Radon and its daughters	0.01–0.1
Radionuclides in the cryostat	0.003–0.03
Detector activation by cosmic rays	0.0004–0.002

A proper design of a low-level gamma-spectrometer is an important prerequisite for later applications in low-level measurements. High energy cosmic rays can initiate a large number of physical processes leading to background induction. Analytic solutions for describing these processes are not available, therefore statistical technique such as Monte Carlo simulation is necessary. The development of a simulation code for background induction is useful for the optimization of a counting system in respect to its background characteristics. It enables to assert the background before the system is built, and also to perform systematic investigation of the influence of various parameters on the background of the detector. The GEANT code has been used for the simulation of the passage of particles through matter as it meets the requirements for simulation with high-energy muons [2, 42, 161,162,163,164]. Monte Carlo predicts that a thickness of 15 cm of lead is the optimum shielding thickness for large volume HPGe detectors situated at sea level or at shallow depths underground. This is much more than usually used 10 cm of lead. If thicker shielding was used in simulations, the background was higher due to interactions of muons with the shield. The simulation also clearly shows that the smallest background is obtained in the smallest shield directly attached to the Ge detector. This has been a surprising result as in previous low-level counting studies opposite recommendations can be found [165]. It is interesting that the background depends only slightly on the shape of the shield if the inner shield dimensions are preserved. Rectangular shields provide only a few percent greater backgrounds than cylindrical ones. The background levels also strongly depend on the internal lining of the shield as the thick layers of low-Z lining increase the detector background (e.g. 1 cm of Cu increase background by factor of two when compared with lead-only shield), explained by smaller self-absorption coefficients for lower-Z materials. Therefore, the shield dimensions should be kept as small as possible (depending on the largest sample size intended for analysis), as the smallest background is obtained in the smallest shield (without any lining, i.e. only with lead walls). For example, it is not worth leaving an extra space in the shield of the well-type detector if samples are only analysed inside the detector well. If the lead X-rays are disturbing, a thin descending Z-lining is superior. It is advisable to design the shield so that the lining can be easily removed. The anticosmic shielding represent an important technology for improving the detection limits of gamma-spectrometers operating at the surface laboratories. A simple plastic scintillation detector placed on the top of the Ge detector which operates as a veto may decrease its background by about a factor of two [161]. The anticosmic shielding is advantageous to use also in shallow underground laboratories operating at depths down to about 100 m w.e. [161].

Multidetector gamma-spectrometry

In some applications, a better sensitivity can be obtained by operating a Ge spectrometer with Compton suppression, or as a gamma–gamma coincidence spectrometer. The antiCompton spectrometer combines both the anticosmic and the antiCompton suppression of the background as the principal Ge detector is surrounded by sufficiently large NaI(Tl) or BGO crystals. The Compton suppression factor should reach values around 40, thus improving detection limits significantly [2]. However, high cosmic-muon rejection factors can be reached only if construction materials with negligible radionuclide contamination have been used for the construction of the Ge detector's cryostat, the surrounding NaI(Tl) detectors and the passive shield. We should also operate the Ge detector in a place with low radon concentration. In specific applications, e.g. analysis of ^{210}Pb or ^{137}Cs in sea sediments in the presence of a high content of ^{40}K it is advantageous to use antiCompton gamma-spectrometer [2, 42, 161, 162]. Due to the relatively high energy of this gamma-emitter (1.46 MeV), radionuclides emitting lower energies of gamma-rays are covered by a Compton continuum from this source. Therefore, the suppression of this continuum (as well as the cosmic-ray induced background) has a great impact on the background of the spectrometer.

AntiCompton gamma-spectrometer with n-type Ge detector (ORTEC, 100% relative efficiency, the peak/Compton ratio 64) has been in operation at the IAEA Monaco laboratory (Fig. 6). The detector arrangement is of a U-type with a preamplifier situated outside of the lead shield, however the FET is mounted on Cu plate connected with cooling finger. The detector cryostat is made of electrolytic copper, the window is made of high purity aluminium. The Ge detector is surrounded by NaI(Tl) shielding. All detectors are housed in a shield made of 10 cm of lead. The background with antiCompton shielding decreased in the energy interval 30–2500 keV by a factor of 5. However, several gamma-lines from natural radionuclides, probably due to a contamination of the detector components and the lead shielding were observed in the Ge-detector background (Fig. 7) [42].

Figure 6: Construction of the anti-Compton gamma-ray spectrometer in the IAEA-EL´s Monaco underground laboratory (100% HPGe detector surrounded by NaI(Tl) detectors in a lead shield).

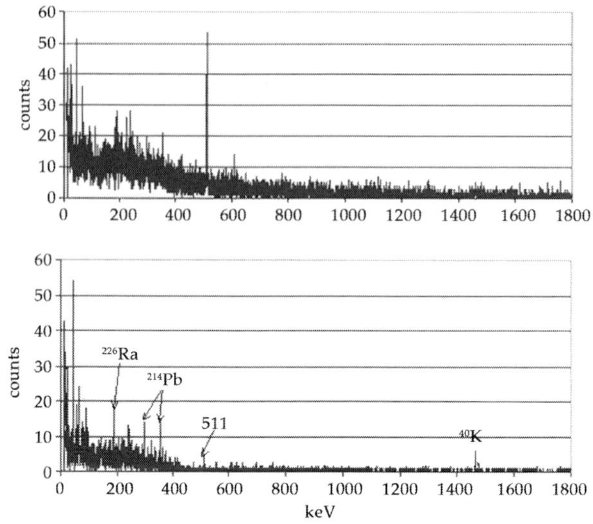

Figure 7: Background spectrum of the antiCompton spectrometer with 100% HPGe detector measured for 60,000 s in the IAEA-EL´s Monaco underground laboratory at 35 m w.e. (top—single spectrum; bottom—anticoincidence spectrum) [2].

A radioactive contamination of construction materials therefore plays a crucial role in low-level antiCompton spectrometers. As for the operation of a semiconductor detector the diode material, in our case Ge, must be very clean to keep loses of charge collection on the detector electrodes as low as possible, the most probably the detector contamination is caused by a cryostat material. However, in the case of the antiCompton spectrometer the main contaminated components may originate in the NaI(Tl) detectors and photomultipliers. The dominant contaminant there is ^{40}K, which is very difficult to remove from the NaI(Tl) detectors. Another possibility would be to use a BGO scintillation detector, which because of higher detection efficiency could have smaller dimensions. However, the radioactive contamination in this case is even worse due to presence of radioactive bismuth and lead. AntiCompton gamma-spectrometers because of contamination problems will not reach therefore a lowest possible background when compared with well-designed single Ge spectrometers. On the other hand, a cosmic-muon rejection factor of at least 40 (at around 1 MeV) is obtained when the antiCompton rejection is operational (Fig. 8) [2]. In such situation, the cosmic-muon background is reduced to such a level that other background components should prevail, like those from the residual contamination of the detector and antiCompton construction materials, or from radon progenies.

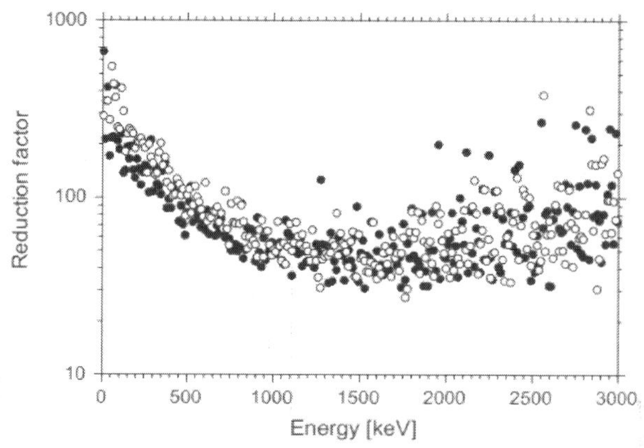

Figure 8: Monte Carlo simulated background reduction factors for an anti-Compton gamma-spectrometer operating at sea level (open circles—vertical positions, dots—horizontal positions).

Except of single and antiCompton gamma-spectrometer several other detection modes can be used in the coincidence-anticoincidence gamma-spectrometry: (i) Gamma-gamma coincidence mode (e.g. [44, 47]; (ii) triple coincidence mode (e.g. [45, 47]; (iii) summing coincidence mode (e.g. [45, 47]; (iv) beta-gamma coincidence mode (e.g. [47]; (v) beta-gamma-gamma coincidence mode (e.g. [47]; (vi) multidimensional gamma-spectrometer [47, 52, 53]; (vii) gamma–gamma spectrometer for measuring angular distribution of gamma-quanta. Coincidence gamma-spectrometers may decrease a background by about two orders of magnitude, what makes these spectrometers superior for very low-level gamma-spectrometry of positron or cascade gamma-emitters (e.g. ^{22}Na, ^{26}Al, ^{60}Co) [48]. Three-dimensional gamma-spectra (volumetric peaks with better identification) can be obtained from multidimensional spectrometers, which can register both the coincidence and non-coincidence peaks simultaneously. Analysing electronics, if two Ge detectors are used in coincidence, require 8000 × 8000 channels, which, with present state of the art computer electronics, is not a problem. The background of multidimensional gamma-spectrometers can be reduced by about two orders of magnitude.

Figure 9 compares gamma-spectra obtained by a single Ge detector (70% relative efficiency) with Ge–NaI(Tl) (10 cm in diameter and 10 cm long NaI(Tl) crystal) coincidence spectrometer installed in the Comenius University laboratory [50]. The sensitivity of the coincidence mode is clearly visible, as the background and the Compton continuum from ^{40}K has been decreased more than an order of magnitude. The detection limit for ^{60}Co in the IAEA-414 reference material (a mixture of Irish Sea and North Sea fish) of 80 mBq kg^{-1} dry weight was obtained.

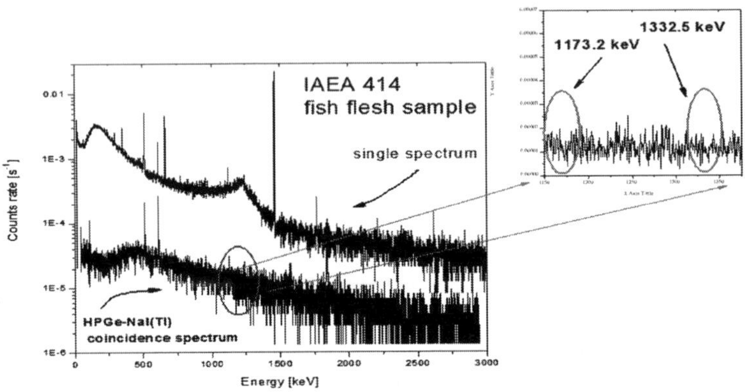

Figure 9: Single and coincidence gamma–gamma spectra of fish flesh sample [50].

Underground Gamma-spectrometry Laboratories

We have seen that the Monte Carlo simulation tools based on the GEANT proved to be capable to model background characteristics of Ge detectors, and thus to optimise the design of low-level gamma-spectrometry systems. Therefore, the influence of various parameters on the detector background can be studied well in advance, and the cosmic-muon induced background can be estimated before a low-level detector system is constructed. Given the detector set-up, its background spectra induced by cosmic-ray muons can be scaled down by a factor corresponding to the shielding depth. Generally, a radioactive contamination of construction parts of the shield and Ge detectors itself is still dominating factor as the obtained background was always higher than the predicted one by Monte Carlo simulations [174, 175].

Shallow-depth laboratories

It has been well known from cosmic-ray physics that fluxes of secondary particles have different behaviour in rocks. While for neutrons the underground depth of 10 m w.e. causes a decrease in their intensity by almost two orders magnitude, for muons an equivalent decrease is reached only at 100 m w.e. (Fig. 10). A low-level Ge spectrometer operating at shallow depth of about 100 m w.e. would benefit therefore at least from a partial suppression of the hard component (muons) of cosmic rays.

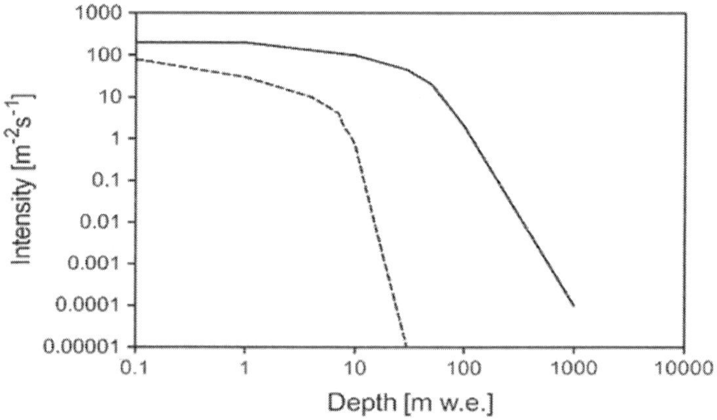

Figure 10: Intensity of secondary cosmic-ray particles with depth underground (the broken line is for neutrons and the full line is for muons).

When designing a new underground gamma-spectrometry laboratory, a single Ge spectrometer of about 100% efficiency (especially if the sample size permits the use of a well detector) is the usual choice for majority of environmental applications. We shall illustrate the applications of the Monte Carlo simulations in designing the underground laboratory constructed at the IAEA Monaco [39, 40, 57]. The laboratory is situated in an underground cellar in a car parking area at a depth of 35 m w.e. The laboratory is equipped with a common lead shield housing four large volume Ge detectors (Fig. 11). An anticosmic veto was made of plastic scintillation detectors surrounding the lead shield. Such a novel design, supported by Monte Carlo simulations, when several Ge detectors are placed in the same lead shield with a common anticosmic guard has been used for the first time in low-level gamma-spectrometry. Its big advantages can be summarized as follows: (i) it reduces the mass of expensive lead/copper shield around the Ge crystals; (ii) it reduces the mass of the outer lead/iron shielding; (iii) it reduces the size of the anticosmic shielding protecting the Ge detectors against cosmic-ray muons.

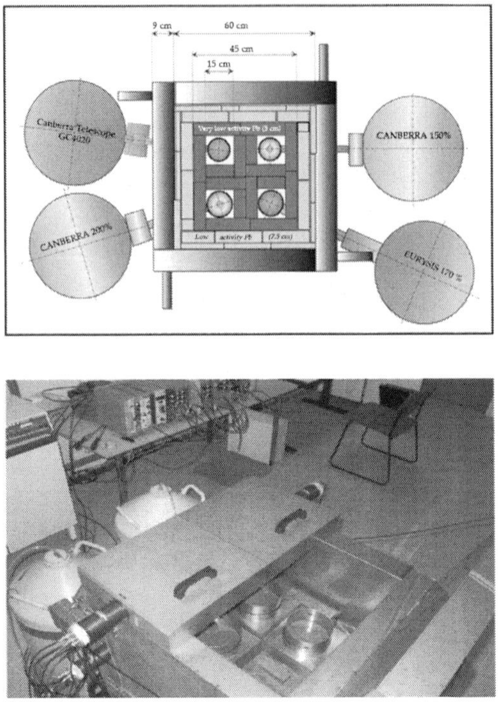

Figure 11: Lead shield with four Ge detectors operating in a shallow underground laboratory in IAEA Monaco with anticosmic shielding made of plastic scintillators [39].

The lead shield housing 4 Ge detectors is made of the outer layer with ordinary lead of 7.5 cm thick and the internal layer of 5 cm thick, made of very low activity lead, which was specially ordered for the underground laboratory (^{210}Pb activity is below 0.1 Bq kg^{-1}). It should be stressed that this detection system is used only for analysis of low-activity samples. The lead shield is surrounded on all sides and from the top by plastic scintillation sheets 7 cm thick, which are viewed by 5 cm diameter photomultipliers. The Ge detectors (coaxial p-type), specifically designed for low-level gamma-spectrometry in an underground laboratory are of U-type with preamplifiers housed outside of the lead shield, however, the FET is mounted on a Cu plate connected to the cooling finger. Only materials with minimum radionuclide contamination were used for the detector construction. Four types of Ge detectors, with parameters listed in Table 3 are housed in the common lead shield.

Table 3: Characteristics of HPGe detectors in the IAEA Monaco underground laboratory

Detector efficiency	**100%**	**150%**	**170%**	**200%**
Type	Coaxial	Well	Coaxial	Well
Cryostat material	Cu	Cu	Al	Cu
End-cap material	C	Cu	Al	Cu
Ge mass (kg)	2.15	2.50	3.29	4.18
Background at 40–2000 keV (h^{-1} $\mathbf{kg^{-1}}$)	290 ± 30	240 ± 20	230 ± 20	200 ± 20
Background at 40–2000 keV with anticosmic veto (h^{-1} $\mathbf{kg^{-1}}$)	40 ± 3	27 ± 2	59 ± 4	33 ± 3
Reduction factor	7.2	8.9	3.9	6.2

ORTEC NIM modular electronics have been used for signal processing and data acquisition. During all measurements radon is expelled from the detector chambers by the evaporation of nitrogen from the detector's Dewar containers, thus keeping stable background during measurements. As the volumes of the detectors differ significantly, it is necessary to compare their background characteristics per kg of Ge. It is interesting to notice that the total detector background per kg of Ge, in the energy window 40–2000 keV, is decreasing with increasing detector volume. However, the background with the anticosmic veto does not follow this rule, but clearly shows a larger contribution of radioactive contamination of construction materials (cryostat plus lead shielding) to its background. The lowest total background with the anticosmic veto was obtained for the 150% efficiency detector, the highest one for the 170% efficiency detector. The 170% efficiency detector have the cryostat made of "pure" aluminium, however, this has been clearly contaminated by U and Th decay products, as well as by ^{40}K. It is important therefore that construction materials are carefully checked by the manufacturers for the presence of radionuclides before the detector construction, especially for the detector's cryostat and its window. However, it has also been a surprise that three detectors with copper cryostats ordered as low background detectors for an underground laboratory from the same company (CANBERRA) had very different background characteristics. The highest background reduction with the anticosmic veto was obtained for the 150% efficiency detector (factor 9), and the lowest for the 170% efficiency detector (factor 4). The 100 and 200% efficiency detectors have the reduction factors within these limits (factors 7 and 6, respectively). The obtained background reduction factors are considerably lower than we would expect from Monte Carlo simulations, which may be due to several reasons: (i) Leakage of muons through the anticosmic shielding, either due to the shielding geometry or energy off-set in the scintillation detectors; (ii) cosmic-ray secondaries produced by muons passing the lead shield which were not discriminated by the anticoincidence circuit. A contamination of the detector's window may also be important, e.g. it may be advantage to made it from thin copper if the threshold energy need not to be very low. If a low energy window is required, an ultra-clean aluminium or a carbon fibre may be a good choice. There are other construction materials that could affect the detector background as well. Preamplifier is usually situated outside of the lead shield, however the FET transistor is connected directly with the Ge diode. The detector holder, copper cooling finger and soldering contacts may be therefore crucial for obtaining a low detector background.

Because of shallow operating depth, the annihilation peak at 511 keV is still dominant in background gamma-spectra, as a result of annihilation of electrons and positrons in the shield and in the Ge detector, which are products of the interaction of secondary cosmic rays with materials surrounding the detector. All these drawbacks would be overcome in deep underground laboratories where fluxes of cosmic-ray secondaries are negligible (Fig. 10).

The background of the described underground Ge detectors (operating only at 35 m w.e. but with anticosmic veto) when compared with other underground laboratories (normalized to the Ge detector mass) is similar to underground laboratories operating with a passive shielding at 250 m w.e. depth [166,167,168]. This is even better seen in Fig. 12 where integral backgrounds of Ge detectors divided by the mass of the Ge crystal, operating in different underground laboratories are compared [169]. Therefore, an anticosmic shielding in an underground laboratory operating at a shallow depth is extremely important for reducing the detector background, and should be widely used. For example, in the case of analysis of ^{137}Cs in seawater samples it has been possible to decrease the sample volume by about a factor of 10, which greatly reduces sampling time. Another advantage is that the same seawater volumes could be used for gamma-spectrometry as well as for mass spectrometry measurements, therefore large samples of about 200 L are not required [1, 2, 107,108,109].

DEVELOPMENTS IN MASS SPECTROMETRY TECHNOLOGIES

As we already mentioned, the most important recent developments in the radioanalytical technologies for analysis of long-lived radionuclides have been applications of mass spectrometric systems, especially the AMS and the ICPMS. When using measurement techniques with the extreme sensitivity at levels below 10^{-15} g per sample, it is very important to minimize the effects of airborne dust particles, reagents, glassware, etc., which can all contribute significantly to the sample blank, and as a result they will modify analytical detection limits which are actually defined by the contamination of the blank sample. For this reason, a successful operation of mass spectrometric systems requires more careful sampling, very clean chemical processing and a higher level of instrumental expertise than in any other radioanalytical methods. Because of limited space available we shall focus in this paper only on ICPMS and AMS technologies.

Inductively Coupled Plasma Mass Spectrometry (ICPMS)

ICPMS has developed into a powerful technique for the analysis of elements, stable and long-lived radioactive isotopes in the environment. With the introduction of the present state of the art high resolution ICPMS machines, this technique competes with TIMS in many respects. The principal advantages of ICPMS are its capability to determine long-lived radioisotopes of metallic elements down to fg levels, to analyse aqueous samples directly and rapidly (in a few minutes) at the low cost per analysis and with small sample size. However, ICPMS is not free of matrix and isotopic effects, therefore careful purification procedures (using e.g. extraction chromatography) are required. A steady increase over the past decade in radioanalytical applications using ICMPS has resulted in a decrease in both the price of instruments and detection limits. New generation of sector field instruments with double-focusing and even multi-collector systems have improved sensitivity (by about an order of magnitude) and precision over traditional quadrupole machines [81, 85, 123]. ICPMS has been used in both higher-resolution and lower-resolution modes. The higher-resolution mode has the advantage of addressing polyatomic interferences, although it cannot solve all the problems with isobaric interferences, which may be caused by incomplete separation chemistry. On the other hand, maximum sensitivity can be reached in the lower-resolution mode. Thus, a combination of the two modes appears to be the best compromise for reaching maximum sensitivity and controlling interferences. The higher count rates under lower-resolution mode give better analytical peaks with lower uncertainties and optimal data quality. Analytes with a relatively strong probability of polyatomic interferences on the isotopes of interest (e.g. ^{238}U which produces a hydride peak that would interfere with ^{239}Pu) should always be scanned. There are problems, however, with the relatively poor abundance sensitivity of sector field ICPMS in the measurement of isotopes with one mass below an abundance peak (e.g. ^{237}Np in the presence of high ^{238}U), and two mass units below (e.g. ^{230}Th in the presence of ^{232}Th, or ^{236}U in the presence of high ^{238}U content). Even when sample matrices are reasonably clean, and care has been taken to minimize oxides during tuning, measurements made near detection limits are sensitive to overestimations due to polyatomic interferences. The use of chromatographic resins [97, 110, 111] has been found to be a suitable technique for processing small volume samples, removing possible interferences by additional cleaning, as well as for cleaning leached plutonium samples electrodeposited on stainless steel disks, previously analysed by alpha-spectrometry.

Although molecular, isobaric and isotopic interferences remain crucial for successful operation of ICPMS, this technique has a large potential for automation by direct coupling with new generation of chromatography instruments. With the introduction of high resolution and high sensitive ICPMS it has been possible to analyse some of the long-lived radionuclides like ^{99}Tc, ^{129}I, ^{236}U, ^{239}Pu and ^{240}Pu at very low levels [84, 85]. This was especially advantageous for reporting separate data for ^{239}Pu and ^{240}Pu, and using their ratio for tracing the origin of plutonium in the environment. ^{238}Pu is difficult to analyse by ICPMS as tracers of ^{238}U (even after careful separation) may be present in the sample. ICPMS has also been widely applied on analysis of long-lived ^{99}Tc and ^{129}I in seawater and seaweeds [85]. Figure 16 shows typical example of ^{239}Pu and ^{240}Pu water profile activity concentrations (together with $^{240}Pu/^{239}Pu$ mass ratios) taken close to the Enewetak Atoll [22]. The medium depth peak, located at 500 m water depth, is clearly visible for both radionuclides, as well as the higher concentrations measured in the bottom sample. The $^{240}Pu/^{239}Pu$ ratio is higher than expected from global fallout (0.186), indicating the influence of high-yield nuclear weapons tests conducted in Bikini and Enewetak atolls on Pu concentrations in the water column of the western North Pacific subtropical gyre [1, 4, 19, 20]. ICPMS has also been effectively applied in certification of reference materials. As an independent method (important for production of certified reference materials) has been used for measurement of U, Th, Pu isotopes in terrestrial and marine environment (sediment, seawater, biota).

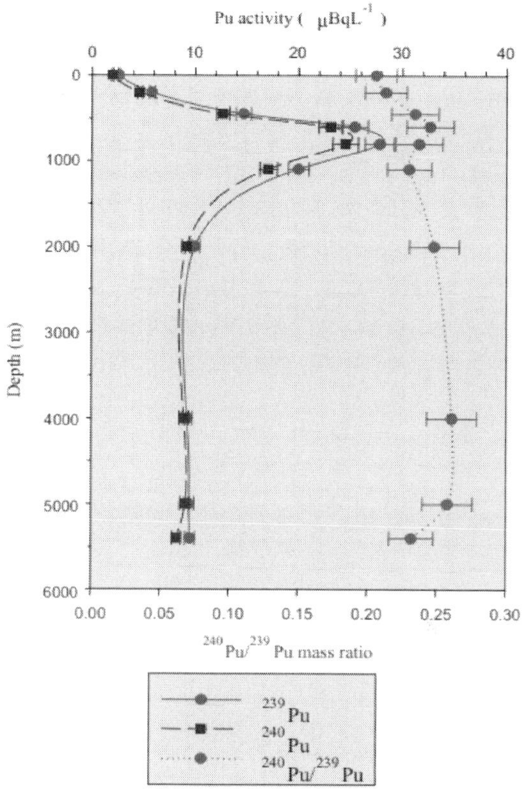

Figure 16: Plutonium isotope seawater profiles in the North-West Pacific Ocean measured by ICPMS [22].

Accelerator Mass Spectrometry (AMS)

AMS represents the most important breakthrough in the analysis of some of the long-lived radionuclides. Development of dedicated tandem accelerators for AMS, firstly for ^{14}C, and then for other radionuclides as well (^{3}H, ^{10}Be, ^{26}Al, ^{32}Si, ^{36}Cl, ^{39}Ar, ^{41}Ca, ^{53}Mn, ^{99}Tc, ^{129}I, ^{135}Cs, ^{237}Np, U and Pu isotopes) expanded AMS applications in all sciences dealing with radionuclide analyses. As in tandem accelerators only negative ions can be used for acceleration, the AMS technique can only be applied for those elements (the great majority) forming negative ions, however, recently there has been research going on using positive ions as well. The most important recent developments in the AMS sector are based on small AMS machines (200–500 kV only) which eliminated the need for a costly pressure tank. This interesting design is a new competitor to larger AMS machines for ^{14}C dating. However, as the operating terminal voltage used for AMS decreases, the technical engineering difficulties increase. The small AMS machines, especially those operating at 1 MV have also a considerable potential for applications for other already mentioned long-lived radionuclides. We may summarize that the AMS has become the most sensitive technique at present for the analysis of long-lived radionuclides [91,92,93,94,95,96].

We shall illustrate the high analytical sensitivity offered by AMS in tracer amount studies of ^{14}C and ^{129}I in seawater around dumped radioactive wastes in the NE Atlantic dumping sites. The ^{14}C results depicted in Fig. 17 show a remarkable peak at medium depths (between 2000 and 3000 m) which is due to the subduction mode water regime when water masses from the surface are transported to medium depths [93]. The $^{129}I/^{127}I$ ratio measured in the same set of seawater samples also showed a medium depth peak, although it has moved to 1250–1500 m water depth. The higher bottom $\Delta^{14}C$ values may indicate a leakage from the wastes, as the $^{129}I/^{127}I$ ratios below 3000 m have shown constant values.

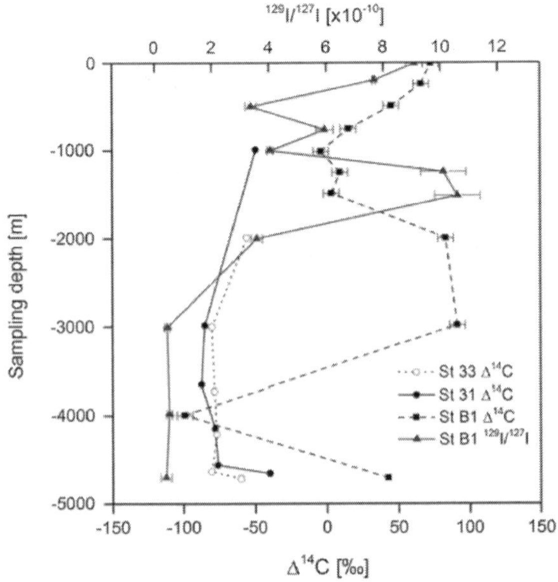

Figure 17: ^{14}C and $^{129}I/^{127}I$ water profiles of NE Atlantic Ocean [93].

In the framework of marine radioactivity assessment of radioactive waste dumping sites (Atlantic, Arctic and Pacific Oceans), radiological assessment of nuclear bomb tests sites (French and USA testing sites in the Pacific Ocean), as well as a wide range of marine radioactivity surveys we established background levels of key radionuclides (^{3}H, ^{14}C, ^{90}Sr, ^{129}I, ^{137}Cs, U, Pu and Am isotopes) in the World Ocean so any recent releases of these radionuclides would be easily recognized over the global fallout background [1,2,3,4, 93, 99, 176,177,178,179].

A COMPARISON OF RADIOMETRIC AND MASS SPECTROMETRY TECHNIQUES

In the case of ^{3}H there is not an alternative method to in-growth ^{3}He mass spectrometry method for ^{3}H analyses down to 0.01 TU. Similarly, for ^{14}C and ^{129}I analysis of environmental samples, especially for water column investigations, the AMS technique is the dominant one. In the case of analysis of ^{137}Cs in water column samples the underground gamma-ray spectrometry is the suitable technique which can produce high density data sets. A comparison of Pu results obtained by AMS, ICPMS and semiconductor alpha-spectrometry (SAS) shows that a reasonably good agreement (within quoted uncertainties) for wide range of activities and different sample matrices analysed has been obtained for all three methods. From the point of view of sensitivity, AMS and ICPMS have detection limits at least by a factor of 10 lower than SAS. Sample chemistry, injection and the physics of the analysis make the ICPMS a simpler technique than the AMS, however, because of a higher sensitivity to possible interferences (e.g. a production of hydrates) the AMS is the most suitable technique for analysis of ^{239}Pu and ^{240}Pu in environmental samples. A comparison of Pu results obtained by ICPMS, AMS and SAS is presented in Table 5. It can be seen that a reasonably good agreement has been obtained between all three methods. The relative precision of Pu results as obtained until now by SAS, AMS and ICP-MS was 5, 10 and 7%, respectively (for $^{239,240}Pu$).

Table 5: A comparison of results of analysis of Pu and U isotopes in IAEA reference materials

Method	^{238}Pu Bq kg^{-1}	^{239}Pu Bq kg^{-1}	^{240}Pu Bq kg^{-1}	$^{239,240}Pu$ Bq kg^{-1}	^{241}Pu Bq kg^{-1}	^{242}Pu Bq kg^{-1}	^{238}U Bq kg^{-}	^{236}U Bq kg^{-1}
IAEA-134 Irish Sea cockle flesh								
SAS	3.0 ± 0.2			16.3 ± 0.8				
ICPMS		9.8 ± 0.8	7.7 ± 0.6	17.5 ± 1.0				
IAEA-135 Irish Sea sediment								
SAS	42 ± 1			222 ± 8				
AMS		129 ± 13	92 ± 9	221 ± 16				(11.2 ± 2.2) 10^{-3}
ICPMS		127 ± 10	98 ± 8	221 ± 13	3970 ± 200	0.051 ± 0.001		(10.3 ± 3.0) 10^{-3}
LSS					3000 ± 650	0.056 ± 0.006		
IAEA381 Irish Sea water								
SAS	3.3 ± 2			14.0 ± 1.0			0.042 ± 0.004	
AMS		8.2 ± 0.3	7.0 ± 0.2	15.7 ± 0.4		0.0046 ± 0.0005		(0.019 ± 0.003) 10^{-3}
ICPMS		8.1 ± 0.8	7.0 ± 0.7	15.1 ± 1.1		0.0047 ± 0.0002		(0.023 ± 0.006) 10^{-3}
LSS					200 ± 60			
IAEA-384 Fangataufa lagoon sediment								
SAS	40 ± 2			111 ± 3			35.3 ± 1.7	
AMS		109 ± 11	14 ± 1	123 ± 11		2.5 ± 0.4		(8.4 ± 1.7) 10^{-6}
ICPMS		102 ± 8	18 ± 2	120 ± 8	240 ± 20	1.9 ± 0.5		
LSS					100 ± 70			
IAEA-385 Irish Sea sediment								
SAS	0.45 ± 0.03			2.94 ± 0.09			29 ± 1	
AMS		1.79 ± 0.28	1.15 ± 0.14					< 0.0001
LSS					30 ± 2			
IAEA-414 Irish and North Sea fish								

SAS	0.025 ± 0.002			0.120 ± 0.005		0.044 ± 0.004	1.11 ± 0.07	
AMS		0.087 ± 0.003	0.053 ± 0.005	0.14 ± 0.01		0.068 ± 0.010		(0.024 ± 0.007) 10^{-3}
0 ICPMS		0.063 ± 0.007	0.046 ± 0.005	0.11 ± 0.02	2.7 ± 0.6			

A general comparison of radiometric and mass spectrometry techniques is presented in Table 6. It can be seen that the most sensitive technique available at present for analysis of environmental samples is AMS which gives the lowest detection limits, three to eight orders of magnitude lower than the radiometric methods (with the exception of ^{3}H, where the ^{3}He ingrowth mass spectrometry method dominates).

Table 6: A comparison of detection limits for frequently analysed long-lived radionuclides in the environment (in Bq)

Method	**^{3}H**	**^{14}C**	**^{99}Tc**	**^{129}I**	**^{236}U**	**^{237}Np**	**^{240}Pu**
RM	10^{-3a}	10^{-4}	10^{-2}	10^{-2}	10^{-3}	10^{-4}	10^{-5b}
^{3}He MS	10^{-3}						
NAA			1	10^{-7}		5×10^{-4}	
ICPMS			10^{-5}	10^{-3}	5×10^{-9}	5×10^{-6}	5×10^{-6}
TIMS			8×10^{-6}	10^{-8}	10^{-10}	10^{-9}	0.5×10^{-6}
RIMS			10^{-5}				$\sim 10^{-6}$
AMS	10	10^{-7}	6×10^{-6}	10^{-10}	10^{-10}	10^{-10}	0.4×10^{-6}

^{3}He in growth mass spectrometry method

[a]After electrolytic enrichment

^{b239,240}Pu

RM **radiometrics,** *ICPMS* **inductively coupled plasma mass spectrometry,** *TIMS* **thermal ionization mass spectrometry,** *RIMS* **resonance ionization mass spectrometry,** *AMS* **accelerator mass spectrometry**

CENTRE FOR NUCLEAR AND ACCELERATOR TECHNOLOGIES (CENTA)

Recent developments in AMS technologies and their applications in nuclear sciences (physics and chemistry), and in environmental, space, biomedical and material researches have been an inspiration for creation of many new tandem accelerator laboratories. Except for AMS, tandem accelerators have been widely used for the ion beam analysis (IBA) and the nuclear reaction analysis (NRA) of environmental and material samples, as well as for ion beam modification (IBM) of materials. All these new technologies represent the most successful developments in small accelerators and their applications in various branches of science. A Centre for Nuclear and Accelerator Technologies (CENTA) has been established recently (2013) at the Comenius University in Bratislava comprising of a state-of-the art tandem accelerator laboratory designed for: (i) AMS studies of long-lived radionuclides in environmental, space and life sciences; (ii) IBA applications in environmental, life and material research, including cultural heritage studies; (iii) NRA studies with charged particles for new generations of fission reactors, for thermonuclear reactors, and for astrophysics; (iv) and for IBM of materials used for construction of new generation of fission reactors and for thermonuclear reactors, and for research in nanotechnology. This orientation of the laboratory has been driven by general needs to establish in Slovakia a national laboratory devoted to ion beam studies and applications, and to assure for the future wide-range research capabilities for successful participation in international and funding programs. Because of financial constrains it has not been possible yet to install in the CENTA laboratory all equipment which will be necessary for carrying out research in the planned topics. Therefore, the present laboratory design was due to limited financial support restricted to two ion sources (Alphatross for gas, and MC-SNICS for solid targets), the injection system, the 3 MV tandem accelerator, and simple high energy analysers with two ion beam end stations. All available equipment was supplied by the National Electrostatics Corporation (NEC, Middleton, USA). The near future installation will include a fully equipped AMS line with 90° magnet, two electrostatic spectrometers and the end of the line detector (Fig. 18). Later installations will include a nuclear microscope, a raster station for IBM studies of materials, and a station for biomedical research. A dedicated hall to accommodate the tandem accelerator laboratory has been built at the Comenius University campus at Mlynska dolina. The hall design separates the ion beam channels (placed in a bunker covered by soil) from the AMS line, enabling thus work in different radiation environments [179, 180].

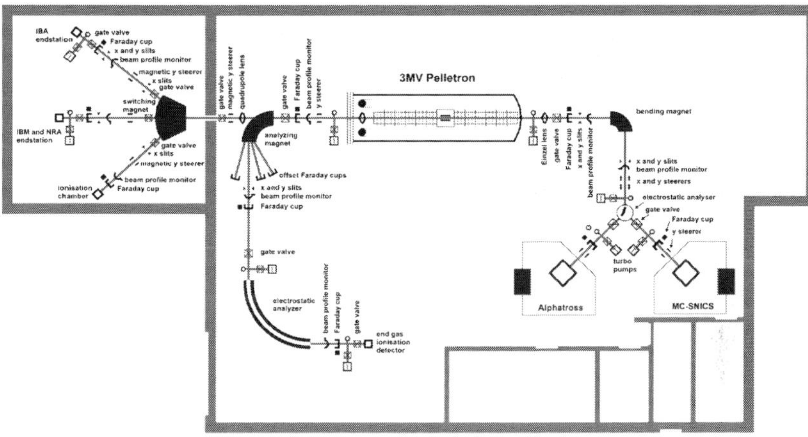

Figure 18: Floor scheme of the CENTA tandem accelerator laboratory with complete AMS line [180].

The first experiments included optimization of parameters of both ion sources (Alphatross and MC-SNICS) for AMS and IBA studies. Studies of transmission characteristics of accelerated ions with different energy and charge states have also been carried out. The nitrogen gas pressure in the gas stripper for acceleration of ions with various charges and energies has been optimized for different elements. Transmission efficiencies of $^{9}Be^{2+}$ and $^{9}Be^{3+}$ were determined for AMS measurements of ^{10}Be. In the first case the transmission efficiency was by more than a factor of 10 greater, therefore the $^{9}Be^{2+}$ beam was used for ^{10}Be measurements [181].

Analysis of ^{10}Be by AMS

As the AMS line at the CENTA laboratory does not yet include a fully capable analysing system, the possibility to measure ^{10}Be using only a small switching magnet as the ion analyser was tested. The ^{10}Be analysis by AMS is mainly limited by the stable isobar ^{10}B, while the requirements for mass separation are the least stringent of all standard radionuclides analysed by AMS. The method for suppression of ^{10}B ions, based on a silicon nitride foil stack used as a passive absorber was developed earlier at the VERA laboratory (Fig. 19). The MC-SNICS was used for the production of $^{10}BeO^{-}$ ions, which were mass separated and injected into the 9SDH-2 Pelletron, operating at 3 MV terminal voltage. The $^{10}Be^{+2}$ ions were selected, and ^{10}B

ions, as well as of most background ions from heavier masses were absorbed in the silicon nitride stack introduced in front of the ionization chamber which was used for the ion detection (Fig. 20). A good separation of $^{10}Be^{2+}$ ions from background (formed mainly by $^{9}Be^{+2}$) has been obtained (Fig. 21). The standard ^{10}Be source (S555 developed at ETH Zürich) with the $^{10}Be/^{9}Be$ mass ratio of $(8.71 \pm 0.24) \times 10^{-11}$ was used in these measurements. Using this setup, a detection limit for $^{10}Be/^{9}Be$ of 10^{-12} was achieved, which was mainly determined by scattering of $^{9}Be^{+2}$ ions on residual gas inside the switching magnet [181].

Figure 19: Tandem accelerator laboratory (from the top: Alphatros and SNICS ion sources, injection line, Pelletron, analyzing line with switching magnet, end of the line PIXE/PIGE chamber).

Figure 20: End of the line detectors for ^{10}Be analysis.

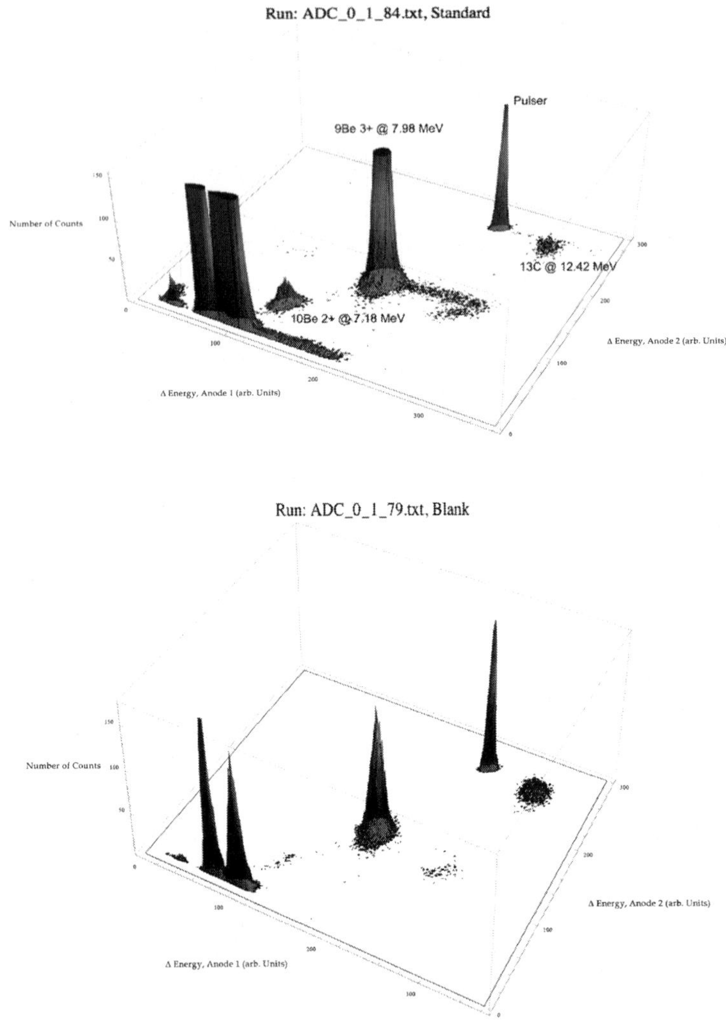

Figure 21: Multidimensional mass spectra of ^{10}Be ions (top) and background (bottom) measured in the ionization chamber with silicon nitride stack absorber [181].

Analysis of Uranium Impurities in Materials

One of the main tasks of the CENTA facility is investigation of U and Th impurities (at levels below 1 nBq g^{-1}) in construction materials of detectors designed for underground physics experiments such as double beta-decay of ^{82}Se and ^{76}Ge (experiments SuperNEMO and LEGEND, respectively), as well as a search for dark matter (experiment EURECA) [182,183,184]. We have been focusing on radiopurity measurements of copper as this material is usually closest to the detector, and therefore its radiopurity has the dominant impact on the detector background. Copper samples can be directly accommodated as targets in the ion source of the tandem accelerator, therefore no pre-concentration chemistry is required. We did preliminary investigations with analysis of uranium in copper wire targets made of OFHC. Uranium and thorium ions extracted from the copper produce in the ion source negative ions either as uranium/thorium oxides, or as uranium/ thorium compounds with copper (Fig. 22) [185]).The ion clusters of ^{63}Cu and ^{65}Cu ($^{63}Cu^{65}{}_{3}Cu$, etc.) with masses of 254 (256), 319 (325) and 374 (388) were observed after the injection magnet, however, the UO and ThO_2 ions with masses of 254 and 264, respectively, should be expected in the first mass peak as well. This has been expected as U and Th oxides are the most frequently observed compounds of these two elements in the environment, and therefore they will make the most influential background contributions during AMS measurements. A more favourable case should be therefore a formation of negative molecules of UCu^- or $ThCu^-$ which would fall into the mass windows of 301 and 295, respectively, where they would be free of copper cluster interferences (Fig. 23; the U_3O_8 and UF_4 targets were prepared at the Czech Technical University in Prague [185]). Further investigations are going on with optimization of ion production/acceleration and post-acceleration ion analyses, including analysis of U and Th in enriched ^{82}Se which will be used as an isotope source in the SuperNEMO experiment.

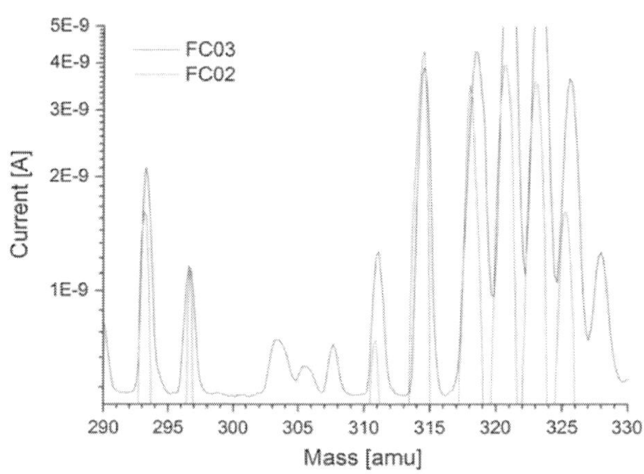

Figure 22: Uranium ions extracted from the copper were either as uranium oxides, or as uranium compounds with copper.

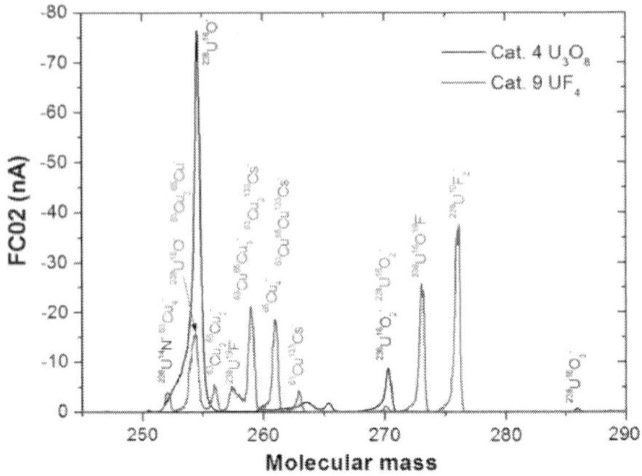

Figure 23: Production of copper clusters after extraction of ions from U_3O_8 and UF_4 targets placed in copper holders.

Al_2O_3 versus AlN Targets for ^{26}Al Analysis by AMS

Al_2O_3 targets have been mostly used in ion sources for the production of aluminium negative ions for AMS analysis of low-level concentrations of ^{26}Al in environmental, biological and space samples with detection limits down to 0.01 fg. Al_2O_3 targets have high-temperature stability, non-toxicity, in-air stability and relatively easy production procedures. It is well known, however, that aluminium does not yield high-intensity negative ion beams, like other elements used in tandem AMS accelerators. An alternative solution could be to use as the target aluminum nitride (AlN) material, which can yield higher Al^- currents. On the other hand, the AlN targets are more difficult to synthesize, and they decompose with water in the air to form $Al(OH)_3$ and ammonia. Commercially available compounds of Al_2O_3 and AlN were mixed with copper, silver and iron high purity powders and sputtered in MC-SNICS ion source for studying ionization yields. Since the production of magnesium and nitrogen negative ions is negligible, a production of MgN^- molecules has been questionable. Obtained results indicate that the AlN matrix could be a suitable material for AMS measurements as the production of $^{27}Al^-$ is higher by a factor of 2 than from the aluminium oxide matrix, while aluminium sulphate and aluminium fluoride showed a very low sputtering efficiency [186]. Materials such as Al_2O_3, AlN, pure aluminium wire and Mg_3N_2 were also tested for $^{24-26}Mg^{14}N^-$ formations on isobaric molecule interference, focusing on searching for interferences at 40 amu ($^{26}Mg^{14}N^-$ creation) as the main isobaric interference for $^{26}Al^{14}N^-$. In our conditions, none of the isobaric interferences was detected. As can be seen in Fig. 24, there is one small peak that can correspond to the isobaric interference $^{26}Mg^{2+}$ (all samples were ^{26}Al blank materials) from injected MgN^- (mass 40 amu) ions. The AMS investigations, with the same Al_2O_3, AlN and Mg_3N_2 matrices, were done in the VERA laboratory using 3 MV terminal voltage. The measurements indicate that the MgN^- creation from AlN matrix is unfortunately suppressed only about 3.3 times than the MgO^- creation from Al_2O_3 matrix. The suppression of MgN^- formation from Mg_3N_2 matrix was, however, 5×10^5 times lower than AlO^- creation from Al_2O_3.

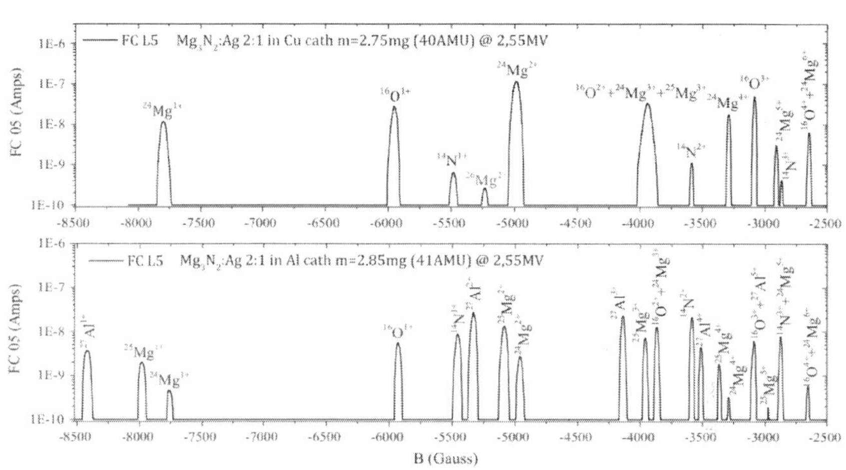

Figure 24: Mass scans of ions from the Mg_3N_2 sample after acceleration and dissociation [186].

Analysis of ^{14}C in Annual Tree Rings by AMS

One of the main goals of the CENTA facility is investigation of ^{14}C variations in the atmosphere and biosphere as a continuation of previous research on ^{14}C variations in atmospheric CO_2 of Bratislava (having 50 y tradition and representing the second longest continuing ^{14}C record in Europe) [187, 188], as well as around Slovak nuclear power plants in Jaslovské Bohunice and Mochovce [189, 190]. Recently we have been focusing on the development of an integral method of ^{14}C variations studies in annually growth tree rings [191]. Such type of sampling enables to reconstruct ^{14}C chronology in a given locality, usually for several decades. To reach one-year time resolution, an analytical method based on AMS measurement is necessary, in which a sufficient sample of carbon quantity is only a few milligrams. Another advantage of the AMS based method of ^{14}C activity determination is a smaller uncertainty of measurements (reduced by about a factor of two).

Preparation of graphite targets for ^{14}C analysis by AMS

Wood samples after extraction from the centre of selected tree rings were treated by the ABA method (1 M HCl 60 °C, 0.1 M NaOH 60 °C, 1 M HCl 60 °C). Dried cellulose separated from the wood was burned in oxygen atmosphere at 900 °C. Cold traps cooled to − 40°C were used for for purification of the CO_2 (Fig. 25). Small graphitization reactors (0.9 cm^3) were used for graphite production (Fig. 25) using a modified Bosh reaction [192]. The iron powder was first purified with oxidation at 900 °C before filling the reactors, and then reduced by pure H_2 at 600 °C. After about 30 min of iron reduction, the reactors were filled with H_2 and CO_2 in the ratio 3:1. A part of the reactor was then cooled by the dry ice to − 78 °C. Iron was heated up to 900 °C for the CO production, and after about 30 min, the temperature was lowered to 550 °C for graphite production. Temperature and pressure were monitored during the whole reaction. In addition to the tree-ring samples, chemical blank was prepared by using high purity graphite with the same procedure as the tree-ring samples. Graphite samples produced in the reactors were then pressed into aluminium cathodes and mounted on the MC-SNICS ion source wheel. For calibration purposes, ^{14}C cellulose standard IAEA-C3 was combusted, graphitized and mounted together with tree-ring samples and blanks on the wheel. The $^{14}C/^{12}C$ ratios were measured in the VERA laboratory using NEC Pelletron accelerator at 2.7 MV terminal voltage [191, 192].

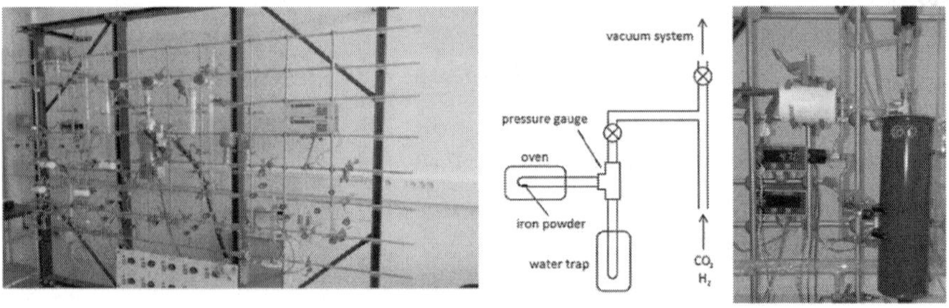

Figure 25: Vacuum-cryogenic apparatus for combustion of tree-ring samples and preparation of graphite samples [191].

Comparison of ^{14}C levels in tree rings from Slovakia

Results of ^{14}C activity in tree rings sequence lime tree collected at the Žlkovce monitoring station (situated close to the Bohunice NPP, about 60 km north of Bratislava) are compared with Vysoka pri Morave (a village situated 30 km north-west of Bratislava) and Bratislava tree rings (Fig. 26). Due to wood biomass accumulation from May to September, a small decrease in the comparison with maximum activities of atmospheric $^{14}CO_2$ (during the summer period) can be expected in tree ring samples. Nevertheless, possible local variations of ^{14}C activities caused by microclimatic differences (a time shift in biomass ingrowth) can be reduced for tree rings compared to leaves and other biota samples. The tree-ring method is limited by the age of the tree, but ^{14}C data over several decades can be reconstructed from old trees. The ^{14}C tree-ring data can provide information on average ^{14}C concentrations during spring and early summer (when new wood, light in colour and usually softer, is formed—so called softwoods), and during a late summer and autumn growing season (when new cells formed are smaller, but of a higher density, and have darker thicker walls—so called hardwoods). The tree-rings during the growing season can integrate impacts from both nuclear and fossil CO_2 sources on the local environment, indicating long-term trends in ^{14}C biospheric concentrations, and can be used to assess radiation doses to the public. Radiocarbon concentrations in tree-ring samples from Vysoká pri Morave show an expected exponential decrease during the last 40 years with decay constant of 14.5 ± 1.2 years ($R^2 = 0.989$), in agreement with similar results obtained at other European ^{14}C stations. The Suess effect, represented by a dilution in ^{14}C levels by fossil fuel CO_2 emissions was observed in both tree-ring data sets. The Vysoká pri Morave ^{14}C data during 1974–1995 were systematically lower by about 50‰ than the Schauinsland (Germany) clean air reference values due to a regional fossil-fuel impact. However, after 1996 the Vysoká pri Morave ^{14}C data were closer to the Schauinsland data [193, 194] due to lower CO_2 emissions as a result of closing some of the heavy industry technologies in the region. The observed interannual changes in $\Delta^{14}C$ with considerably larger variances in the Žlkovce than in Vysoká pri Morave $\Delta^{14}C$ tree-ring data could be caused by small ^{14}C releases from the Jaslovské Bohunice NPP. On the other hand, the Bratislava tree-ring ^{14}C data have been heavily influenced by fossil fuel emissions (a local Suess effect).

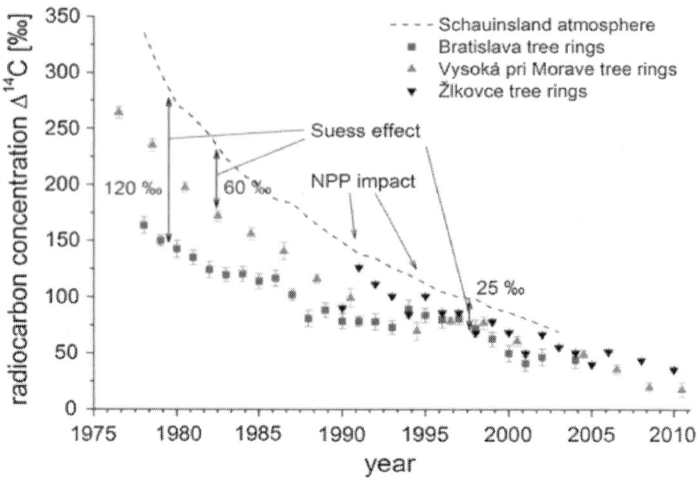

Figure 26: Comparison of ^{14}C levels in tree-ring samples from Slovakia with Schauisland atmosphere [192, 197, 198].

PIXE Beam-line Analyses

We shall present first results obtained with the PIXE beam line installed at the Bratislava CENTA tandem accelerator facility. The PIXE reaction chamber is equipped with a vertically movable sample holder for positioning of up to eight samples depending on their dimensions. The holder is capable of rotation around its vertical axis to adjust the angle how the incident beam should hit the sample (Fig. 27). Proton and ^4He ion beams produced in the Alphatross ion source and accelerated in the 3 MV Pelletron were used in the investigations. Results obtained with the ^4He ion beam showed better detection limits when compared with protons of the same energy and beam intensity. For detection of produced X-rays, a BEGe detector has been used, covering the energy range from 3 keV to 3 MeV. Analyses of PIXE laboratory standards and old silver coins with ^4He ions of 3.5 MeV energy showed reproducible results [195].

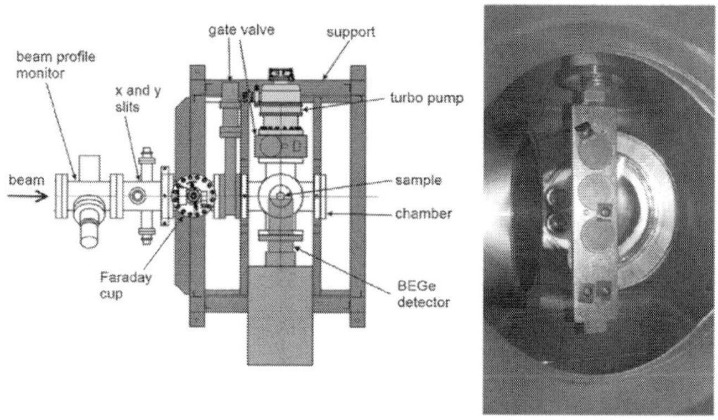

Figure 27: Scheme of the PIXE/PIGE chamber with sample holder and metallic samples. Characteristic X-rays were detected at the 45° angle by the BEGe detector (the endcap with carbon window is visible on the left side of the image) [195].

PIXE measurements of two Slovak coins using 4.5 MeV helium beam were carried out performed (Fig. 28). A silver coin of nominal value 20 crowns which was issued in 1941 during the Slovak state, and a coin of nominal value 1 crown which was used as a former currency in the Slovak Republic (between 1993 and 2004) were analysed. GUPIX-WIN was used for the composition determination of these coins. The 1941 Slovak silver coin was measured using 50 pA beam intensity for 10 min showed presence of silver (65 ± 5%), and the rest was copper. The copper characteristic X-ray lines (K_a and K_b) were sufficiently resolved. The second coin measured using higher beam intensity (200 pA) for 20 min showed a dominating copper concentration 85 ± 5%, and the rest was tin (14%) and iron (< 1%) [195].

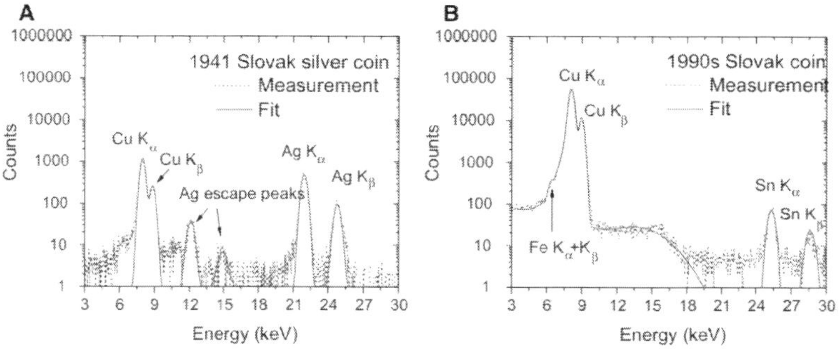

Figure 28: PIXE analysis of two Slovak coins (modified from [195].

As the second example of PIXE analysis we present measurements of Fe in a section of brain rabbits after electromagnetic irradiation with similar energy and intensity as of mobile phones. There have been long discussions about side effects of frequently used mobile phones for human health, especially possible impacts on children. Figure 29 compares abundance of Fe in the sections of rabbit brains after electromagnetic irradiation and without the irradiation. While the background Fe concentrations were below 2 ppm, the irradiated brain contains aggregations with Fe concentrations up to 55 ppm. A further research is in progress to measure other metals in brains after electromagnetic irradiation including human brains (a paper under preparation).

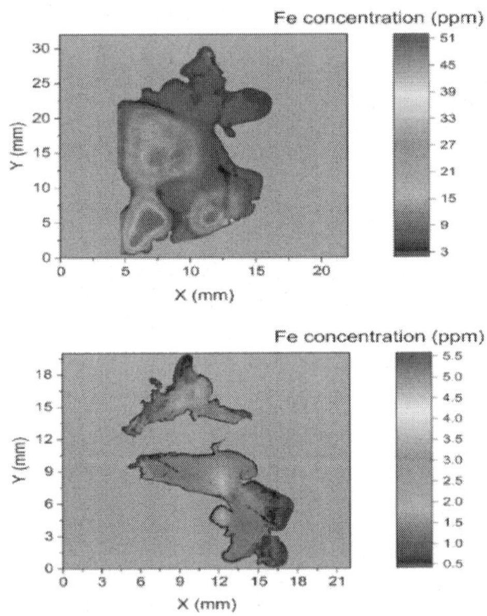

Figure 29: Distribution of Fe in a thin section of rabbit brains after (top) and without (bottom) electromagnetic irradiation.

EXAMPLES OF LARGE-SCALE PROJECTS CARRIED OUT WITH NEW RADIOANALYTICAL TECHNOLOGIES

We have only a limited space available to show where new radioanalytical technologies made important breakthroughs in new science. We may just state that in all sciences (e.g. space, environmental, marine, nanotechnologies, biomedical, pharmaceutical, agricultural, etc.), which require radionuclide analyses, the impact of new technologies have been significant as they have enabled to carry out investigations which were not possible before either because of small sample size or the required sensitivity. We shall illustrate the recent developments in the radioanalytical technologies on our recent research activities carried out in the framework of international collaboration, e.g. on investigations of rare decays (SuperNEMO), on atmospheric radioactivity, on isotope groundwater studies and on marine radioactivity studies (SHOTS, ANTARES and FUKUSHIMA projects).

SuperNEMO and Rare Nuclear Processes and Decays

Background is playing an essential role in underground experiments investigating rare nuclear processes and decays such as neutrinoless double beta-decay experiments (e.g. in SuperNEMO [182] and LEGEND [183]), as well as in search for dark matter (EURECA experiment [184]). As this is a very complex topic, we shall focus here only on the SuperNEMO experiment, which will search for neutrinoless double beta-decay of ^{82}Se, and its first modul (called Demonstrator) is presently under construction in the Modane underground laboratory [35, 36, 182]. It is expected that it will be operational at the second half of 2018. The two-neutrino double beta-decay process was observed for several isotopes (e.g. ^{48}Ca, ^{96}Zr, ^{100}Mo, and others, e.g. [196]). The double beta-decay is a second order process (allowed by the Standard Model), and half-lives of the order of 10^{20} years have been reported. However, no positive result has been obtained till now for a neutrinoless double beta-decay which would violate the lepton number conservation, requiring a Majorana neutrino (the identical neutrino with antineutrino), and giving unique information on the neutrino mass hierarchy, representing thus a new physics behind the Standard Model. With more than ten experiments which have been going on in several underground laboratories, only upper half-live limits of the order of 10^{24} years have been reported. The SuperNEMO experiment is further developing an underground physics technology used by its predecessor, the NEMO-3 experiment. It combines tracking and calorimetry techniques for the reconstruction of the final state topology (including timing and kinematics of the double beta-decay transition events), offering thus a powerful tool for background rejection. The SuperNEMO experiment in its full scale will consist of 20 modules (8 m high, 7 m long and 6 m wide) having the source foil situated in the centre of the module, tracking chambers at each side and finally calorimeters at each side for energy and time of flight measurements (Fig. 30). The selenium powder (^{82}Se enriched source) is distributed on the foil with surface density of 53 mg cm^{-2}. After the enrichment, it has been chemically purified using extra pure materials at super clean conditions. All the materials entering the source production have been carefully selected and prepared to have ultra-low radionuclide contamination levels.

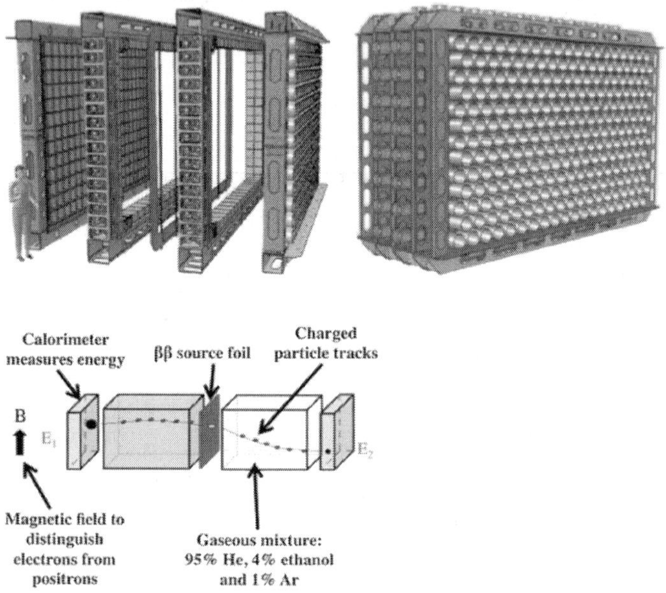

Figure 30: Construction view of the SuperNEMO module with description of its functions (modified from [35]).

The background constrains of the SuperNEMO detector are determined by external and internal radiation sources. The external ones are mainly due to cosmic rays (muons, neutrons and gamma-rays), neutrons from the underground laboratory walls, gamma-rays from the construction parts of the detector, and radon contamination of the laboratory air, which can also infiltrate inside the detector. The background from internal sources will be mostly composed of: (i) Double beta-decay events, which may imitate neutrinoless beta-decay; (ii) radioactive contamination from the internal construction parts, mainly in the tracker; (iii) radon inside the tracker. While the first constrain can only diminish by high quality spatial, timing and energy resolution capabilities of the SuperNEMO detection system, the last two constrains require using as clean as possible radioactivity free construction materials. Radon (^{222}Rn) and its decay products (^{214}Bi and

^{214}Po), as well as decay product of thoron, ^{220}Rn (^{208}Tl) have been identified as the most dangerous contaminants for the SuperNEMO detector. The most sensitive parts of the SuperNEMO detector from the point of view of radioactive contamination are: (i) isotope source (^{82}Se); (ii) supporting foil for the isotope source, and (iii) tracker (mainly the components which are inside the tracker, i.e. anode wires, cathodes, front/end insulators). The radon and thoron inside the tracker will decay to ^{214}Bi and ^{208}Tl, respectively, and their decay products will deposit on source/foil surfaces and on anode wires. They can also infiltrate in other parts of the detector, and their decay products due to emitted beta-electrons with high energy (up to 3.3 MeV and 4.9 MeV for ^{214}Bi and ^{208}Tl, respectively) and gamma-rays (up to 2.2 and 2.6 MeV for ^{214}Bi and ^{208}Tl, respectively) may imitate double beta-decays. The tracker amelioration mainly relies on the reduction of the radon presence inside the tracker volume. To achieve this reduction, it is necessary to perform an exhaustive control of the radon-tightness and radon-emanation of the different materials conforming the tracker. After the installation of the tracker modules, the remaining radon activity inside the tracker should < 0.15 mBq m^{-3}, requiring a continuous purification of the counting gas in the tracking chambers from radon. Radon adsorption materials have also been tested to improve the purification systems, as well as to build at the LSM a radon-free air factory that will flush the air around the SuperNEMO detector. Following the experience from the NEMO-3 experiment, the radiopurity limits defined for the SuperNEMO experiment are based on the goal to reach for the neutrinoless double beta-decay the half– life of 1×10^{26} y. The stringent limits are set for ^{222}Rn present in the tracker (< 0.15 mBq m^{-3}), for the radioactive contamination of the internal parts of the detector (the isotope source, supporting foil, wires and walls of the tracker) they are < 2 nBq/g and < 10 nBq/g for ^{208}Tl and ^{214}Bi, respectively. Although ^{40}K also emits high-energy gamma-rays (1.46 MeV), and it is frequently present mainly in glass of photomultipliers, it is usually outside of the sensitive volume of the tracker, and therefore radiopurity limits for ^{40}K are not so strict (< 100 mBq kg^{-1}). When compared with the NEMO-3 detector, the radio-contamination limits for the SuperNEMO are lower by about a factor of 30, 50 and 30 for ^{222}Rn, ^{208}Tl and ^{214}Bi contamination, respectively.

Low-level Ge gamma-spectrometry facilities were developed underground in LSM (Modane, France) and in Boulby (Cleveland, UK), and surface ones in Bordeaux and Bratislava to check radiopurity of construction materials. Table 7 lists as an example a few results of recently carried radiopurity measurements of construction materials.

Table 7: Recent radiopurity measurements of construction materials.

Radio-nuclide	Activity (nBq/g)								
	Mylar foil[a]	Plastic scintil-lator[b]	Ultra-clean copper[c]	Clean copper	M-copper[d]	Electrolytic copper[f]	Signal cables[b]	Tracker wires[b]	PMT glass[b]
^{232}Th	0.9 ± 0.2	< 100	0.034 ± 0.008	60 ± 20[d] < 19[e]	80 ± 30	1600 ± 200	< 1000	850 ± 180	65,000 ± 7000
^{238}U	1.0 ± 0.3	< 300	0.13 ± 0.04	1.0 ± 0.5[d] < 16[e]	2.1 ± 1.1	< 1600	< 1000	410 ± 90	4000 ± 500
^{40}K		2200 ± 1000					25,000 ± 4000	5600 ± 2700	150,000 ± 11000

[a]BiPo-3 [197]

[b]HPGe [36]

[c]ICPMS [198]

[d]AMS [199]

[e]HPGe [56]

[f]NAA [200]

The SuperNEMO experiment is the new generation of tracking-calorimetry neutrinoless double beta-decay experiments. No background event is expected in the neutrinoless double beta-decay region in 2.5 years of Demonstrator operation using 7 kg of ^{82}Se. The half-life sensitivity is expected to be $T^{0\nu}_{\beta\beta} > 6.5 \times 10^{24}$ years, corresponding to an effective neutrino mass sensitivity of |0.2–0.4 eV| (90% C.L.). The full SuperNEMO experiment comprising of 20 modules with 100 kg of ^{82}Se source should reach a neutrino mass sensitivity of |0.04–0.1 eV| (90% C.L.), and a half-life limit of (1×10^{26} years).

New generations of underground experiments will require, however, further developments of ultra-sensitive radioanalytical technologies. The big problem of the gamma-ray spectrometry for ultra-low-level radioactivity screening is the fact that the same radionuclides, which should be analysed in construction materials are also found in the detector background. Therefore, alternative methods for analysis of primordial radionuclides (mainly ^{238}U and ^{232}Th and their decay products) in construction materials have been suggested. The revolutionary approach in the radiometric sector has been the BiPo coincidence spectrometer, where already a third generation of this ultra-sensitive detector system has been developed [197]. The detector can measure ultra-low levels of alpha- and beta-emitters on large surface materials (e.g. supporting foils, ^{82}Se isotope source on the foil) used in the SuperNEMO experiment. To gain in sensitivity, the principle is to detect the delayed beta-alpha coincidences of the ^{214}Bi–^{214}Po cascades (^{214}Bi is the main contributor to the SuperNEMO detector background). The high-energy gamma-emitter ^{208}Tl is qualified in the BiPo detector through its mother product, the ^{212}Bi. The detector consists of 2 face-to-face planar calorimeters made of pure aluminized polystyrene scintillators coupled to 5″ low radioactivity PMTs to detect the beta- and alpha-particles, and measure the time delay to distinguish the two isotopes. The total surface area of the detector is 3.6 m^2. The BiPo-3 detector is operating since 2013 in the Canfranc underground laboratory in Spain.

The present state-of-the-art detection limits of the ultrasensitive detection methods are compared in Table 8. It can be seen that the leaders are AMS, ICPMS and BiPo-3. It would be advantageous if no pre-concentration treatment of samples would be carried out, as this process could add radioactive contamination from chemicals used during sample processing. From this point of view the AMS could be the preferably technique as samples such as copper, steel, etc. can be directly used as targets in ion sources, while ICPMS analyses would always require a pre-concentration chemistry. Similarly, NAA could be used without pre-irradiation chemistry as well, having also advantages in post-irradiation chemistry, which could improve detection limits.

Table 8: A comparison of detection limits for ^{238}U and ^{232}Th

Radio-nuclide	Detection limits (μBq)					
	Alpha-spectrometry[a,h]	Underground gamma-spectrometry[i]	BiPo-3[d,i]	ICPMS[h]	AMS[f,i]	NAA[g,i]
^{232}Th	100	5000[b] 2400[c]	1.4	0.003[e]	0.0002	0.08
^{238}U	100	4000[b] 2000[c]	1.6	0.010[e]	0.0001	0.2

[a]Estimated from [111]

[b]Estimated from [201]

[c]Estimated from [202]

[d]Estimated from [197]

[e]Estimated from [198]

[f]Estimated from [199]

[g]Estimated from [200]

[h]With pre-concentration chemistry

[i]Without pre-concentration chemistry

Radioanalytical Impacts on Environmental Studies

New ultra-sensitive radioanalytical technologies made important breakthroughs in environmental sciences, however, we can present in this review only a few examples of successful studies tracing radionuclides in the atmosphere, groundwater and in the marine environment.

Atmospheric radioactivity

We shall focus on investigations of ^{137}Cs from global fallout, Chernobyl and Fukushima accidents as ^{137}Cs has been considered to be the most important radionuclide for the long-term radiological impact after nuclear accidents mainly because of large releases, its relatively long physical half-life and its bioavailability. Thus because of its high abundance and accumulation in human tissue, it has been important for delivering radiation doses to the public [203, 204]. The ^{137}Cs in the terrestrial environment is present in atmospheric aerosols, soil and vegetation, and it has been used worldwide as a tracer of environmental processes, specifically for studying transport process in atmospheric, aquatic and terrestrial ecosystems. The ^{137}Cs after its release to the atmosphere is rapidly associated with aerosol particles, which represent a major reservoir of pollutants in the atmosphere. A new generation of high-sensitive Ge spectrometers, operating under optimum shielding conditions, have been able to carry out ^{137}Cs activity concentration measurements down to 0.1 μBq m^{-3}, decreasing thus atmospheric aerosol sampling time to several hours. The ^{137}Cs released from all sources (mainly global fallout and the Chernobyl accident) has primarily been deposited on the Earth's surface by dry and wet deposition. Before the Chernobyl accident, temporal variations of ^{137}Cs in the surface air were strongly related to global fallout, which was stratospherically controlled [205], i.e. having maxima in late spring and early summer months (Fig. 31) due to transport of stratospheric air to the troposphere. Concentrations of ^{137}Cs in the atmosphere had a decreasing trend after the moratorium on testing nuclear weapons in the atmosphere signed in 1963.

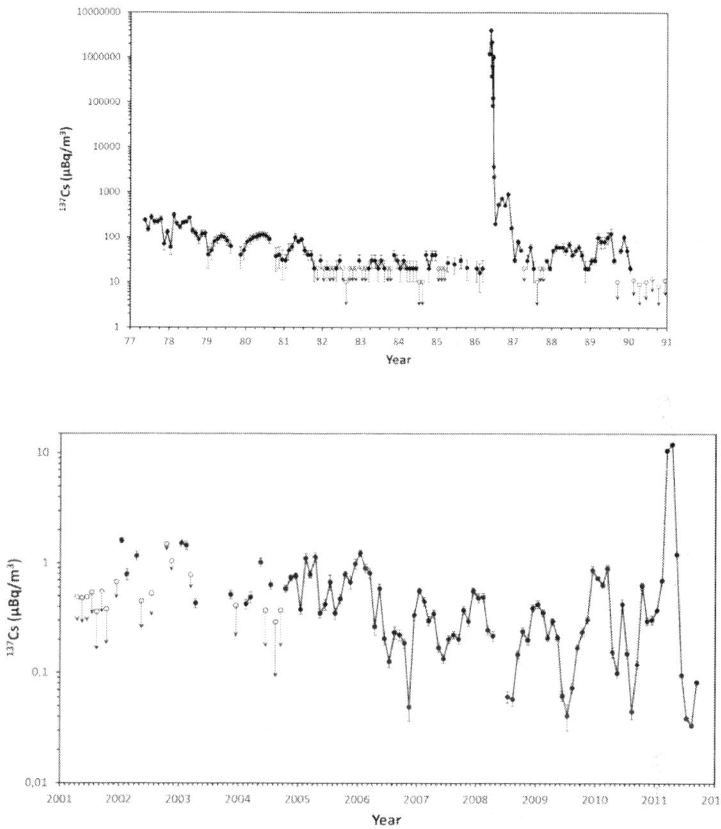

Figure 31: ^{137}Cs activity concentrations in atmospheric aerosols in the Bratislava air (Slovakia, Central Europe) (modified from [207].

The situation has changed in 1986 when due to the Chernobyl accident a high peak appeared in ^{137}Cs activity concentration in the surface air (Fig. 31) [206]. The Chernobyl accident released about 85 PBq of ^{137}Cs to the atmosphere [4], representing thus the largest single release of ^{137}Cs which occurred till now. The maximum ^{137}Cs levels observed in the beginning of May 1986 in the Bratislava air were 4 Bq m^{-3}, about 200 000 higher than the pre-Chernobyl value (Fig. 31). A few years after the Chernobyl accident the summer maxima, characteristic for the transport of ^{137}Cs from the stratosphere to the troposphere were still visible, later, during the 1990 s the main source of atmospheric ^{137}Cs became the soil resuspension and its transport in the ground-level air [192].

The Fukushima accident caused in Europe only a negligible increase in ^{137}Cs activity concentration in the ground level air. The pre-Fukushima ^{137}Cs

levels measured during the 2010–2011 in Bratislava ranged from 0.04 to 1.0 μBq m^{-3} (Fig. 31) [207]. Radionuclide activity concentrations observed in the Bratislava air showed three concentration maxima which were associated with different air masses present in the central Europe. When compared with the Chernobyl data (4 Bq m^{-3}) we see that they were at least by four orders of magnitude lower. This was mainly due to a long transport of air masses from Japan to central Europe, as well as by about a factor of 5 lower release rates of ^{137}Cs to the atmosphere during the Fukushima accident [4]. Therefore, estimated radiation doses to the public of Europe were negligible, around six orders of magnitude lower than the accepted limit of 1 mSv $year^{-1}$.

Radiocarbon in groundwater of the Danube River Basin

Isotope tracers (^{14}C, ^{13}C and ^{18}O) have been applied for better understanding of the groundwater origin at the Žitný Island situated in the south-western Slovakia between the Danube and Small Danube rivers. The spatial distribution of ^{14}C in shallow groundwater (in the form of dissolved inorganic carbon) of the Žitný Island (down to 20 m depth) is presented in Fig. 32 [208]. The observed values are mostly > 80 pMC (percent of modern carbon), except for two wells (depths of 10.9 and 15.2 m) on the east side of the island where the ^{14}C values of 31.5 and 69.9 pMC, respectively, were obtained, indicating that in this region we are dealing with a groundwater reservoir which has been outside of the direct influence of the Danube River. The core of the ^{14}C profiles represents, however, modern groundwater as the majority of groundwater has ^{14}C content above 80 pMC. The groundwater levels at Žitný Island have been depending therefore on water levels (and flow volumes) in the Danube River. This has been well manifested in the central and western part of the island where the thickness of Quaternary sediment is tens (even hundreds) of meters. However, in the eastern part of the island (where groundwater with lower ^{14}C content was observed), Neogene clays were found a few meters below the surface, which prevent a direct infiltration of groundwater of Danube origin to deeper layers. This would indicate the existence of a confined aquifer formed below the layer of Neogene clay sediments. This is not the case in the central and western parts of the island, where the Quaternary sediments are much thicker, and where mainly the Danube River system is directly influencing the groundwater regime of the Žitný Island, even at depths down to 90 m. The obtained results on spatial and vertical variability of ^{14}C, suggest isotopic heterogeneity in the groundwater of Žitný Island. This has been a first attempt in the isotope hydrology to construct isotope maps and to study

gyre on a time scale of several decades. The gyre acts as a reservoir, maintaining higher radionuclide concentrations in the region. The ^{14}C water profile data clearly shows a penetration of bomb ^{14}C around 40°S (transect at 60°E) down to almost 5000 m (similarly as we could see it in the case of tritium), what documents that the south Indian Ocean is important for sink of anthropogenic carbon. The south Indian Ocean have been acting on a time scale of several decades as a final reservoir of contaminants transported from the northern Indian and Pacific Oceans, which will cause that all garbage present in the world ocean will be finally stored in the Subtropical gyre of the Indian Ocean. The observed distribution of isotopic tracers in the Crozet Basin reflects the complex dynamics and advection of different water masses, which makes the basin one of the most interesting oceanographic places in the world ocean. The world ocean generally control the climate on the Earth, but the Indian Ocean in this respect is playing there a crucial role.

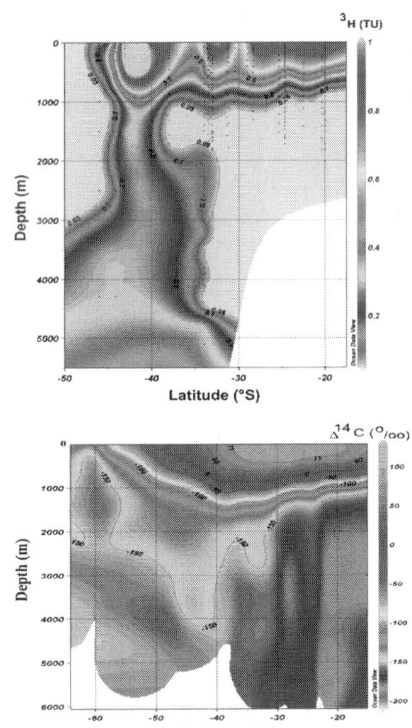

Figure 34: Spatial distribution of ^{3}H and ^{14}C in waters of the South Indian Ocean (ANTARES data were combined with nearest WOCE stations). Downward tritium and radiocarbon transport around 40°S (at 60°E) can be seen in these profiles (modified from [32].

Fukushima impact on the Pacific Ocean waters

From many results, we published recently on Fukushima radionuclides in the Pacific Ocean we present here at least a few data on ^{137}Cs, ^{3}H and ^{14}C in the water column of the NW Pacific. It has been of great interest how Fukushima accident has changed the distribution of these tracers in waters of the open Pacific Ocean, important for climate change studies. The measured ^{137}Cs activity concentrations in surface waters collected 30-600 km offshore the Fukushima NPP in June 2011 [4, 102] were up to three orders of magnitude higher than the global fallout background, although the cruise track did not go closer than 30 km from the coast. The geographical distribution of ^{137}Cs levels in the water column indicates that an atmospheric deposition of ^{137}Cs occurred within the 600 km zone offshore Fukushima. Figure 35 shows as an example the observed ^{3}H and ^{14}C water profiles with typical higher concentrations at the sub-surface layers. All Δ^{14}C values observed offshore Fukushima were negative, much lower when compared with our previous investigations [215, 216], as they were influenced (similarly as the ^{3}H levels) by Oyashio Intrusion, which brought low radioactivity waters from the north. From the full set of ^{3}H/^{137}Cs activity ratios observed in the water column, and the previously estimated ^{137}Cs releases to the sea [104], we may do the first estimation of the total ^{3}H activity released and deposited offshore of the Fukushima coast to be (0.3 ± 0.2) PBq. Measured ^{3}H, ^{14}C and ^{137}Cs concentrations in the NW Pacific Ocean confirm that their distribution has been influenced by the Fukushima accident. Their pre-Fukushima levels, e.g. those previously gathered in the framework of the WOCE and WOMARS projects [217], do not represent anymore the global fallout distribution of these radionuclides in the NW Pacific Ocean.

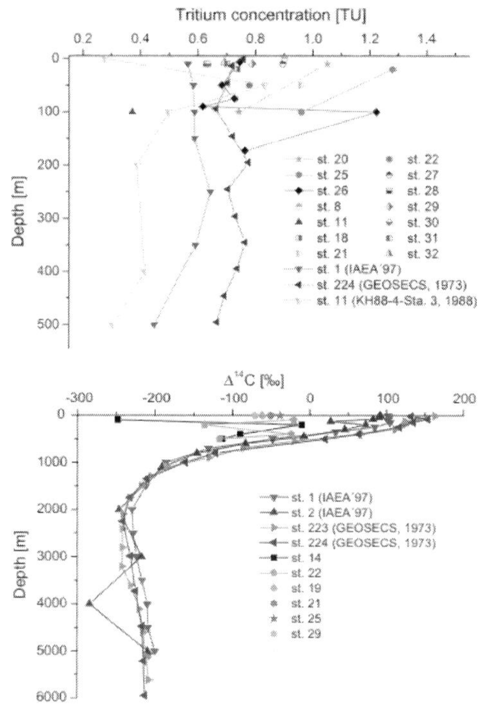

Figure 35: ^{3}H and ^{14}C seawater profiles measured offshore Fukushima NPP during the KoK epedition in June 2011 [103].

CONCLUSIONS AND OUTLOOK

We may summarize that the new developments in ultra-sensitive radioanalytical technologies have had great impact on nuclear and environmental investigations. New applications of radionuclides as tracers of environmental processes were presented in this review, which were not possible before either because of large samples required for radionuclide analyses, or because of limited sensitivities. We illustrated these developments on several topics, however, we could not cover in this review all our previous developments, e.g., on coincidence and multidimensional gamma-spectrometry, on background sources (cosmic rays, cosmogenic, radiogenic and anthropogenic radionuclides) in underground laboratories, on rare nuclear processes, on Monte Carlo simulations, on management of data quality and production of reference materials, etc. Similarly we did not discuss investigations of radionuclides in extra-terrestrial objects (from Luna and Apollo missions to Martian and Lunar meteorites and recently fallen chondrites), cosmic rays and solar variations, radioactive waste dumping sites in the Arctic, Atlantic and Pacific Oceans and adjacent seas, assessment of impacts of nuclear weapons test sites at Mururoa/Fangataufa, Bikini/Enevetak, Algeria and Australia, assessment of environmental impacts from using depleted uranium in warheads (Mediterranean Sea, Adriatic Sea, former Yugoslavia), environmental impacts due to operation of nuclear power plants, global fallout, Chernobyl and Fukushima impacts, submarine groundwater studies, isotope manifestations of climate change studies, to mention at least a few of them. Readers may found references to the corresponding papers in WOS or SCOPUS databases.

We may conclude that the most important development in the radiometric sector has been associated with underground operation of large volume Ge detectors, which have been dominating in analysis of short- and medium-lived radionuclides emitting gamma-rays. This has been mainly because of their high efficiency and excellent resolution, which helped to improve detection limits down to ~ 1 μBq kg^{-1}, and to decrease a sample size by about a factor of ten. Specific attention was given to Monte Carlo simulations of background of Ge detectors, which if carried out in advance of the construction of a low-level spectrometry system could predict its parameters. The mass spectrometry sector, represented mainly by AMS and ICPMS revolutionized ultra-sensitive analyses of long-lived radionuclides with detection limits < 1 nBq g^{-1}. The ICPMS has proved to be powerful tool because of its high sensitivity, multi-isotopic capability and the low cost per analysis. Nevertheless, there could be problems with molecular, isobaric and isotopic interferences even if careful purification procedures are used. The most exciting breakthrough in the analysis of many long-lived radionuclides has been made in the AMS sector. The AMS systems have been operating at the highest sensitivities with minimum sample size, and small matrix and interference effects. All these new developments in radiometric and mass spectrometry sectors for ultra-sensitive radionuclide analyses have had great impact on investigations of rare nuclear processes and applications in environmental, life and space sciences.

However, there is still a room for new exciting developments which can be documented by the development of new ultra-sensitive laser based analytical techniques, ultra-trace isotope detection of noble gases, development of positive ion sources for tandem accelerators, development of ion traps technologies, to mention at least a few of them. They may include new developments in low energy accelerator/ion trap technologies, or a combination of different analytical technologies including laser based systems. Developments in analytical technologies will further support a transfer from bulk sample analyses to compound specific isotope analyses with on-line coupling of analytical instruments. These new developments would make a single atom counting technology available for many radionuclides, what would be the major achievement in ultra-sensitive analysis of radionuclides, and their wide applications in all sciences based on radionuclide studies. The new technologies will, however, bring new problems, which will require additional attention: (i) A danger of contamination will considerably increase—therefore clean laboratories of class 10–100 will be required; (ii) super-clean chemistry during sample preparation will become even more important; (iii) sample inhomogeneities could mask investigated effects—therefore compound specific isotope analysis of samples will be widely required.

Ultra-sensitive radioanalytical technologies have always been crucial for realization of new scientific ideas in all branches of science. We believe that further developments in new single atom counting technologies will open windows for new even more exciting scientific investigations in the future.

ACKNOWLEDGEMENTS

My scientific carrier was strongly influenced mainly by three past events. The first one was associated with my stay during 1969–1970 at the Tata Institute of Fundamental Research in Bombay where I had a privilege to work under the leadership of late Prof. D. Lal, who had heavily influenced my further research orientation on radionuclides (both cosmogenic and radiogenic) as tracers of processes in terrestrial and extra-terrestrial bodies. The second story was associated with my stay during 1993–2005 at the International Atomic Energy Agency's Marine Environment Laboratories in Monaco where I was leading research on development of radioanalytical methods and their applications in marine radioactivity studies. Thanks to Deputy Director General of IAEA, late Dr. S. Machi, I got strong extra-budgetary financial support from Japan to establish in Monaco modern radioanalytical laboratories including an underground laboratory. And the third story, the most recent one (after my return from Monaco in 2006) has been associated with creation of the Centre for Nuclear and Accelerator Technologies (CENTA) at my home Comenius University in Bratislava, which would not be possible without generous support from European Structural Funds, and from International Atomic Energy Agency. However, I have also been very lucky to work with excellent people worldwide. It has been great pleasure to work on recent developments of radioanalytical techniques and their applications in nuclear, space and environmental sciences with many colleagues from the Comenius University and other Slovak institutions (M. Bagínová, E. Bédi, R. Breier, M. Chudý, Š. Cimbák, L. Ďurana, P. Emrich, M. Grgula, V. Hlinka†, K. Holý, M. Ješkovský, V. Jurina, J. Kaizer, I. Kontuľ, M. Kopáni, A. Kováčik, K. Kozlíková, Š. Krnáč, J. Masarik, J. Morávek, V. Palušová, J. Pánik, V. Porubčan, M. Richtáriková, O. Slávik, J. Staníček, J. Szarka, I. Sýkora, Š. Šáro, M. Šeliga, F. Šimkovic, A. Šivo, S. Tokár, J. Tóth, J. Vanko, P. Vojtyla, J. Zahoran, J. Zeman, and others). During and after my stay at the International Atomic Energy Agency's Marine Environment Laboratories in Monaco I have the privilege to work with many colleagues from the Department of Nuclear Sciences and Applications (P.K. Aggarwal, L. Araguas—Araguas, M.S. Baxter, M. Betti, J.-F. Comanducci, C. Dovlete, M. Eriksson, S.W. Fowler, K. Froehlich†, J. Gastaud†, M. Groening, T. Honda†, L. Huynh-Ngoc, Y. Ikeuchi, T. Ito, R. Kaiser, J. La Rosa, S.-H. Lee, I. Levy, L. Liong Wee Kwong, H.D. Livingston, F.R. Mantoura, J.C. Miquel, M. Nakano, H. Nies, B. Oregioni, I. Osvath, H. Petersson, M.K. Pham, J.A. Sanchez-Cabeza, J. Scholten, S. Steyskal, O. Togawa, E.J. Wyse and others). It has also been a great pleasure to take part in many international projects and collaborate with distinguished scientists from all over the world on various aspects of environmental radioactivity and rare nuclear processes (to mention at least the leaders of the groups participating in joint projects: A. Aarkrog (Riso), V.A. Alexeev (Moscow), M. Aoyama (Tsukuba, Fukushima), T. Aramaki (Mutsu), P. Bailly de Bois (Octeville), J.A., Barci-Funel (Nice), M. Baskaran (Detroit), L. Benedik (Ljubljana), H. Bokuniewicz (Stony Brook), K. Buesseler (Woods Hole), A.A. Burchuladze† (Tbilisi), W.C. Burnett (Tallahassee), G.S. Burr (Tucson), F. Bréchignac (Saint Paul Lez Durance), E. Chamizo (Sevilla), M.A. Charette (Woods Hole), A. Chatt (Halifax), J.R. Chisholm (Monaco), L. Coppola (Toulouse), R. Delfanti (La Spezia), L. Ferrière (Vienna), L.K. Fifield (Canberra), E. Fiorini (Milano), D. Frekers (Münster), M. Fukasawa (Yokosuka), R. Golser (Vienna), I. Hajdas (Zürich), Y. Hamajima (Kanazawa), S. Hauser (Palermo), K. Hirose (Tsukuba, Tokyo), G.-H. Hong (Seoul), M.A.C. Hotchkis (Sydney), X. Hou (Riso), M. Hult (Geel), Y. Igarashi (Tsukuba), Y. Ikeuchi (Chiba), K.G.W. Inn (Gaithersburg), T. Ito (Tokai), J.M. Jaubert (Monaco), P. Jean-Baptist (Gif-sur-Yvette), C. Jeandel (Toulouse), A. Ioannidou (Thessaloniki), J. John (Prague), A.J.T. Jull (Tucson), P.J. Kershaw (Lowestoft), W. E. Kieser (Toronto, Ottawa), C. Koeberl (Vienna), M. Koehler (Dresden), K.

Komura† (Kanazawa), E. Kontar (Moscow, Findlay), J. Kučera (Prague), Y. Kumamoto (Yokosuka), W. Kutschera (Vienna), M. Laubenstein (Gran Sasso), H.-C. Li (Taipei), A. E. Litherland (Toronto), H.H. Loosli (Bern), G. Lujaniené (Vilnius), R.J. Macke (Vatican), O. Masson (Saint Paul Lez Durance), M.M.M. Meier (Zürich), M. Molnár (Debrecen), W.S. Moore (Columbia), U. Morgenstern (Lower Hutt), C. Musa (Paris), M. Nakano (Tokai), H. Nies (Hamburg), S. Nisi (Gran Sasso), H. Oeschger† (Bern), N. Ogrinc (Ljubljana), J. de Oliveira (Sao Paulo), Y.R. Ozorovich (Moscow), S.V. Pagava (Tbilisi), L. Palcsu (Debrecen), S. M. Pan (Nanjing), W. Plastino (Roma), A. Priller (Vienna), K.F. von Reden (Woods Hole), P. Roos (Riso), T.P. Ryan (Dublin), S. Shima (Mutsu), H. Simgen (Heidelberg), B.L.K. Somayajulu† (Ahmedabad), P. Steier (Vienna), T. Stieglitz (Townsville), I. Svetlik (Prague), M. Taniguchi (Kyoto), F. Terrasi (Caserta), O. Togawa (Tokai), Z. Top (Miami), Ch. Tsabaris (Athens), G.K. Ustinova (Moscow, Dubna), P. Vojtyla (Geneva), R. Wieler (Zürich), E.M. Wild (Vienna), D. Woodhead (Lowestoft), I. Zvara (Dubna), and with more than two hundred colleagues participating in international underground physics research projects NEMO-3, SuperNEMO, LEGEND and EURECA. Also I would like to thank Prof. K. Hirose for valuable comments which helped to improve this paper. The developments of ultra-sensitive radioanalytical technologies and their applications at the Comenius University in Bratislava have recently been supported by the EU Research and Development Operational Program funded by the ERDF (projects 26240120012, 26240120026 and 26240220004), by the International Atomic Energy Agency (projects SLR 0/008, SLR 0/009, SLR1001 and RER 7008), by the Slovak Research and Development Agency (Projects APVV 15 – 0576 and APVV-0516-10) and by the VEGA Scientific Granting Agency of Slovakia (Projects 1/0891/17 and 1/0783/14).

REFERENCES

1. Povinec PP (2004) In: Livingston HD (ed) Marine Radioactivity. Elsevier, New York
2. Povinec PP, Betti M, Jull AJT, Vojtyla P (2008) Acta Phys Slov 58:1–154
3. Povinec PP, Scholten J, Erikson M, Betti M (2012) In: L'Annunziata MF (ed) Handbook of Radioactivity Analysis. Elsevier, New York
4. Povinec PP, Hirose K, Aoyama M (2013) Fukushima Accident: Radioactivity Impact on the Environment. Elsevier, New York
5. Povinec, PP et al (1995) In: Proc. monitoring of nuclear contamination in Arctic Seas. NRL/MR/6610-95-7674. NRL, Washington D.C. IV 2-17
6. Povinec PP, Osvath I, Baxter MS (1996) Int J Appl Rad Isot 47:1127–1133
7. Povinec PP et al (1997) Mar Pollut Bull 35:235–241
8. Osvath I, Povinec PP, Baxter MS (1999) Sci Total Environ 237(238):167–179
9. Osvath I, Povinec PP, Huynh-Ngoc L, Comanducci J-F (1999) Sci Total Environ 237(238):277–286
10. Osvath I, Povinec PP, Baxter MS, Huynh-Ngoc L (2001) J Radioanal Nucl Chem 248:735–739
11. Povinec PP et al (2001) J Radioanal Nucl Chem 248:713–718
12. Osvath I, Povinec PP (2001) J Environ Radioact 53:335–349
13. Osvath I, Povinec PP, Livingston HD, Ryan TP, Comanducci J-F, Mulsow S (2005) J Radioanal Nucl Chem 263:437–440
14. Povinec PP et al (2006) Cont Shelf Res 26:874–884
15. Povinec PP et al (2006) Radioact Environ 8:373–381
16. Povinec PP et al (2008) Coast Shelf Sci 76:522–542
17. Povinec PP, Osvath I, Comaducci J-F (2008) Radioact Environ 11:449–579
18. Livingston HD, Povinec PP (2000) Ocean Coast Manag 43:689–712
19. Livingston HD, Povinec PP, Ito T, Togawa O (2001) In: Kudo A (ed) Plutonium in the Environment. Elsevier, Amsterdam
20. Livingston HD, Povinec PP (2002) Health Phys 82:656–668
21. Povinec PP et al (2001) J Radioanal Nucl Chem 248:713–718

22. Povinec PP et al (2003) Deep Sea Res II 50:2607–2638
23. Povinec PP et al (2003) Deep Sea Res II 50:2751–2760
24. Povinec PP (2005) J Radioanal Nucl Chem 263:413–417
25. Povinec PP (2008) J Radioanal Nucl Chem 276:771–776
26. Povinec PP (2010) J Radioanal Nucl Chem 286:401–407
27. Povinec PP et al (2003) Deep Sea Res II 50:2785–2801
28. Povinec PP et al (2003) Deep Sea Res II 50:2835–2846
29. Povinec PP et al (2004) J Environ Radioact 76:113–117
30. Povinec PP et al (2005) J Environ Radioact 81:63–87
31. Povinec PP et al (2006) In: Povinec PP (ed) Isotopes in Environmental Studies. IAEA, Vienna
32. Povinec PP et al (2011) Earth Planet Sci Lett 302:14–26
33. Povinec PP et al (2012) Progr Oceanogr 89:17–30
34. Povinec PP et al (2012) J Environ Radioact 104:24–45
35. Povinec PP (2017) Nucl Instrum Methods Phys Res A 845:398–403
36. Povinec PP (2017) Appl Radiat Isot 126:26–30
37. Burchuladze AA, Pagava SV, Povinec P, Togonidze GI, Usačev S (1980) Nature 287:320–322
38. Povinec PP et al (2006) Radioact Environ 8:538–553
39. Povinec PP, Comanducci JF, Levy-Palomo I (2004) Appl Radiat Isot 61:85–93
40. Povinec PP, Comanducci JF, Levy-Palomo I (2005) J Radioanal Nucl Chem 263:441–445
41. Povinec PP et al (2006) In: Povinec PP (ed) Radioactivity in the environment. Elsevier, New York
42. Povinec PP, Vojtyla P, Comanducci JF (2008) Analysis of environmental radionuclides. IAEA, Vienna, pp 163–208
43. Povinec P, Jurina V, Emrich P (1984) Acta Phys Slovaca 34:58–64
44. Staníček J, Povinec PP (1986) Nucl Instrum Methods B17:462–466
45. Sýkora I, Povinec PP (1986) Nucl Instrum Methods B17:467–471
46. Sýkora I, Ďurčík M, Staníček J, Povinec PP (1992) In: Povinec PP (ed) Rare Nuclear Processes. World Scientific, Singapore
47. Zvara I, Povinec PP, Sykora I (1994) Pure Appl Chem 66:2537–2586

48. Povinec PP, Sykora I, Porubcan V, Jeskovsky M (2007) J Radioanal Nucl Chem 282:805–810
49. Povinec PP (2008) J Radioanal Nucl Chem 276:771–775
50. Sýkora I et al (2008) J Radioanal Nucl Chem 276:779–787
51. Kovacik A, Sykora I, Povinec PP, Porubcan V (2012) J Radioanal Nucl Chem 293:339–343
52. Povinec PP (1982) Isotopenpraxis 18:92–97
53. Burnett JL, Davies AV (2011) J Radioanal Nucl Chem 288:699–703
54. Heusser G (2015) Eur Phys J C75:531–539
55. Hirose K, Aoyama M, Igarashi Y, Komura K (2005) J Radioanal Nucl Chem 263:349–354
56. Heusser G, Laubenstein M, Neder H (2006) In: Povinec PP, Sanchez-Cabeza JA (eds) Radionuclides in the environment. Elsevier, Amsterdam
57. Povinec PP (2012) J Anal Sci Technol 3:42–71
58. Povinec PP, Pham MK (2001) J Radioanal Nucl Chem 248:211–216
59. Povinec PP (2013) J Radioanal Nucl Chem 295:537–544
60. Levy I et al (2011) Progr Oceanogr 89:120–133
61. Povinec P, Chudý M, Seliga M, Saro S, Szarka J (1972) J Radioanal Chem 12:513–518
62. Chudý M, Povinec P, Šeliga M (1972) Acta Phys Slov 22:179–185
63. Šeliga M, Povinec P, Chudý M (1972) Coll Czechoslov Chem Comm 37:448–452
64. Povinec P, Chudý M, Seliga M, Saro S (1971) Isotopenpraxis 7:54–57
65. Povinec P, Šáro Š, Chudý M, Šeliga M (1968) Int J Appl Radiat Isot 19:877–881
66. Šeliga M, Povinec P, Chudý M (1970) Coll Czechoslov Chem Comm 35:1278–1282
67. Usacev S, Povinec P, Chudý M, Saro S (1973) Radiocarbon 15:443–450
68. Povinec P (1978) Nucl Instr Methods 156:441–446
69. Povinec P et al (1978) Radiochem Radioanal Lett 33:67–76
70. Povinec P (1975) Int J Appl Radiat Isot 26:465–469
71. Povinec P (1979) Nucl Instr Methods 163:363–368

72. Povinec P, Szarka J, Usacev S (1979) Nucl Instrum Methods 163:369–375
73. Povinec P (1980) Nucl Instrum Methods 176:111–117
74. Povinec P (1980) Nucl Instrum Methods 177:465–469
75. Cimbák S, Povinec P (1985) Environ Intern 11:65–69
76. Cimbák S, Čechová A, Grgula M, Povinec P, Šivo A (1986) Nucl Instrum Methods B17:560–563
77. Csongor E, Wilhelmová L, Dvorák Z, Povinec P, Grgula M (1988) Appl Radiat Isot 39:401–405
78. Chudý M, Janik R, Povinec P (1979) J Radioanal Chem 51:167–172
79. Liong Wee Kwong L, Povinec PP (1999) In: Povinec PP (ed) Marine pollution. IAEA, Vienna
80. Liong Wee Kwong L et al (2004) J Radioanal Nucl Chem 261:283–289
81. Wyse E, Lee SH, La Rosa J, Povinec PP, de Mora SJ (2001) J Anal At Spectrom 16:1107–1110
82. Povinec PP et al (2002) J Radioanal Nucl Chem 251:369–374
83. Povinec PP (2005) J Radioanal Nucl Chem 263:413–417
84. Hou X, Roos P (2008) Anal Chim Acta 608:105–109
85. Ross P (2008) In: Povinec PP (ed) Radionuclides in the Environment. Elsevier, Amsterdam
86. Wendt K, Trautmann N, Bushaw BA (2000) Nucl Instrum Methods Phys Res B172:162–167
87. Erdmann N et al (2008) In: Povinec PP (ed) Radionuclides in the Environment. Elsevier, Amsterdam
88. Dai MH et al (2001) J Environ Radioact 53:9–15
89. Lewis D et al (2001) J Radioanal Nucl Chem 249:115–119
90. Betti M et al (2008) In: Povinec PP (ed) Radionuclides in the environment. Elsevier, Amsterdam
91. Tuniz C, Bird JR, Fink D, Herzog GF (1998) Accelerator mass spectrometry: ultrasensitive analysis for global science. CRC Press, Boca Raton
92. Raisbeck G, Yiou F (1999) Sci Total Environ 237(238):31–38
93. Povinec PP, Oregioni B, Jull AJT, Kieser WE, Zhao XL (2000) Nucl Instrum Methods Phys Res B 172:672–678
94. Kutschera W (2005) Int J Mass Spectrom 242:145–152

95. Jull AJT et al (2008) In: Povinec PP (ed) Radionuclides in the environment. Elsevier, Amsterdam
96. Fifield LK (2008) In: Povinec PP (ed) Radionuclides in the environment. Elsevier, Amsterdam
97. Lee SH et al (2009) J Radioanal Nucl Chem 282:831
98. Povinec PP, Litherland AE, von Reden KF (2009) Radiocarbon 51:45–58
99. Lujaniene G et al (2015) Radiocarbon 57:481–492
100. Lujaniene G et al (2016) J Radioanal Nucl Chem 307:2231–2237
101. Lujaniene G et al (2017) Appl Radiat Isot 126:49–53
102. Kaizer J et al (2018) J Environ Radioact 184(185):83–94
103. Povinec PP et al (2017) J Environ Radioact 166:56–66
104. Povinec PP et al (2013) Biogeosciences 10:5481–5496
105. Hou X et al (2013) Environ Sci Technol 47:3091–3098
106. Aoyama M, Hirose K, Miyao T, Igarashi Y, Povinec PP (2001) J Radioanal Nucl Chem 248:789–796
107. Aoyama M et al (2011) Prog Oceanogr 89:7–15
108. Sanchez-Cabeza JA et al (2011) Prog Oceanogr 89:31–37
109. Sachez-Cabeza JA et al (2007) In: Kumamoto Y, Watanabe S (eds) Blue earth global expedition, vol. 3. YAMSTEC, Yokosuka
110. La Rosa J et al (2001) J Radioanal Nucl Chem 248:765–771
111. Lee SH et al (2001) J Radioanal Nucl Chem 248:757–764
112. Povinec PP (ed) (2008) Analysis of environmental radionuclides. Elsevier, Amsterdam
113. Povinec PP, Sanchez-Cabeza JA (eds) (2006) Radionuclides in the Environment. Elsevier, New York
114. Povinec PP, Hirose K (2012) Encyclopaedia of sustainability science and technology. Springer, New York
115. Povinec PP (2010) Encyclopaedia of inorganic chemistry. Radionuclides in the environment. Wiley, New York
116. Theodorson P (1996) Measurement of weak radioactivity. World Scientific, Singapore
117. Debertin K, Helmer RG (1998) Gamma and X-ray spectrometry with semiconductor detectors. Elsevier, Amsterdam

118. Scott EM, Dixon M (2008) In: Povinec PP (ed) Analysis of environmental radionuclides. Elsevier, New York
119. Macášek F (2008) In: Povinec PP (ed) Analysis of environmental radionuclides. Elsevier, New York
120. Aoyama M, Hirose K (2008) In: Povinec PP (ed) Analysis of environmental radionuclides. Elsevier, New York
121. Currie LA (2008) In: Povinec PP (ed) Analysis of environmental radionuclides. Elsevier, New York
122. Hou X (2008) In: Povinec PP (ed) Analysis of environmental radionuclides. Elsevier, New York
123. Nisi S et al (2009) Appl Radiat Isot 67:828–832
124. Levine J et al (2009) Int J Mass Spectr 208:36–43
125. Povinec P, Usacev S (eds) (1977) Low-radioactivity measurements and applications. SPN, Bratislava
126. Povinec P (ed) (1982) Low-level counting. VEDA, Bratislava
127. Povinec, P. (ed) (1986) Nucl Instrum Methods Phys Res B17:377–588
128. Povinec P (ed) (1987) Low-level counting and spectrometry. VEDA, Bratislava
129. Povinec P (ed) (1991) J Phys G 17:1–543
130. Povinec P (ed) (1992) Rare nuclear processes. World Scientific, Singapore
131. Povinec PP (ed) (1999) Sci Total Environ 237/238:1–526
132. Libby WF (1955) Radiocarbon dating. Univ Chicago Press, Chicago
133. de Vries H, Barendsen GW (1953) Physica 19:987–1003
134. Suess HE (1954) Science 120:5–7
135. Burke WH Jr, Meinschein WG (1955) Rev Sci Instrum 26:1137–1140
136. Arnold JR (1954) Science 119:155–157
137. Tamers MA, Pearson FJ (1965) Nature 205:1205–1207
138. Muller RA (1977) Science 196:489–494
139. Purser KH et al (1977) Rev Phys Appl 12:1487–1492
140. Nelson DE, Korteling RG, Stott WR (1977) Science 198:507–508
141. Blackett PMS, Occhialini GPS (1933) Proc R Soc A 139:699–726
142. Povinec P (1972) Radiochem Radioanal Lett 9:127–135
143. Lal D (1965) In: Chatter RM (ed) Proc radiocarbon and tritium dating.

Pullman, Washington
144. Mace EK et al (2016) J Environ Radioact 155(156):122–129
145. Oeschger H, Wahlen M (1975) Ann Rev Nucl Sci 25:423–463
146. Povinec P (1972) Nucl Instrum Methods 101:613–614
147. Povinec P (1981) Nucl Energy 27:207–209
148. Szarka J, Povinec P (1979) Nucl Instrum Methods 164:463–468
149. Kubinec P, Masarik J, Melo I, Povinec P (1986) Nucl Instrum Methods B17:454–457
150. Kuzminov VV et al (1986) Nucl Instrum Methods B17:452–453
151. Bellotti E et al (1992) Nucl Instrum Methods A323:125–134
152. Povinec P et al (1990) Nucl Instrum Methods A293:562–568
153. Povinec P (1981) Int J Appl Radiat Isot 32:729–732
154. Krajcar-Bronić, Obelić B, Srdoć D (1986) Nucl Instrum Methods B17:498–500
155. La Rosa J et al (2005) J Radioanal Nucl Chem 263:427–433
156. Povinec PP et al (1996) In: Liquid scintillation spectrometry. Radiocarbon, pp 149–155
157. Korinková T et al (2016) J Radioanal Nucl Chem 307:2295–2299
158. Simek P et al (2016) J Environ Radioact 166:83–90
159. Palcsu L, Major Z, Kollo Z, Papp L (2010) Rapid Commun Mass Spectrom 24:698–704
160. Jean-Baptiste P et al (2010) J Environ Radioact 101:185–189
161. Povinec PP (2012) J Anal Sci Technol 3:42–71
162. Vojtyla P, Povinec PP (2000) Appl Radiat Isot 53:185–189
163. Breier R, Povinec PP (2009) J Radioanal Nucl Chem 282:799
164. Breier R, Povinec PP (2010) Appl Radiat Isot 68:1231–1235
165. Watt DE, Ramsden D (1964) High sensitivity counting techniques. Pergamon Press, Oxford
166. Pagava SV et al (1992) In: Povinec PP (ed) Rare nuclear processes. World Scientific, Singapore
167. Niese S et al (1998) In: Povinec PP (ed) Analysis of environmental radionuclides. Elsevier, New York
168. Komura K, Hamajima Y (2004) Appl Radiat Isot 61:185–190
169. Laubenstein M et al (2004) Appl Radiat Isot 61:167–172

170. Bellini G et al (2012) JCAP 05:015
171. Armengaud E et al (2016) Edelweiss collaboration JCAP 05:019
172. SNOLAB User's Handbook (2013) www.snolab.ca
173. Breier R, Hamajima Y, Povinec PP (2016) J Radioanal Nucl Chem 307:1957–1960
174. Breier R, Laubenstein M, Povinec PP (2017) Appl Radiat Isot 126:188–190
175. Breier R, Brudanin VB, Loiaza, P, Piquemal F, Povinec PP, Štekl I (2017) J Environ Radioact (submitted)
176. Baxter MS, Harms I, Osvath I, Povinec PP, Scott EM (1998) J Environ Radioact 39:161–181
177. Hirose K et al (1999) J Environ Radioact 43:1–13
178. Povinec PP et al (1999) Sci Total Environ 237(238):249–267
179. Povinec PP et al (2015) Nucl Instrum Methods Phys Res B 342:321–326
180. Povinec PP (2015) Nucl Instrum Methods Phys Res B 361:87–94
181. Ješkovský M et al (2015) Nucl Instrum Methods Phys Res B 361:139–142
182. Arnold R et al (2010) Eur Phys J C 70:927–943
183. Abgrall N et al (2017) AIP Conf Proc 1894, 020027
184. Angloher G et al (2014) Phys Dark Univ 3:41–74
185. Povinec PP et al (2015) J Radioanal Nucl Chem 304:67–73
186. Povinec PP et al (2009) J Environ Radioact 100:125–130
187. Povinec PP et al (2012) J Environ Radioact 108:33–40
188. Povinec PP et al (2008) Appl Radiat Isot 66:1686–1690
189. Povinec PP et al (2015) Radiocarbon 57:355–362
190. Ješkovský M et al (2015) Nucl Instrum Methods Phys Res B 361:129–132
191. Kontuľ I et al (2017) Appl Radiat Isot 126:58–60
192. Levin I, Kromer B (2004) Radiocarbon 46:1261–1272
193. Levin I, Kromer B, Hammer S (2013) Tellus B 65:20092
194. Zeman J et al (2017) J Radioanal Nucl Chem 311:1409–1415
195. Arnold R et al (2017) Phys Rev Lett 119:041801
196. Barabash AS et al (2017) JINST 12 P06002

197. LaFerriere BD et al (2015) Nucl Instrum Methods Phys Res A 775:93–98
198. Famulok N et al (2015) Nucl Instrum Meth Phys Res B 361:193–196
199. Kučera J, Kameník J, Povinec PP (2016) J Radioanal Nucl Chem. https://doi.org/10.1007/s10967-016-4930-2
200. Loaiza P et al (2017) Appl Radiat Isot 123:54–59
201. Budjáš D et al (2009) Appl Radiat Isot 67:755–758
202. Aarkrog A et al (1997) J Environ Radioact 34:69–90
203. Povinec PP, Hirose K (2016) Sci Rep 5:9016. https://doi.org/10.1038/srep09016
204. Hirose K, Povinec PP (2015) Sci Rep 5:15707. https://doi.org/10.1038/srep15707
205. Povinec PP et al (1988) J Radioanal Nucl Chem Lett 126:467–478
206. Povinec PP et al (2012) J Environ Radioact 114:81–88
207. Povinec PP et al (2013) Radiocarbon 55:1017–1028
208. Aoyama M, Hirose K, Igarashi Y (2006) J Environ Monit 8:431–438
209. Povinec et al PP (2004) J Environ Radioact 76:113–137
210. Povinec PP et al (2006) In: Povinec PP, Sanchez-Cabeza JA (eds) Isotopes in Environmental Studies IAEA, Vienna
211. Aoyama M, Hirose K (1995) J Environ Radioact 29:57–74
212. Kumamoto Y et al (2011) Prog Oceanog 89:49–60
213. Aoyama M, Povinec PP, Sanchez-Cabeza JA (2011) Prog Oceanog 89:1–6
214. Aramaki T, Mizushima T, Kuji T, Povinec PP, Togawa O (2001) Radiocarbon 43:857–867
215. Povinec PP et al (2004) Radiocarbon 46:583–594
216. Povinec PP et al (2005) J Environ Radioact 81:63–87

17

Radiochemistry on Electrodes: Synthesis of an ^{18}F-labelled and In Vivo Stable COX-2 Inhibitor

Artem Lebedev[1,2], Jing Jiao[1] , Jason Lee[1,3], Fan Yang[4] , Nathanael Allison[5], Harvey Herschman[1,3,6], Saman Sadeghi[1,3,5,6]

[1] UCLA Department of Molecular and Medical Pharmacology, David Geffen School of Medicine, Los Angeles, California, United States of America
[2] Traceability, Culver City, California, United States of America
[3] UCLA Crump Institute for Molecular Imaging, Los Angeles, California, United States of America
[4] The Department of Chemistry at The Scripps Research Institute, La Jolla, California, United States of America
[5] UCLA Biomedical Physics Interdepartmental Program, Los Angeles, California, United States of America
[6] UCLA Ahmanson Translational Imagining Division, Los Angeles, California, United States of America

ABSTRACT

New radiochemistry techniques can yield novel PET tracers for COX-2 and address the shortcomings in in vivo stability and specificity, which have held back clinical translation of tracers to image COX-2 expression. Current techniques limit radiosynthesis to analogs

Citation: Lebedev A, Jiao J, Lee J, Yang F, Allison N, Herschman H, et al. (2017) "Radiochemistry on electrodes: Synthesis of an 18F-labelled and in vivo stable COX-2 inhibitor". PLoS ONE 12(5): e0176606. https://doi.org/10.1371/journal.pone.0176606

of the COX-2 inhibitors with fluorine-18 added via a carbon chain, or on an aromatic position which renders the radiolabeled analog less specific towards COX-2, resulting in tracers with low *in vivo* stability or specificity. To solve this problem, we have developed a new high affinity, ^{18}F-labelled COX-2 inhibitor that is radiolabeled directly on a heteroaromatic ring. This molecule exhibits favorable biodistribution and increased metabolic stability. Synthesis of this molecule cannot be achieved by traditional means; consequently, we have developed an automated electrochemical radiosynthesis platform to synthesize up to 5 mCi of radiochemically pure ^{18}F-COX-2ib in 4 hours (2% decay-corrected radiochemical yield). *In vitro* studies demonstrated clear correlation between COX-2 expression and uptake of the tracer. PET imaging of healthy animals confirmed that the molecule is excreted from blood within an hour, mainly through the hepatobiliary excretion pathway. *In vivo* metabolism data demonstrated that > 95% of the injected radioactivity remains in the form of the parent molecule 1 hour after injection.

INTRODUCTION

There is strong evidence suggesting a relationship between inflammation and carcinogenesis, as well as neuroinflammation and CNS disease progression. Several processes that are involved in carcinogenesis, including apoptosis, angiogenesis, cell proliferation, invasiveness and metastasis, are correlated with COX-2 overexpression. Epidemiological data support the correlation of COX-2 overexpression with cancer, since aspirin or other NSAIDs lower incidence of deaths from various types of cancer[1]. Genetic studies have provided further correlation between carcinogenesis and COX-2 overexpression. For example targeted COX-2 deletion led to decreased intestinal polyps in female mice [2] and enhanced COX-2 expression is sufficient to induce mammary gland tumorigenesis[3]. COX-2 expression has also been shown to have a direct role in modulating breast cancer progression [4].

Cyclooxygenase-2 (COX-2), located on the luminal side of the endoplasmic reticulum and nuclear membrane, plays a major role in regulating the rate of conversion of arachidonic acid to the various prostanoids and their downstream products[5]. COX-2 overexpression is a characteristic feature of many premalignant neoplasms[6] and appears to be both a marker and an effector of neural damage, both after a variety of acquired brain injuries and in natural or pathological aging of the brain[7]. While co-expression of COX-2 with tumor metastatic phenotype has been observed in certain types of cancer[8], evidence of a direct role for COX-2 in carcinogenesis and neurodegenerative processes remains controversial, and, in the absence of a viable COX-2 *in vivo* imaging agent, hypotheses either way cannot be confirmed without the availability for non-invasive longitudinal studies. Developing a non-invasive COX-2 imaging agent will be of great value, contributing to our understanding of the molecular mechanisms associated with inflammatory processes, by monitoring COX-2 levels throughout the progression of diseases such as neurodegenerative Alzheimer disease and Parkinson's disease. Furthermore, early detection of this inflammation related process, i.e., induction of COX-2 expression, can potentially stratify patients and provide a rationale for selective therapies and their optimization in treatment of CNS disorders and cancers. One such example is a clinical study demonstrating stratification of patients with the presence of COX-2 in premalignant cancer lesions as an important determinant of their response to adjuvant celecoxib therapy [9]. Currently, only *ex vivo* analysis can provide quantitative information on COX-2 expression. However, *ex vivo* analysis is laborious, will not provide localization and biodistribution, and can be inaccurate, since COX-2 mRNA and protein are not stable *ex vivo* [10].

Positron emission tomography (PET) is a real-time, *in vivo* three dimensional imaging technique that has unparalleled specificity and sensitivity for visualizing biochemical processes[11]. It is uniquely suited to provide data on *in vivo* expression of COX-2 and its involvement in disease development and progression. This modality is already widely used in the clinic and clinical translation of novel PET tracers has recently yielded an array of newly approved tracers. [12]. The unparalleled sensitivity of this method makes it the only viable candidate for visualization of low abundance targets, such as COX-2. PET relies on the administration of an exogenous tracer–a radiolabeled molecule with a known biodistribution, administered at sub-pharmacological levels for visualizing and quantifying molecular processes *in vivo*. While a variety of positron-emitting radionuclides can be used for this purpose, ^{18}F remains the staple for clinical PET tracers due to the favorable 109.8 min radioactive half-life, 97% low energy positron emission and the fact that addition of a small fluorine atom minimally perturbs the parent molecule. Desirable properties of ^{18}F label have given rise to a whole research field of ^{18}F radiolabeling and continuous advances in this field have yielded an impressive set of tools for labeling of a wide range of molecules.[13]

Despite these advances, a clinical translatable tracer for visualization of COX-2 expression through PET imaging still awaits the development of an adequate probe. A large set of potent COX-2 inhibitors have been developed for therapeutic applications; these molecules can potentially be used as vectors for radioactive labels. However, low tolerance of these molecules to structural modifications has left very few options for labeling using traditional methodologies. Most known attempts have yielded tracers suffering from either low metabolic stability or low affinity to the enzyme. Previous efforts to address this problem are summarized in a recent review. [14] However, despite attracting substantial attention, finding a suitable COX-2 PET tracer remains an unresolved challenge.

Among COX-2 inhibitors reported to date, the celecoxib analog **1**[15] (Fig 1) stands out because of two major advantages for PET imaging: 1) low nanomolar affinity to the enzyme and 2) presence of a fluorine atom directly attached to heteroaromatic moiety. Radiolabeling of this molecule can be performed, with no structural perturbation, by replacing the natural fluorine isotope with ^{18}F-fluorine. The introduced radioactive label will be present at a metabolically stable position. The importance of minimal structural perturbation is highlighted by extensive structure–activity studies that suggest bulky substituents on COX-2 inhibitors are not tolerated *in vivo*.[15] Tracers having a fluorine atom attached to the aromatic ring generally exhibit better metabolic stability compared to agents with the ^{18}F label attached to aliphatic chain, which often experience rapid defluorination, especially if excreted through hepatobiliary pathway.[16][17]. Working within the confines of traditional radiochemistry, several groups have attempted to place an ^{18}F-fluorine on an aromatic moiety of various analogs of COX-2 inhibitors. Unfortunately, traditional aromatic radiolabeling methods are restricted in the placement of the ^{18}F label to only a few positions on the COX-2 inhibitor scaffold. However, substituents on the aromatic ring alter the binding affinity of the radiolabeled COX-2 inhibitors. Such compounds may show probe uptake in COX-2–expressing cells, but were ineffective *in vivo* [18–20]. Placing the ^{18}F-fluorine label on the five membered ring thus emerges as an attractive approach, but there are very few examples of either late stage "cold" 19F-fluorination or ^{18}F radiolabeling of electron rich five-membered heteroaromatic rings.[21],[22]

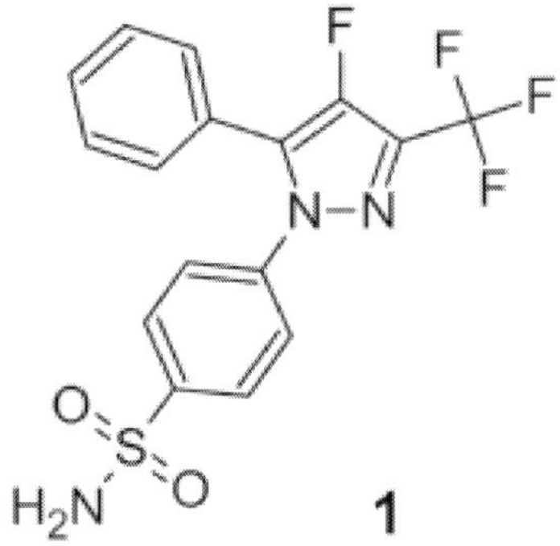

1

Figure 1: Celecoxib analog 1.

IC_{50} = 1.7 nM.

https://doi.org/10.1371/journal.pone.0176606.g001

Radiofluorination of aromatic molecules in general has so far remained a highly desirable but elusive goal. Unfortunately, ^{18}F-fluoride, the most accessible form of radioactive fluorine, exhibits low reactivity in classic nucleophilic substitution. Historically, electrophilic reactions involving ^{18}F-F_2 gas have been a solution for this problem, but these reactions suffer from low specific activity, lack of availability of ^{18}F-F_2 and have experimental challenges associated with the handling of corrosive F_2 gas. Precursors bearing a strongly electron withdrawing group can react with fluoride directly, but this method severely restricts the scope of target tracers [23,24]. A number of methods have been developed to address these challenges. These methods focus on purely chemical methodologies, employing various catalysts,[25] organometallic[26] and iodonium[27] leaving groups or strong oxidizing reagents[28], to facilitate unfavorable interaction between the negatively charged fluoride anion and the electron rich aromatic moiety. Recent reviews have summarized these approaches.[29,30]

Electrochemical fluorination stands out from the late stage radiofluorination methods described above.[31],[32] This method relies on the oxidation of the precursor on an electrode surface maintained at a potential above the oxidation potential of the precursor, to generate a cationic species reactive toward fluoride anion. In the context of selective fluorination of organic molecules, this method has been well studied and applied to a variety of molecules.[33] However, utilization of this methodology in fluorination of aromatics and radiosynthesis has been lagging.

MATERIALS AND METHODS

All animal experiments were performed according to protocol approved by the UCLA Office of Animal Research Oversight. Isoflurane anesthesia was used for the invasive procedures, animals were sacrificed by carbon dioxide asphyxiation.

Electrochemical radiolabeling was performed using an in-house built radioelectrochemical synthesizer. A detailed description and evaluation of the apparatus is under preparation for publication. Fluidic schematic of the platform can be found in (Fig 1). No-carrier added [^{18}F]fluoride was produced via $^{18}O(p,n)^{18}F$ nuclear reaction in a RDS-112 (Siemens, USA) cyclotron by 11 MeV proton bombardment of 95% oxygen-18 enriched water ([$^{18}O]H_2O$, Rotem, Inc., Israel) in a tantalum target with a beam strike volume of 860μL. A beam current of 35 μA for 60 min yielded ~1100 mCi of [^{18}F]fluoride, which was directly delivered to the ^{18}F vial of the radioelectrochemical synthesizer.

After delivery of radioactivity from the cyclotron, [^{18}F] fluoride in target water was pushed through a cartridge containing 10 mg of anion exchange resin (Bio-Rad AG MP-1M), followed by 6 ml of dry acetonitrile. The cartridge was further dried by blowing dry nitrogen at 10 psi for 5 min.

After the cartridge was dried, 4 ml of solution of precursor (50 mM), Bu_4NClO_4 (50 mM) and Et_4NF*4HF (20 mM) in dry acetonitrile was pushed through the MP-1 cartridge (controlled flow rate of 40 uL/s) and directed into the cylindrical Teflon reactor. The use of the fluoride salt is primarily dictated by the need for an electrochemically inert anion capable of replacing fluoride on anion exchange cartridge. While this limits the achievable specific activity, the added carrier fluoride is not a fundamental prerequisite for the electrochemical reaction.

The solution was then electrolyzed at ambient temperature using Ø1.0 mm Pt wire electrodes (working electrode ~800 mm^2, auxiliary electrode ~1300 mm^2) and Ø1.5 mm Ag wire as a pseudo reference electrode. Electrolysis and concurrent fluorination in the cell was conducted for 70 min in 700 cycles using an Autolab PGSTAT204 driven by Nova 1.9 software (Metrohm USA). Each cycle consisted of 5 sec of working phase at 2.7 V and 1 sec of recovery phase at 0.3 V. A controlled stream of dry nitrogen (0.5 to 2 ml/min) was blown across the surface of the solution to remove hydrogen gas formed at the auxiliary electrode. Due to incomplete conversion of the precursor, chemical yield of the reaction can be improved by extending the electrolysis process, although this will not improve radiochemical yield because of the competing radioactive decay.

After hydrolysis, the dark yellow reaction mixture was transferred into a pre-heated reactor charged with 1 ml of 37% HCl_{aq}. The mixture was stirred for 15 min and 12 ml of water was added. The resulting slurry was passed through a C-18 SPE cartridge (Waters Sep-Pak Classic Short). The product adsorbed on the cartridge was washed off with 1 ml of EtOH followed with 1 ml of water.

The resulting suspension of crude product in 50% EtOH was transferred into the loading loop of the prep-HPLC part of the system and purified using a Phenomenex Gemini column (21x250 mm) at 20 ml/min of 47% MeCN in water as an eluent. The radioactive peak at approx. 30 min was collected in approx. 20 ml of eluent.

20 ml of water was added to the collected fraction and the solution was passed through a C-18 SPE cartridge (Waters Sep-Pak Light). The product was washed off the cartridge using 400 uL of EtOH followed by 400 uL of water.

After radioactive decay and losses in the system, typically, 2 to 5 mCi of the radiochemically pure final product was obtained. The entire process takes approx. 4 hours (0.8 to 2% DCY). ^{1}H, ^{13}C, ^{19}F NMR, ESI MS as well as analytical HPLC of non-radioactive carrier **^{19}F-1** and radioactive **^{18}F-1** obtained in this reaction are provided i. Specific activity at the end of synthesis was approximately 3 Ci/mmol. (Fig 2). This is in line with typical carrier added syntheses and approximately 1000 times lower than a typical no-carrier added synthesis.

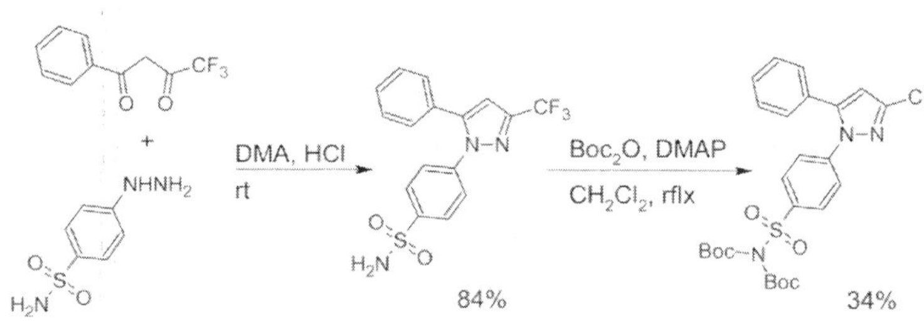

Figure 2: Synthesis of radiolabeling precursor.

https://doi.org/10.1371/journal.pone.0176606.g002

RESULTS

Radioelectrochemical Synthesis

Here we present a new method of radiofluorination of the pyrazole moiety based on electrochemical oxidation. We demonstrate the method by synthesis of an isotopomer of **1**. The resulting molecule ^{18}F-**1**, shows both *in vivo* stability not previously demonstrated for ^{18}F-labelled COX-2 inhibitors and *in vitro* specificity to COX-2 expression, making it a promising candidate as a molecular probe for visualization of COX-2 expression. We also demonstrate short blood half-life and fast excretion in a healthy mouse model. In its current form, the method only yields a marginal specific activity of 3 Ci/mmol (S2 Protocol, Fig 2), which needs to be improved before the tracer can be used as a sensitive *in vivo* imaging agent for COX-2 expression.

We prepared **^{18}F-1**, a radiolabeled version of COX-2 inhibitor **1**, using the electrochemical radiofluorination approach presented in Fig 4. The final precursor for the electrochemical radiolabeling was synthesized via a two-step process presented in Fig 2. Electrolysis was performed under pulsating potentiostatic conditions on platinum wire electrodes. Oscillating potential on the working electrode was used to prevent electrode surface fouling. Potential was kept at 2.7 V for 4 seconds, followed by a 1 sec 0.3 V pulse to regenerate the surface. Fig 3a shows a typical current vs time curve for this experiment. Inset in the figure demonstrates the section of the curve that corresponds to two consecutive voltage oscillations. Fig 3b represents cyclic voltammograms (CV) of the reaction mixture before and after electrolysis, along with the CV of the background solution (electrolyte + $Et_4NF*4HF$), which guided the selection of the oxidation potential for the reaction. Onset potential for the oxidation of the precursor was ~2.2 V and at the 2.7 V operating voltage of the cell, ~10 mA of current was due to precursor oxidation above the background current at this potential (Fig 3b). As expected, after electrolysis, oxidation current was found to be lower due to both precursor and electrolyte consumption.

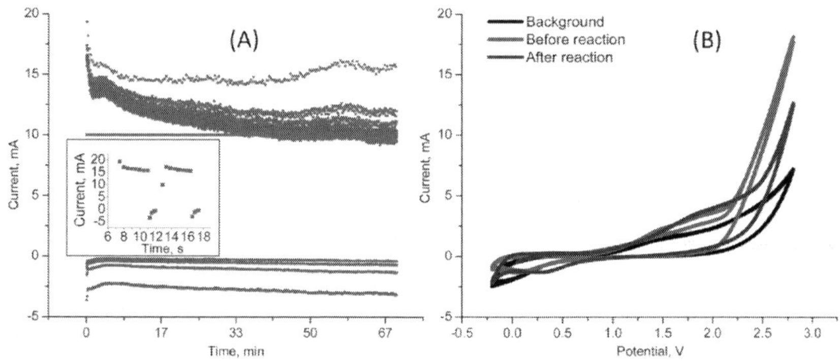

Figure 3: Electrochemical characteristics of radiolabeling process.

A) is a typical current vs time plot for the duration of electrolysis and B) shows the cyclic voltammogram of the background and the reaction mixture before and after electrolysis.

https://doi.org/10.1371/journal.pone.0176606.g003

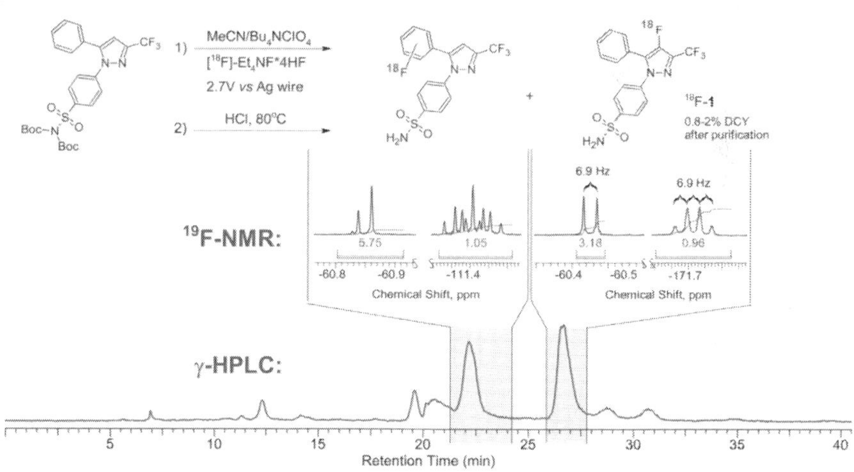

Figure 4: The reaction scheme of the electrochemical radiosynthesis of COX-2 inhibitor ^{18}F-1.

Green trace is the preparative radiochromatogram of the reaction mixture. Areas shaded green correspond to the collected fractions, the key areas of the ^{19}F-NMR spectra of the collected fractions are shown above.

https://doi.org/10.1371/journal.pone.0176606.g004

The structure of the final product was confirmed using NMR and AP-ESI mass spectroscopy. Due to the carrier added mode of the synthesis, the amount of the product produced during radiosynthesis was sufficient for NMR characterization; therefore, all analysis was done on the decayed samples obtained in the radiolabeling runs. Typically, semipreparative HPLC trace (Fig 4, green trace) of the reaction mixture contains two major radioactive peaks along with smaller unidentified byproducts. Both compounds were isolated and analyzed using ^{1}H, ^{13}C, ^{19}F NMR. ^{19}F NMR (Fig 4, red trace) clearly shows that the compounds of the earlier fraction do not contain any ^{19}F nuclei coupled with $-CF_3$ group (singlets at ~-60.9 ppm) while the compound in the later fraction exhibits a spectral pattern consistent with–F substituent (q, 6.9 Hz, -171.7 ppm) coupled with $-CF_3$ group (d, 6.9 Hz, -60.4 ppm). ^{1}H and ^{13}C NMR as well as AP-ESI mass spectra are also consistent with the structure of 1.

Up to 5 mCi of the final purified product was obtained using this procedure (2% DCY). After HPLC purification and subsequent reformulation, chemically and radiochemically pure doses of ^{18}F-**1** were produced and used in subsequent studies. Analytical HPLC traces of the pure injectable dose produced following this procedure is presented. Due to low solubility in aqueous ethanol, the compound was formulated in 40% EtOH prior to *in vitro* or *in vivo* administration.

In Vitro Cell Uptake Studies

To demonstrate the potential of **^{18}F-1** for quantification of COX-2 expression, *in vitro* COX-2 dependent uptake of the probe in LPS-treated RAW264.7 mactrophage-like cells was demonstrated.

A clear dependence of the ^{18}F-1 uptake on the concentration of LPS was observed in these experiments. Fig 5 shows progressive increase in the uptake with increasing LPS concentration and COX-2 levels up to 400 ng/ml, followed by uptake saturation beyond LPS concentrations exceeding 500 ng/ml. Analysis of the western blot data overlaid in Fig 5 illustrates that COX-2 expression with LPS stimulation also starts to saturate above 400ng/ml and probe uptake and expression data closely parallel one another.

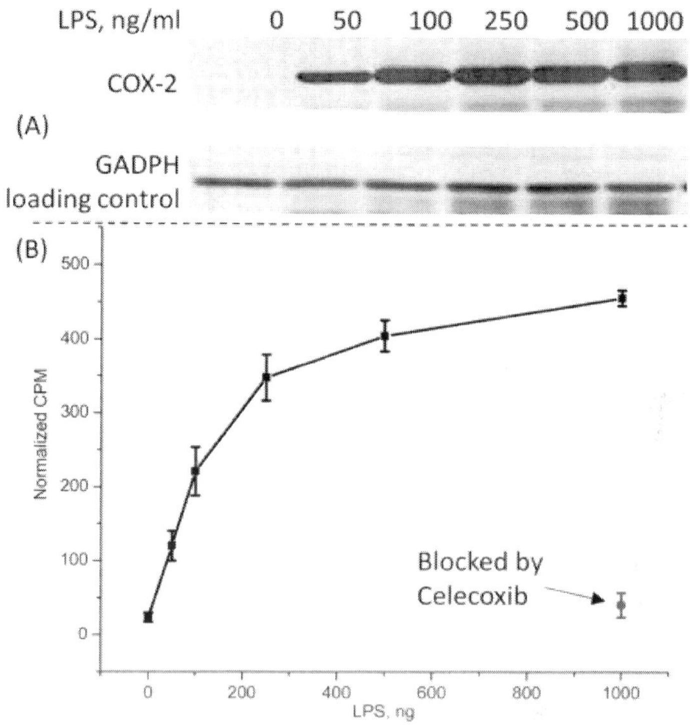

Figure 5: In vitro experiments.

(A) Western blot analysis of LPS treated cells. The upper panel is COX-2 expression and the lower blot is GAPDH loading control, (B) In vitro COX-2 dependent uptake of the probe in LPS-treated RAW264.7 macrophage-like cells, normalized to 1000 ng/ml of LPS.

https://doi.org/10.1371/journal.pone.0176606.g005

Additionally, LPS induced uptake of **^{18}F-1** was blocked by 32 μg/ml of Celecoxib, a selective COX-2 inhibitor (Fig 5), supporting the conclusion that COX-2 expression is responsible for probe retention in the cells.

In Vivo Metabolism and PET Imaging

In vivo metabolic stability of the probe was investigated. 1h after bolus injection of ^{18}F-**1**, urine, kidney, brain, liver, small intestines and blood plasma were collected from 3 healthy female mice and the radioactivity was extracted and analyzed by HPLC. Experimental protocols are detailed.

HPLC traces show over 95% intact probe and no discernable metabolites from the radio-HPLC analysis of the organic extracts (Fig 6A). The results of radio-TLC analysis of the same fractions agree with the HPLC traces (see SI). Metabolized radiolabelled molecules were observed in the urine where polar metabolites are excreted, and in the aqueous extracts of the major organs. However, these degradation products accounted for less than 1% of the total activity in all major organs. This is illustrated in (Fig 6B), which presents the chromatograms of the aqueous and organic extracts of small intestines as an example. While the organic fraction is composed of >98% of the parent compound, the aqueous fraction composition includes only approx. 20% of parent in the mixture. However, the amount of radioactivity in the aqueous solution is approximately 1% of the radioactivity in the organic phase, indicating that the metabolites comprise only an insignificant fraction of the total radioactivity.

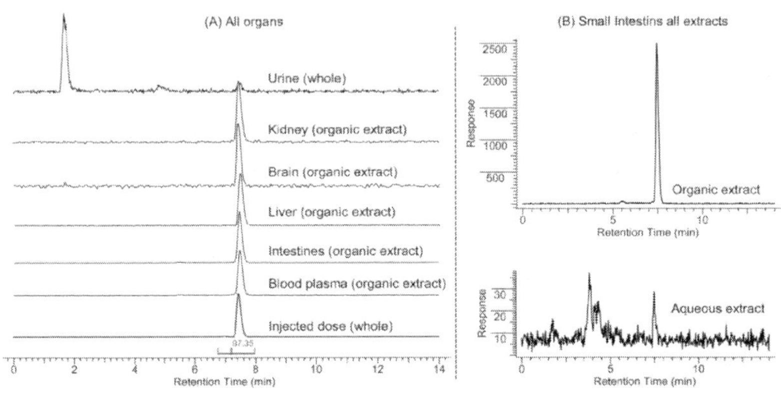

Figure 6: (A) Analytical HPLC traces of organic extracts of major organs 1h after administration of 50 uCi of 18F-1. (B) Analytical HPLC traces of all extracts of small intestines represented on the same scale.

https://doi.org/10.1371/journal.pone.0176606.g006

Ex-vivo biodistribution of the tracer was also studied following a 1 h uptake after administration of the probe. The same organs were collected and radioactivity accumulation was measured and normalized by weight and injected dose. The data presented in Fig 7 show that liver, kidney, and intestines accumulated the most radioactivity. Blood, muscle and brain each accumulated only 1% ID/g. The lowest amount of radioactivity was found in urine and bones.

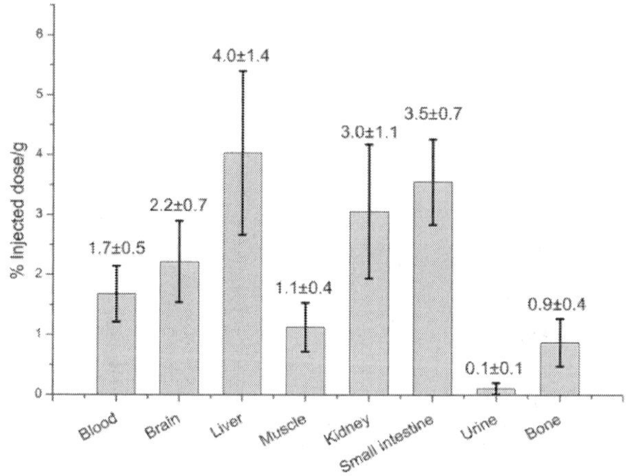

Figure 7: Ex-vivo biodistribution of ^{18}F-1.

Healthy mice (n = 5) 1 hour after bolus injection.

https://doi.org/10.1371/journal.pone.0176606.g007

Static and dynamic PET imaging studies were performed following bolus injection of ^{18}F-**1** in healthy wild-type mice. The time series PET images are presented and the region of interest (ROI) analysis from the 60 minutes post injection image is in agreement with the ex-vivo biodistribution data.

DISCUSSION

Radioelectrochemical Synthesis of COX-2 Inhibitor 1

The mechanism of the electrochemical fluorination of aromatic compounds has been extensively studied over the past 40 years.[34] The current consensus is that the reaction follows the ECEC sequence presented in Fig 8 and that the mechanism involves four distinct steps: electrochemical oxidation leading to radical cation I in Fig 8, followed by nucleophilic attack of fluoride (II in Fig 8), then another oxidation and the production of the cation III in Fig 8 and subsequent deprotonation and synthesis of the product.

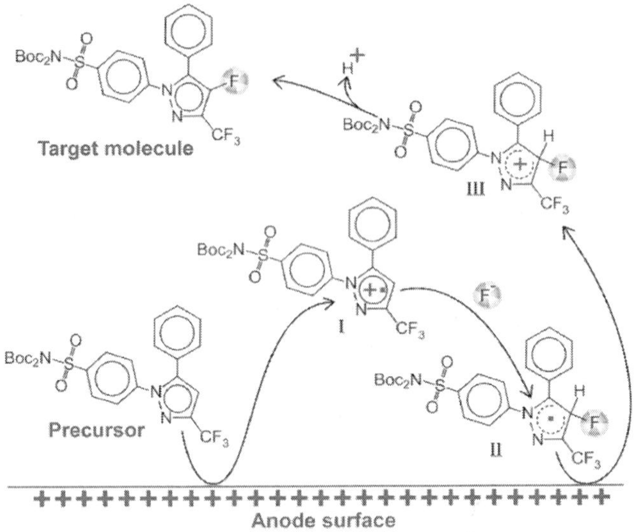

Figure 8: Mechanism of product formation.

https://doi.org/10.1371/journal.pone.0176606.g008

The details of this mechanism dictated the choice of the starting material. Cationic species (**I** in Fig 8) originating at the electrode surface must participate in a bimolecular reaction with fluoride anion in order to form the target product. Monomolecular decomposition of **I** however, can occur immediately after its formation, thus slowing conversion to the target molecule. This reasoning guided selection of protecting groups for the sulfonamide moiety, in that both hydrogen atoms have to be replaced by protecting groups stable toward monomolecular decomposition. Indeed, our attempts to use a precursor having only one Boc protecting group failed to yield any radiofluorinated products (fluoride conversion 0%). Upon substitution of a protecting group for a trityl, 24% fluoride conversion was achieved, but the only fluorinated product was found to be trityl fluoride, most likely due to the high stability of trityl cation that was in turn formed via the same unimolecular decomposition of the cation-radical intermediate.

Development of an automated electrochemical radiosynthesizer platform was a critical element for successful radiosynthesis of the target compound ^{18}F-**1**. This was due to large quantities (up to 1000 mCi) of radioactive precursor used in the synthesis, the need to tightly control the electrolysis parameters, and sensitivity of the reaction to moisture and other impurities. Using the in-house built automated platform, we achieved reliable production of the target molecule. More than 20 runs, needed for the studies discussed below, were performed and resulted in up to 5 mCi of the final purified product (2% DCY).

In Vitro Studies

Reliable supply of the radiolabeled ^{18}F-**1** enabled us to probe the factors affecting the *in vitro* cell uptake of the probe. Studies in macrophages demonstrated that the uptake of the tracer increases proportionately to LPS stimulation and can be blocked by excess Celecoxib. These observations illustrate the specificity of **^{18}F-1** as a potential imaging agent.

Multiple processes can drive accumulation of the probe in the cells. Depending on the cell type, the COX-2 enzyme is localized in the nuclear envelope and/or endoplasmic reticulum.[35],[36] The enzyme itself is a homodimer tethered to the membrane surface via the intermembrane domain.[37] Therefore, before associating with the target protein, ^{18}F-**1** has to diffuse through two membranes and cross two cell compartments filled with protein solution. Considering that >95% protein binding of celecoxib in plasma,[38] it is reasonable to expect the tracer, **^{18}F-1**, to be bound to one of the cytosolic proteins. Moreover, it has been recently suggested that COX-2 inhibitors have to first enter through the membrane binding domain, before passing through an opening, made exclusively of hydrophobic residues, before entering the active site of the enzyme.[39] Therefore, the diffusion process from the medium to the COX-2 molecule involves many intermediate steps (Fig 9). While available data do not provide any information on the relative rates of these steps, it is clear that dissociation of ^{18}F-**1** from the COX-2 molecule is the slowest process. The fact that LPS treatment modulates the uptake of the tracer (Fig 5) is likely due to the increased expression of COX-2 in response to LPS treatment.[40] Close correlation of COX-2 expression measured with western blot and the radioactivity uptake indicates that the uptake saturation is likely due to saturation of COX-2 enzyme expression with increasing LPS, and not because of loss of ^{18}F-**1** binding affinity to COX-2. Reduction of uptake in presence of excess celecoxib also points to the fact that association of the probe with the catalytic site of COX-2 enzyme is the driving force in tracer uptake.

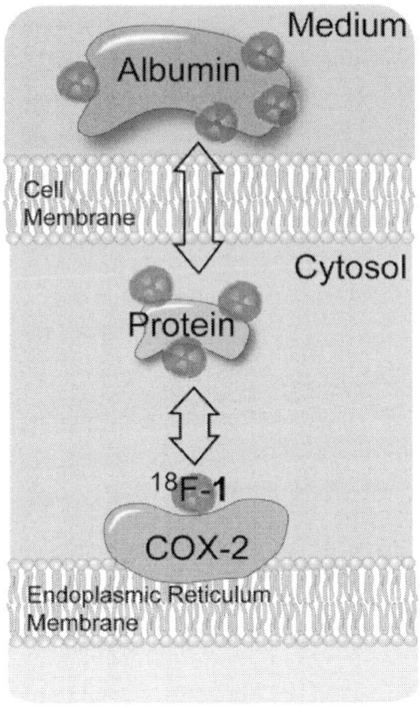

Figure 9: Schematic for association of 18F-1 with COX-2 in cell culture.

https://doi.org/10.1371/journal.pone.0176606.g009

In Vivo Metabolism and PET Imaging

Study of the metabolic fate of the compound following *in vivo* administration in mice confirmed that the aromatic position of the fluorine atom confers high *in vivo* stability. This is an improvement over other attempts at creating COX-2 imaging agents. Previous use of the known celecoxib analogs as PET imaging tracers was crippled by low metabolic stability. Up to 83% of radioactivity was found to be metabolized 1 hour after injection in baboons [18]. Our data demonstrates that ^{18}F-**1** is free of this drawback.

HPLC analyses of the radioactivity accumulated in all major organs estimated the amount of metabolites 1h post injection to be on the scale of 1% of the amount of the radioactivity present. The great majority of the radioisotope was still present in the form of unchanged parent compound ^{18}F-**1**. TLC trace, a method much more sensitive to the presence of polar metabolites, also found no polar metabolites in all major organs.

Analysis of PET images also supports the notion of high metabolic stability. Fig 10 presents a PET/CT image of the head of a healthy mouse inoculated with ^{18}F-**1** 90min prior to the scan. Noticeably, there is no significant increase in ^{18}F-**1** retention in the skull or vertebras, where bone uptake if present, would be easily visualized. Due to the small size of the structure, quantitative analysis of the image is not possible, but a characteristic pattern of fluoride uptake in the bone structures,[18] resulting from de-fluorination of the tracer, is not apparent on this image.

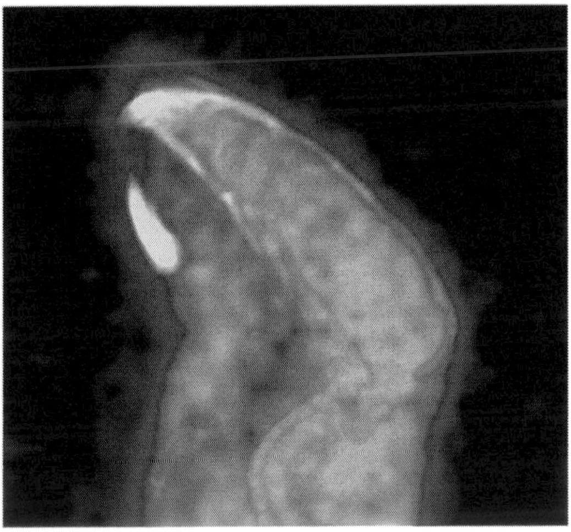

Figure 10: PET/CT image of the mouse head.

90 min after administration of ^{18}F-**1**.

https://doi.org/10.1371/journal.pone.0176606.g010

Accumulation patterns observed in ex-vivo organ counting agrees with the notion of high metabolic stability as well. Bone uptake, the main indicator of de-fluorination, was at the same level as muscle uptake and below that of blood and brain. Furthermore, there was no free ^{18}F-fluoride present in the ex-vivo TLC analysis of the blood.

It is important to note that the metabolic behavior of the tracer is very different from that of celecoxib.[41] While celecoxib is almost completely metabolized through oxidation of the methyl group, compound **1** remains virtually unchanged due to the lack of this substituent. This observation highlights an advantage of ^{18}F-**1**, as compared to drugs such as celecoxib, for imaging applications.

Dynamic PET/CT scans of healthy mice demonstrated favorable pharmacokinetics of ^{18}F-**1** for imaging studies with background subsiding within 1 hour and no noticeable defluorination. Fig 11 shows the relative time-activity profiles of all major organs. Within 1 hour after injection concentration of the tracer in all major organs, with the exception of the small intestine and the right side of the liver discussed below, stabilized at a level not exceeding twice that of muscle non-specific uptake. This indicates that there is potential of reaching a good image contrast within 1 hour after injection. Imaging data are in line with ex-vivo biodistribution (Fig 7) with kidney being a notable exception. While the dynamic scan indicates little difference in uptake in the kidney, blood and brain, ex-vivo counting indicates approximately 50% more accumulation in kidney as compared to the brain and blood. The apparent discrepancy between imaging and ex vivo counting data may be due to accumulation of activity in the thin renal cortex excluded from the conservatively defined ROI.[42] Uptake of ^{18}F-1 in the kidney may be attributed to known high levels of COX-2 expression in this organ.[43]

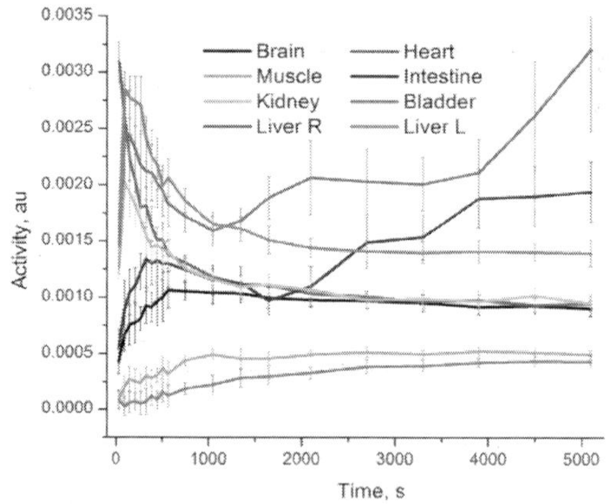

Figure 11: Time-activity profile of major organs.

Healthy mouse after bolus iv administration of ^{18}F-**1**.

https://doi.org/10.1371/journal.pone.0176606.g011

Gallbladder is likely the main excretion organ for the tracer. Increase in activity with time in the intestine and low activity in the bladder rule out urinary excretion. Unexpectedly, the left and right sides of the liver exhibit a 2 fold difference of the radioactivity uptake after 90 min. Furthermore, activity in one side of the liver increased with time while the other side of the liver demonstrated a washout rate similar to that of other organs. Additionally, static images reveal boluses of radioactivity distributed along the intestines, likely due to accumulation in feces. A similar accumulation pattern has been reported previously in human subjects for other 18F labeled tracers, where larger image sizes allowed for positive identification of gallbladder as the main excretion organ. [44]

Of particular interest is the time-activity curve of the brain ROI. Following the initial ramp up, brain radioactivity concertation decreases and is then stabilized at a level similar to other organs and twice the uptake of muscle. This is likely due to initial diffusion across the blood brain barrier and subsequent washout of the tracer from brain tissue. This indicates the potential applicability of this molecule for brain imaging of COX-2 expression, which is of particular interest for imaging of neurodegenerative diseases.[45]

CONCLUSION

Early detection and imaging of COX-2 overexpression would be an important clinical tool for monitoring disease progression, therapy intervention, evaluation of therapeutic treatment efficacy, and identification of patients who can be selected for COX-2 inhibitor or NSAID treatment. A COX-2 PET probe could potentially replace invasive biopsy and should reduce sampling errors due to target heterogeneity, as well as instability of COX-2 protein and mRNA, and should help determine the pharmacokinetics and *in vivo* binding characteristics of new COX-2 inhibitors.

The probe structure of ^{18}F-**1** is based on some of the lessons learned over the last two decades of research in developing a viable COX-2 PET tracer. To address the challenges of *in vivo* stability, we have placed ^{18}F on a stable heteroaromatic ring in a structure that maintains high COX-2 affinity and specificity, using a synthesis that is practical and made possible through the electrochemical platform. Electrochemical synthesis, performed on the automated platform developed in our lab, is readily adaptable for use by imaging researchers without requiring expertise in electrochemistry. Pharmacokinetic studies with ^{18}F-**1** in healthy mice revealed no bone retention or defluorination within 2 hours of injection, significant blood clearance, crossing of the BBB and no significant metabolites in major organs, making the probe ideally suited for imaging studies.

The current decay corrected radiochemical yield of 2%, although enough for preclinical studies and production of a single patient dose, needs improvement. More importantly, the use of $Et_4NF*4HF$ has limited the specific activity to 3 Ci/mmol, severely limiting the application of the probe as a successful imaging agent. We anticipate that the use of microfluidics and electrochemical flow cells for this surface activated reaction will increase the yield and specific activity. To address the specific activity problem, we are investigating the use of alternative quaternary ammonium or phosphonium salts of non-nucleophilic anions, polar and solvating solvents and reducing cation reactivity through cold temperature electrolysis and stabilized cation pools.[46]

REFERENCES

1. Harris RE, Beebe-Donk J, Doss H, Burr Doss D. Aspirin, ibuprofen, and other non-steroidal anti-inflammatory drugs in cancer prevention: a critical review of non-selective COX-2 blockade (review). Oncol Rep. 2005;13: 559–583. pmid:15756426
2. Cherukuri DP, Ishikawa T, Chun P, Catapang A, Elashoff D, Grogan TR, et al. Targeted Cox2 gene deletion in intestinal epithelial cells decreases tumorigenesis in female, but not male, ApcMin/+ mice. Mol Oncol. 2014;8: 169–177. pmid:24268915
3. Liu CH, Chang SH, Narko K, Trifan OC, Wu MT, Smith E, et al. Overexpression of cyclooxygenase-2 is sufficient to induce tumorigenesis in transgenic mice. J Biol Chem. 2001;276: 18563–18569. pmid:11278747
4. Esbona K, Inman D, Saha S, Jeffery J, Schedin P, Wilke L, et al. COX-2 modulates mammary tumor progression in response to collagen density. Breast Cancer Res BCR. 2016;18: 35. pmid:27000374
5. Fujiwaki R, Iida K, Kanasaki H, Ozaki T, Hata K, Miyazaki K. Cyclooxygenase-2 expression in endometrial cancer: correlation with microvessel count and expression of vascular endothelial growth factor and thymidine phosphorylase. Hum Pathol. 2002;33: 213–219. pmid:11957147
6. Knudson AG. Two genetic hits (more or less) to cancer. Nat Rev Cancer. 2001;1: 157–162. pmid:11905807
7. Minghetti L. Cyclooxygenase-2 (COX-2) in inflammatory and degenerative brain diseases. J Neuropathol Exp Neurol. 2004;63: 901–910. pmid:15453089
8. Tsujii M, Kawano S, DuBois RN. Cyclooxygenase-2 expression in human colon cancer cells increases metastatic potential. Proc Natl Acad Sci U S A. 1997;94: 3336–3340. pmid:9096394
9. Edelman MJ, Watson D, Wang X, Morrison C, Kratzke RA, Jewell S, et al. Eicosanoid modulation in advanced lung cancer: cyclooxygenase-2 expression is a positive predictive factor for celecoxib + chemotherapy—Cancer and Leukemia Group B Trial 30203. J Clin Oncol Off J Am Soc Clin Oncol. 2008;26: 848–855.
10. Inoue H, Taba Y, Miwa Y, Yokota C, Miyagi M, Sasaguri T. Transcriptional and posttranscriptional regulation of cyclooxygenase-2

expression by fluid shear stress in vascular endothelial cells. Arterioscler Thromb Vasc Biol. 2002;22: 1415–1420. pmid:12231559

11. Phelps ME. Positron emission tomography provides molecular imaging of biological processes. Proc Natl Acad Sci U S A. 2000;97: 9226–9233. pmid:10922074
12. Mosessian S, Duarte-Vogel SM, Stout DB, Roos KP, Lawson GW, Jordan MC, et al. INDs for PET Molecular Imaging Probes-Approach by an Academic Institution. Mol Imaging Biol MIB Off Publ Acad Mol Imaging. 2014;
13. Cole EL, Stewart MN, Littich R, Hoareau R, Scott PJH. Radiosyntheses using Fluorine-18: the Art and Science of Late Stage Fluorination. Curr Top Med Chem. 2014;14: 875–900. pmid:24484425
14. Laube M, Kniess T, Pietzsch J. Radiolabeled COX-2 Inhibitors for Non-Invasive Visualization of COX-2 Expression and Activity—A Critical Update. Molecules. 2013;18: 6311–6355. pmid:23760031
15. Penning TD, Talley JJ, Bertenshaw SR, Carter JS, Collins PW, Docter S, et al. Synthesis and Biological Evaluation of the 1,5-Diarylpyrazole Class of Cyclooxygenase-2 Inhibitors: Identification of 4-[5-(4-Methylphenyl)-3- (trifluoromethyl)-1H-pyrazol-1-yl] benzenesulfonamide (SC-58635, Celecoxib). J Med Chem. 1997;40: 1347–1365. pmid:9135032
16. Kharasch ED, Thummel KE. Identification of cytochrome P450 2E1 as the predominant enzyme catalyzing human liver microsomal defluorination of sevoflurane, isoflurane, and methoxyflurane. Anesthesiology. 1993;79: 795–807. pmid:8214760
17. Park BK, Kitteringham NR, O'Neill PM. Metabolism of Fluorine-Containing Drugs. Annu Rev Pharmacol Toxicol. 2001;41: 443–470. pmid:11264465
18. Prabhakaran J, Underwood MD, Parsey RV, Arango V, Majo VJ, Simpson NR, et al. Synthesis and in vivo evaluation of [18F]-4-[5-(4-methylphenyl)-3-(trifluoromethyl)-1H-pyrazol-1-yl] benzenesulfonamide as a PET imaging probe for COX-2 expression. Bioorg Med Chem. 2007;15: 1802–1807. pmid:17166726
19. de Vries EFJ, van Waarde A, Buursma AR, Vaalburg W. Synthesis and In Vivo Evaluation of 18F-Desbromo-DuP-697 as a PET Tracer for Cyclooxygenase-2 Expression. J Nucl Med. 2003;44: 1700–1706. pmid:14530489

20. McCarthy TJ, Sheriff AU, Graneto MJ, Talley JJ, Welch MJ. Radiosynthesis, In Vitro Validation, and In Vivo Evaluation of 18F-Labeled COX-1 and COX-2 Inhibitors. J Nucl Med. 2002;43: 117–124. pmid:11801714
21. Makino K, Yoshioka H. Selective fluorination of ethyl 1-methylpyrazole-4-carboxylates with poly(hydrogen fluoride)-amine complex under electrolytic anodic oxidation. J Fluor Chem. 1988;39: 435–440.
22. Siméon FG, Wendahl MT, Pike VW. The [18F]2-fluoro-1,3-thiazolyl moiety—an easily-accessible structural motif for prospective molecular imaging radiotracers. Tetrahedron Lett. 2010;51: 6034–6036. pmid:21057601
23. Kniess T, Laube M, Bergmann R, Sehn F, Graf F, Steinbach J, et al. Radiosynthesis of a 18F-labeled 2,3-diarylsubstituted indole via McMurry coupling for functional characterization of cyclooxygenase-2 (COX-2) in vitro and in vivo. Bioorg Med Chem. 2012;20: 3410–3421. pmid:22560838
24. Tian H, Lee Z. Synthesis of 18F-labeled cyclooxygenase-2 (COX-2) inhibitor as a potential PET imaging agent. J Label Compd Radiopharm. 2006;49: 583–593.
25. Lee E, Kamlet AS, Powers DC, Neumann CN, Boursalian GB, Furuya T, et al. A Fluoride-Derived Electrophilic Late-Stage Fluorination Reagent for PET Imaging. Science. 2011;334: 639–642. pmid:22053044
26. Lee E, Hooker JM, Ritter T. Nickel-Mediated Oxidative Fluorination for PET with Aqueous [18F] Fluoride. J Am Chem Soc. 2012;134: 17456–17458. pmid:23061667
27. Rotstein BH, Stephenson NA, Vasdev N, Liang SH. Spirocyclic hypervalent iodine(III)-mediated radiofluorination of non-activated and hindered aromatics. Nat Commun. 2014;5.
28. Gao Z, Lim YH, Tredwell M, Li L, Verhoog S, Hopkinson M, et al. Metal-Free Oxidative Fluorination of Phenols with [18F]Fluoride. Angew Chem Int Ed. 2012;51: 6733–6737.
29. Tredwell M, Gouverneur V. 18F labeling of arenes. Angew Chem Int Ed Engl. 2012;51: 11426–37. pmid:23086547
30. Preshlock S, Tredwell M, Gouverneur V. 18F-Labeling of Arenes and Heteroarenes for Applications in Positron Emission Tomography. Chem Rev. 2016;116: 719–766. pmid:26751274

31. He Q, Wang Y, Alfeazi I, Sadeghi S. Electrochemical nucleophilic synthesis of di-tert-butyl-(4-[18F]fluoro-1,2-phenylene)-dicarbonate. Appl Radiat Isot. 2014;92: 52–57. pmid:25000498
32. Kienzle GJ, Reischl G, Machulla H-J. Electrochemical radiofluorination. 3. Direct labeling of phenylalanine derivatives with [18F]fluoride after anodic oxidation. J Label Compd Radiopharm. 2005;48: 259–273.
33. Noel M, Suryanarayanan V. Current approaches to the electrochemical synthesis of organo-fluorine compounds. J Appl Electrochem. 2004;34: 357–369.
34. Rozhkov IN. Radical-cation Mechanism of the Anodic Fluorination of Organic Compounds. Russ Chem Rev. 1976;45: 615–629.
35. Parfenova H, Parfenov VN, Shlopov BV, Levine V, Falkos S, Pourcyrous M, et al. Dynamics of nuclear localization sites for COX-2 in vascular endothelial cells. Am J Physiol Cell Physiol. 2001;281: C166–178. pmid:11401839
36. Liou J-Y, Aleksic N, Chen S-F, Han T-J, Shyue S-K, Wu KK. Mitochondrial localization of cyclooxygenase-2 and calcium-independent phospholipase A2 in human cancer cells: implication in apoptosis resistance. Exp Cell Res. 2005;306: 75–84. pmid:15878334
37. Rouzer CA, Marnett LJ. Cyclooxygenases: structural and functional insights. J Lipid Res. 2009;50: S29–S34. pmid:18952571
38. Paulson SK, Kaprak TA, Gresk CJ, Fast DM, Baratta MT, Burton EG, et al. Plasma protein binding of celecoxib in mice, rat, rabbit, dog and human. Biopharm Drug Dispos. 1999;20: 293–299. pmid:10701700
39. Blobaum AL, Xu S, Rowlinson SW, Duggan KC, Banerjee S, Kudalkar SN, et al. Action at a Distance MUTATIONS OF PERIPHERAL RESIDUES TRANSFORM RAPID REVERSIBLE INHIBITORS TO SLOW, TIGHT BINDERS OF CYCLOOXYGENASE-2. J Biol Chem. 2015;290: 12793–12803. pmid:25825493
40. Wadleigh DJ, Reddy ST, Kopp E, Ghosh S, Herschman HR. Transcriptional Activation of the Cyclooxygenase-2 Gene in Endotoxin-treated RAW 264. 7 Macrophages. J Biol Chem. 2000;275: 6259–6266. pmid:10692422
41. Gong L, Thorn CF, Bertagnolli MM, Grosser T, Altman RB, Klein TE. Celecoxib pathways: pharmacokinetics and pharmacodynamics. Pharmacogenet Genomics. 2012;22: 310–318. pmid:22336956

42. Soret M, Bacharach SL, Buvat I. Partial-Volume Effect in PET Tumor Imaging. J Nucl Med. 2007;48: 932–945. pmid:17504879
43. Breyer MD, Harris RC. Cyclooxygenase 2 and the kidney. Curr Opin Nephrol Hypertens. 2001;10: 89–98. pmid:11195058
44. Koole M, Lewis DM, Buckley C, Nelissen N, Vandenbulcke M, Brooks DJ, et al. Whole-Body Biodistribution and Radiation Dosimetry of 18F-GE067: A Radioligand for In Vivo Brain Amyloid Imaging. J Nucl Med. 2009;50: 818–822. pmid:19372469
45. Choi S-H, Aid S, Bosetti F. The distinct roles of cyclooxygenase-1 and -2 in neuroinflammation: implications for translational research. Trends Pharmacol Sci. 2009;30: 174–181. pmid:19269697
46. Suga S, Suzuki S, Yoshida J. Reduction of a "cation pool": a new approach to radical mediated C—C bond formation. J Am Chem Soc. 2002;124: 30–31. pmid:11772058

INDEX

D

E

F

G

H

I